AF787676

Identifying Emerging Issues in Disaster Risk Reduction, Migration, Climate Change and Sustainable Development

Karen Sudmeier-Rieux • Manuela Fernández
Ivanna M. Penna • Michel Jaboyedoff
J.C. Gaillard
Editors

Identifying Emerging Issues in Disaster Risk Reduction, Migration, Climate Change and Sustainable Development

Shaping Debates and Policies

Springer

Editors
Karen Sudmeier-Rieux
Risk-group - Institute of Earth Sciences
Faculty of Geosciences and Environment
University of Lausanne
Lausanne, Switzerland

Ivanna M. Penna
Geological Survey of Norway
Trondheim, Norway

Risk-group - Institute of Earth Sciences
Faculty of Geosciences and Environment
University of Lausanne
Lausanne, Switzerland

J.C. Gaillard
School of Environment
The University of Auckland
Auckland, New Zealand

Manuela Fernández
Institute of Geography and Sustainability
Faculty of Geosciences and Environment
University of Lausanne
Lausanne, Switzerland

Michel Jaboyedoff
Risk-group - Institute of Earth Sciences
Faculty of Geosciences and Environment
University of Lausanne
Lausanne, Switzerland

ISBN 978-3-319-33878-1 ISBN 978-3-319-33880-4 (eBook)
DOI 10.1007/978-3-319-33880-4

Library of Congress Control Number: 2016944357

© Springer International Publishing Switzerland 2017
This work is subject to copyright. All rights are reserved by the Publisher, whether the whole or part of the material is concerned, specifically the rights of translation, reprinting, reuse of illustrations, recitation, broadcasting, reproduction on microfilms or in any other physical way, and transmission or information storage and retrieval, electronic adaptation, computer software, or by similar or dissimilar methodology now known or hereafter developed.
The use of general descriptive names, registered names, trademarks, service marks, etc. in this publication does not imply, even in the absence of a specific statement, that such names are exempt from the relevant protective laws and regulations and therefore free for general use.
The publisher, the authors and the editors are safe to assume that the advice and information in this book are believed to be true and accurate at the date of publication. Neither the publisher nor the authors or the editors give a warranty, express or implied, with respect to the material contained herein or for any errors or omissions that may have been made.

Printed on acid-free paper

This Springer imprint is published by Springer Nature
The registered company is Springer International Publishing AG Switzerland

Foreword by the International Organization for Migration

The year 2015 will be remembered as a key moment in the evolution of the global development architecture. Expectations are that the endorsement of the Sendai Framework on Disaster Risk Reduction 2015–2030 (SFDRR) and of the post-2015 Sustainable Development Goals (SDGs), together with progress under the UN Framework Convention on Climate Change (UNFCCC), will result in a stronger and more cohesive international framework to promote well-being, security and resilience.

All three policy tracks have given significant attention to population movements. UNFCCC and SFDRR 2015–2030 mention displacement as one of the main potential consequences of environmental shocks and stresses, and migration and relocations as options to reduce their impacts. The dialogue on Urban Settlements—culminating in the 2016 Habitat-III conference—has also drawn attention to the vulnerability of migrants in crisis situations, while underlining the role they can play in relief and recovery.

All these processes call for an important change in perspective. The narrative that depicts human mobility as an undesirable (albeit often inevitable) consequence of natural and man-made shocks and stresses, poverty and underdevelopment must give way to a more nuanced view of mobility as a part of the normal, productive life of households and communities. Despite the political sensitivities that are often associated with the management of internal and international population movements, human mobility is clearly an option people and communities exercise in their attempt to achieve well-being, acquire resilience and successfully adapt to environmental change.

This is not to deny that population movements can have negative impacts on the lives of those who move, as well as on those of people in home and host communities. It is simply to acknowledge that mobility is an unavoidable and integral part of our contemporary world. This being the case, international, regional, national and local-level policies and actions must be put into place to maximize the positive outcomes of migration.

This book looks at the various facets of the nexus between human mobility, well-being and environmental hazards through the lens of risk reduction. The wealth of case studies, perspectives and ideas it provides will hopefully help unpack the complexity of this matter, providing useful indications to academics, practitioners and policy-makers alike. We therefore warmly welcome its timely publication, and trust that it will further the understanding of and encourage dialogue on what is a central issue for the sustainable development of our modern, interconnected societies.

William Lacy Swing

Acknowledgments

We are extremely grateful to the Faculty of Geosciences and Environment at the University of Lausanne, Switzerland for encouraging and providing funding for the 2013 workshop which provided the impetus for this book, "Linking Sustainable Development, Global Migration, Climate Change Adaptation and Disaster Risk Reduction – Identifying Emerging Issues". The workshop was also a follow-up activity to a two-year project funded by the Swiss Network for International Studies: "Sustainable Land Management in Mountain Regions of Bolivia and Nepal In the Context of Outmigration, Climate Change and Disaster Risk Reduction 2012-2014". Secondly, we extend a special thanks to Lorenzo Guadagno at the International Organization for Migration, Switzerland for his technical advice and close collaboration with the editors, as well as Alice McSherry at the University of Auckland, New Zealand for English editing assistance.

Finally we thank the many book authors and external reviewers who provided comments to chapters. External reviewers included: David Alexander (Institute for Risk and Disaster Reduction, University College London, UK), Soumyadeep Banerjee (International Centre for Integrated Mountain Development, Nepal), Roberto Barrios (Southern Illinois University Carbondale, USA), William Critchley (Sustainable Land Management Associates, UK), Marc-Henri Derron (University of Lausanne, Switzerland), A.J. Faas (San Jose State University, USA), Victoria Reyes-García (ICREA and Autonomous University of Barcelona, Spain), Devendra Gauchan (Nepal Agricultural Research Council, Nepal), James Lewis (Datum International, UK), Susan Martin (Georgetown University, USA), Ashutosh Mohanty (Mongolia International University, Mongolia), Julien Rebotier (CNRS, Université de Toulouse - Jean Jaurès, France), Roanne van Voorst (University of Amsterdam, Netherlands).

Contents

Contributors

Anu Adhikari International Union for Conservation of Nature, Kathmandu, Nepal

Irasema Alcántara-Ayala Departamento de Geografía Física, Instituto de Geografía, Universidad Nacional Autónoma de México, Circuito Exterior, Ciudad Universitaria, Mexico City, Mexico

Alex Amiguet Institute of Earth Surface Dynamics, Faculty of Geosciences and Environment, University of Lausanne, Lausanne, Switzerland

Valentin Bonnefoy Department of Operations and Emergencies, International Organization for Migration, Geneva, Switzerland

Patrick Bottazzi Institute of Geography and Sustainability, Faculty of Geosciences and Environment, University of Lausanne, Lausanne, Switzerland

Vibeke Brandvold Department of Geology and Mineral Resources Engineering, Norwegian University of Science and Technology, Trondheim, Norway

Andrew E. Collins Department of Geography/Disaster and Development Network, Faculty of Engineering and Environment, Northumbria University, Newcastle upon Tyne, UK

Silwan Daouk Hôpitaux Universitaires de Genève, Geneva, Switzerland

Sanjaya Devkota FEED (P) Ltd., Lalitpur, Nepal

Institute of Engineering, Tribhuvan University, Kathmandu, Nepal

Emilie Edelblutte University of Lyon, CNRS, University of Lyon 3 Jean Moulin, Environnement Ville Société UMR 5600, Lyon, France, Lyon, France

Manuela Fernández Institute of Geography and Sustainability, Faculty of Geosciences and Environment, University of Lausanne, Lausanne, Switzerland

J.C. Gaillard School of Environment, The University of Auckland, Auckland, New Zealand

Ethel Gandia Department of Operations and Emergencies, International Organization for Migration, Geneva, Switzerland

Lorenzo Guadagno Department of Operations and Emergencies, International Organization for Migration, Geneva, Switzerland

Anne Gueye-Girardet Fedevaco, Lausanne, Switzerland

Kenneth Hewitt Department of Geography and Environmental Studies, Wilfrid Laurier University, Waterloo, ON, Canada

Carmen Hitz Stadt Aarau, Umweltfachstelle, Stabsstelle Stadtentwicklung, Aarau, Switzerland

Michel Jaboyedoff Risk-group - Institute of Earth Sciences, Faculty of Geosciences and Environment, University of Lausanne, Lausanne, Switzerland

Johanna Jacobi Centre for Development and Environment, University of Bern, Bern, Switzerland

Ina Rehema Jahn Department of Operations and Emergencies, International Organization for Migration, Geneva, Switzerland

Stéphanie Jaquet Centre for Development and Environment, University of Bern, Bern, Switzerland

Raoul Kaenzig Institute of Geography, University of Neuchâtel, Neuchâtel, Switzerland

Ilan Kelman University College London, London, UK

University of Agder, Kristiansand, Norway

Ajay Chandra Lal Institute of Engineering, Tribhuvan University, Kathmandu, Nepal

Hans-Peter Liniger Centre for Development and Environment CDE, University of Bern, Bern, Switzerland

Keith Morrison Pacific Centre for Environment and Sustainable Development, The University of the South Pacific, Suva, Fiji

Mamadou Lamine Ndiaye Saint-Hyacinthe, QC, Canada

Seydou Niang Laboratoire Lateu, IFAN, Université C.A.Diop, Dakar, Senegal

Tomohito Okuda International Development and Regional Planning Unit, Department of Urban Engineering, Graduate School of Engineering, University of Tokyo, Tokyo, Japan

Anthony Oliver-Smith Department of Anthropology, University of Florida, Gainesville, FL, USA

Ivanna M. Penna Geological Survey of Norway, Trondheim, Norway

Risk-group - Institute of Earth Sciences, Faculty of Geosciences and Environment University of Lausanne, Lausanne, Switzerland

Hans-Rudolf Pfeifer Institute of Earth Surface Dynamics, Faculty of Geosciences and Environment, University of Lausanne, Lausanne, Switzerland

Patrick Pigeon Université de Savoie-CISM, Le Bourget du Lac, Cedex, France

Maria Isabel Pillco Faculty of Agronomy, Universidad Mayor de San Andrés, La Paz, Bolivia

Stephan Rist Centre for Development and Environment, University of Bern, Bern, Switzerland

Jessica Roberts Institute of Earth Surface Dynamics, Faculty of Geosciences and Environment, University of Lausanne, Lausanne, Switzerland

Cyril Royez urbaMonde, Genève, Switzerland

Monika Schneider Research Institute of Organic Agriculture, Frick, Switzerland

Gudrun Schwilch Centre for Development and Environment, University of Bern, Bern, Switzerland

Gitta Shrestha Nepal Centre for Contemporary Research, Kathmandu, Nepal

Karen Sudmeier-Rieux Risk-group - Institute of Earth Sciences, Faculty of Geosciences and Environment, University of Lausanne, Lausanne, Switzerland

Pauline Texier-Teixeira University of Lyon, University of Lyon 3 Jean Moulin, Environnement Ville Société UMR 5600, Lyon, France, Université Lyon 3 Jean Moulin, Lyon, France

Bishnu Raj Upreti Nepal Centre for Contemporary Research, Kathmandu, Nepal

Torsten Vennemann Institute of Earth Surface Dynamics, Faculty of Geosciences and Environment, University of Lausanne, Lausanne, Switzerland

Benoît Zen-Ruffinen Institute of Earth Surface Dynamics, Faculty of Geosciences and Environment, University of Lausanne, Lausanne, Switzerland

Editor Biographies

Karen Sudmeier-Rieux is a senior researcher at the University of Lausanne, Institute of Earth Science, where she manages two research projects on landslide risk, bio-engineering, migration, resilience and vulnerability of mountain communities in Western Nepal. She also works as a training and education consultant for UNEP in Geneva. She is the thematic lead for disaster risk reduction group of the IUCN Commission on Ecosystem Management (IUCN-CEM) and co-editor of a recent book, *The Role of Ecosystems in Disaster Risk Reduction* (UNU-Press), in collaboration between IUCN-CEM, UNEP and the United Nations University. She has also published a number of articles on the topic of resilience for disaster risk reduction, community-based landslide management in Nepal, and the 2005 earthquake in Pakistan. She has also authored a number of IUCN publications on ecosystems, livelihoods and disaster risk reduction, and was a contributor to the IPCC Special Report on Extreme Events (2012).

Manuela Fernández is a political scientist and geographer. She completed her Ph.D. in Environmental studies from University of Lausanne (UNIL) on factors of risk governance and conflict resolution between actors involved in landslide risk management in Guatemala. She is also trained in interdisciplinary risk methods (FORIN) and approaches. She has developed in-depth knowledge on institutions, actors, public policies and models of public management in Latin America and Europe, and has published papers on this topic as well as several risk recommendations reports for the municipalities of Guatemala.

Ivanna M. Penna completed her Ph.D. at the Department of Geological Sciences from the University of Buenos Aires (Argentina) on large landslide occurrence in the Southern Central Andes of Argentina, with a specialization on assessment and management of geological and climate related risk at the University of Geneva (Switzerland). Dr. Penna's publications address the conditioning factors for large landslide occurrence and related primary and secondary hazards, as well as their impact on the long-term evolution of landscapes. She has conducted fieldwork in

Argentina, Nepal, Bolivia and the Swiss Alps, and co-managed an interdisciplinary research project on migration, disaster risk reduction and climate change in Bolivia and Nepal. She previously worked as researcher at the Institute of Earth Sciences, University of Lausanne, and has been working at the Landslide Group at the Geological Survey of Norway since 2015.

Michel Jaboyedoff is a geologist with a degree in physics and a PhD. in clay mineralogy. During his Ph.D., his research focused on natural hazards, especially on rockfall. Since 2005, he has been working as a full professor at the Institute of Earth Science, University of Lausanne conducting research on natural hazards and its related risks. He has been involved in several risk management projects around the world (Argentina, Canada, Nepal, Norway, Switzerland) linking natural hazards to climate change and adaptation, and is part of several European FP7 Projects (Mountain Risks, Safeland, CHANGES). He was also the principal investigator for the SNIS project, "*Sustainable land management in mountain regions of Bolivia and Nepal in the context of outmigration, climate change and disaster risk reduction.*" Many of his studies involve the use of remote sensing techniques such as Lidar his team's research efforts have also concentrated on the development of software for hazard and risk analysis. He leads a team of several PhD students and researchers who are in charge of the project GEOCOVER that aims to provide full digital coverage of Switzerland using 1:25,000 scale geological maps provided by Swisstopo.

J.C. Gaillard is an Associate Professor at the University of Auckland in New Zealand. His present work focuses on developing participatory tools for disaster risk reduction and on involving marginalized groups in disaster-related activities with an emphasis on ethnicity, gender, prisoners and homeless people. J.C. Gaillard also collaborates on participatory mapping and community-based DRR trainings with NGOs, local governments and community-based organizations in Asia and the Pacific. He is author of numerous publications, including the recent *Handbook of Hazards and Disaster Risk Reduction*, Routledge, in partnership with Prof. Ben Wisner and Dr. Ilan Kelman, and is chief editor of the journal *Disaster Prevention and Management*, Emerald.

Acronyms

ADB	Asian Development Bank
AIRDR	Assessment of Integrated Research on Disaster Risk
BNPB	National Disaster Mitigation Agency (Indonesia)
CBS	Central Bureau of Statistics (Nepal)
CCA	Climate Change Adaptation
CIAT	Comité Interministériel d'Aménagement du Territoire (Interministerial Committee for Land use planning) (Haiti)
CNR	Compagnie nationale du Rhône (National Company of Rhone River)
CRED	Centre de Recherche sur l'Epidémiologie des Désastres (Research Center on the Epidemiology of Disasters)
CRM	Climate Risk Management
DDN	Disaster and Development Network
DFID	Department for International Development (United Kingdom)
DHM	Department of Hydrology and Meteorology (Nepal)
DIDR	Development-induced displacement and resettlement
DRM	Disaster risk management
DRR	Disaster risk reduction
EbA	Ecosystem-based adaptation
EC	European Commission
ECHO	European Union Humanitarian Aid and Civil Protection Department
EIA	Environmental impact assessment
EM-DAT	Emergency events database
GCCA	Global Climate Change Alliance (European Union)
EU-GCCA	European Union—Global Climate Change Alliance
EWASs	Early warning articulated systems
EWSs	Early warning systems
FAO	Food and Agriculture Organization of the United Nations
FEMA	Federal Emergency Management Agency (United States)

FGD	Focus group discussion
FiBL	Swiss Research Institute of Organic Agriculture
GDP	Gross domestic product
GFMD	Global Forum for Migration and Development
GII	Gender inequality index
GIS	Geographic information system
GLOFs	Glacial outburst floods
GNDR	Global Network of Civil Organisations for Disaster Reduction
GNS	Gross national savings
GT	Glacier Trust (Nepal)
HDI	Human Development Index
HFA	Hyogo framework for action
HIV/AIDS	Human immunodeficiency virus infection/acquired immune deficiency syndrome
HS	Homeland Security [United States Department of]
ICIMOD	International Centre for Integrated Mountain Development
ICRC	International Committee of the Red Cross
ICT	Information communication technology
IDP	Internally displaced person
IDRM	Integrated disaster risk management
IEE	Initial environmental examination
IEM	Integrated emergency management
IFRC	International Federation of Red Cross and Red Crescent Societies
IAASTD	International Assessment of Agricultural Knowledge, Science and Technology for Development
ICRAF	World Agroforestry Center
IFAD	International Fund for Agricultural Development
IFOAM	International Federation of Organic Agriculture Movements
IMF	International Monetary Fund
IOM	International Organization of Migration
IPCC	Intergovernmental Panel for Climate Change
IRDR	Integrated Research for Disaster Risk
IRGC	International Risk Governance Council
IUCN	International Union for Conservation of Nature
MASL	Metres above sea level
MAT	Mean annual temperature
MCGM	Municipal Corporation of Greater Mumbai (India)
MDGs	Millennium development goals
MINUSTAH	United Nations Stabilization Mission in Haiti
MMRDA	Mumbai Metropolitan Region Development Authority (India)
MoSTE	Ministry of Science, Technology and Environment (Nepal)
MRPDA	Mithi River Protection and Development Authority (India)
NAST	National Academy of Sciences and Technology (Nepal)
NATO	North American Treaty Organization
NGO	Non-governmental Organization

NIC	National Institute of Colonization (Bolivia)
NIDS	Nepal Institute for Development Studies
NLSS	Nepal Living Standard Survey
OCHA	Office for the Coordination of Humanitarian Affairs (UN)
OECD	Organisation for Economic Co-operation and Development
OLTB	Overcoming land tenure barriers (Haiti)
OWGGASDG	Open Working Group on Sustainable Development Goals
PAR	Pressure-and-release
REDD	United Nations Collaborative Programme on Reducing Emissions from Deforestation and Forest Degradation
RIA	Risk interpretation and action
RPP	Risk prevention plans
SASMI	South Asian Seasonal Monsoon Index
SDGs	Sustainable Development Goals
SES	Social–ecological systems
SFDRR	Sendai framework for disaster risk reduction
SIC	Security industrial complex
SLM	Sustainable land management
SSHAC	Senior Seismic Hazard Analysis Committee
STD	Sexually transmitted disease
SWOT	Strengths, weaknesses, opportunities and threats
TLTB	Tauke Land Trust Board (Fiji)
TEPCO	Tokyo Electric Power Company (Japan)
UAE	United Arab Emirates
UN	United Nations
UNCTAD	United Nations Conference on Trade and Development
UNDAF	United Nations Development Assistance Framework
UNDESA	United Nations Department of Economic and Social Affairs
UNDP	United Nations Development Programme
UNFCCC	United Nations Framework Convention for Climate Change
UNHCR	United Nations High Commission on Refugees
UNICEF	United Nations Children's Fund
UNIFEM	United Nations Development Fund for Women
UNISDR	United Nations Office for Disaster Reduction
UNOSEH	United Nations Office of the Special Envoy for Haiti
UNPEC	Urban National Parks in Emerging Countries and Cities
USAID	United States Agency for International Development
USP	University of South Pacific
USOHDACA	United States Overseas Humanitarian, Disaster and Civic Aid
VDC	Village Development Committee (Nepal)
WCDRR	World Conference on Disaster Risk Reduction
WOCAT	World Overview of Conservation Approaches and Technologies
WWAP	World Water Assessment Programme
WWF	World Wildlife Fund

Chapter 1
Introduction: Exploring Linkages Between Disaster Risk Reduction, Climate Change Adaptation, Migration and Sustainable Development

Karen Sudmeier-Rieux, Manuela Fernández, J.C. Gaillard, Lorenzo Guadagno, and Michel Jaboyedoff

1 Tackling Inter-linkages

We live in a world of increasingly complex and inter-linked environmental and societal issues, such as disasters and migrations. Disasters, development, migration and climate change are among some of the most pressing issues facing our societies today. They are closely linked through a myriad of interconnections and causal connections and have recently been given more attention in the scientific literature, primarily from the perspective of development, climate change or so-called *environmental migration* (Foresight 2011; Piguet 2013; UNDP 2010). Large disasters, which took almost 68,000 lives and affected 218 million on average each year in the last decade (1994–2013), and the many small everyday disasters that are often not accounted for, are on the rise (EM-DAT 2015; GNDR 2015; UNISDR 2015). At the same time there is a steep rise in the number of migrants worldwide, up from

K. Sudmeier-Rieux (✉) • M. Fernández • M. Jaboyedoff
Faculty of Geosciences and Environment, University of Lausanne, Quartier Mouline- Batiment Géopolis, 1015 Lausanne, Switzerland
e-mail: karen.sudmeier-rieux@unil.ch; manuelafernandezirujo@gmail.com; Michel.Jaboyedoff@unil.ch

J.C. Gaillard
School of Environment, The University of Auckland, Private Bag 92019, 1142 Auckland, New Zealand
e-mail: jc.gaillard@auckland.ac.nz

L. Guadagno
Department of Operations and Emergencies, International Organization of Migration, 17, route des Morillons, 1200 Grand Saconnex, GE, Switzerland
e-mail: lguadagno@iom.int

© Springer International Publishing Switzerland 2017
K. Sudmeier-Rieux et al. (eds.), *Identifying Emerging Issues in Disaster Risk Reduction, Migration, Climate Change and Sustainable Development*,
DOI 10.1007/978-3-319-33880-4_1

155 million in 1990 to 232 million in 2013 (OECD 2013; UNDESA 2015). Seemingly every day, the media report about a new wave of migrants taking to the sea or land to seek better opportunities elsewhere, often escaping unsustainable livelihood conditions, political upheaval or environmental stress, or a combination of all the above (UN News Centre 2015). The goal of this book is to explore these inter-linkages from a number of different geographical, social and natural science angles and contribute to the debate about how to improve disaster risk reduction (DRR) policies and practices, taking into account migration process from a large perspective where both natural and social factors are crucial and mutually "alloyed".

It is also a contribution to counter the often criticized "silo" approach taken by institutions and academia alike, which do not foster enough integrative policies or transdisciplinary research (Gall et al. 2015). Yet, policies and practices in the fields of DRR, climate change adaptation (CCA), environmental management and migration are highly intertwined. The year 2015 will be remembered as a cross roads for the renegotiation of several relevant international agreements: the post Hyogo Framework for Action (HFA) framework led by the United Nations Office for Disaster Risk Reduction (UNISDR), which became the Sendai Framework for Disaster Risk Reduction (SFDRR) 2015–2030; the renewal of the Kyoto Protocol at the Conference of Parties meeting in Paris under the United Nations Framework Convention on Climate Change (UNFCCC); the Sustainable Development Goals (SDGs) by the SDG working group to replace the Millennium Development Goals (MDGs) and finally the Third Financing for Development Conference which aims to contribute to and support the implementation of the post-2015 development agenda. Migration is an important component in all of these agreements, and adequate migration policies will be key to the achievement of their objectives. However, it is also a politically sensitive topic, as demonstrated by the contentious inclusion of references to migration and displacement in the texts of the SFDRR and the SDGs.

This volume contributes in particular to the discussion on the SFDRR and its potential as an integrative DRR policy framework, some seeing it as a positive step forward in terms of embracing a multitude of issues, others doubting that the agreement will lead to much concrete action towards real DRR on the ground. The overarching guiding theme is succinctly summarized by Hewitt: (Chap. 3, p:45) "Risk reduction in each preventive field has required working with, and the development of improvements for, those most likely to be harmed, usually those living in disadvantaged places and groups", and this must include migrants, who were largely sidetracked in DRR policies and practices until recently. When we refer to DRR, we understand it to encompass the practices and policies that span prevention, mitigation and response through early warning, risk analysis and management, communications and longer-term recovery (Chaps. 8 and 7).

Taking a DRR approach provides a more comprehensive platform to exploring inter-linkages between disasters, migration, climate change and sustainable development, as it adopts a comprehensive perspective. Human mobility has direct relevance for hazard impacts, exposure and vulnerability, and is therefore key to risk management decisions and policies. Mobility choices are part of complex household-level livelihood strategies to minimize risks and optimize economic, social, political gains. However, a household decision to move from a mountain slope to a crowded urban floodplain may simultaneously translate into positive economic outcomes, increased livelihood security, access to infrastructure and education and into increased levels of disaster risk for the people involved. These livelihood decisions will also affect collective risk in the place of origin and the place of destination. It is possible that migration can both reduce and increase exposure, vulnerabilities and capacities. Adopting a DRR perspective to understanding migration allows to identify the variety of effects migration has on risk, potentially revealing dynamics of "redistribution rather than reduction of disaster risk" Guadagno (Chap. 2).

Thus, despite the varying opinions and perspectives reflected in this book, we identify two emerging themes:

- First is that disasters, climate change and migration processes are increasingly complex, cascading and intertwined and often based in skewed or mal-development, including the unsustainable management of natural resources;
- Second, migration can act as a positive and negative factor for adapting to economic, social or environmental stressors (i.e. migration to coastal urban centres can reduce economic risks by improving livelihoods while creating new exposure to hazards).

2 Exploring Emerging Themes

This section explores the above-mentioned themes, while providing a critical analysis of the environmental migration/displacement/climate change discourse from a DRR perspective.

2.1 Trends in Disasters and Migration

In analyzing disaster data for the past decade, we observe that flooding is associated with most middle-size and large disasters, affecting 2.5 billion people and accounting for 43 % of all recorded events (CRED 2015). Although this figure seems high, the average number of affected people has fallen over the past 20 years, from 267 million in 1994 to 214 million in 2015. Death rates however have increased over the same period, from 68,000 deaths per year for the full 20-year period (1994–2013) to 99,700 deaths per year between 2004 and 2013. This trend is partially explained by three mega disasters: the 2004 Asian tsunami, Cyclone Nargis in 2008 and the 2010 Haitian earthquake but also by continued vulnerability of some communities to natural hazards (CRED 2015). Important to note is the number of small, cumulative disasters which often go unreported but are extremely devastating for lives and livelihoods (UNISDR 2015). Although there is a steady increase in the number of climate-related disasters, which increased by 44 % from the 1980–1989 level, a close analysis of these trends points to exposure and vulnerability or people living in unsafe places along coastlines, rivers, steep hillsides or the flanks of volcanoes and compounded by environmental degradation, rather than climate change per se as their main driver (CRED 2015; IPCC 2014a; UNISDR 2015). This is not to deny that climate change certainly is amplifying the magnitude and frequency of certain hazards, primarily sea level rise, heat waves and certain extreme rainfall events (IPCC 2014b).

According to CRED (2015), Asia led by the greatest number of disasters, with 3.3 billion people affected in China and India. Although the largest disaster events occurred in upper-middle income countries (56 %), the highest number of fatalities occurred in low income countries (68 %). Disaster profiles thus differ considerably depending on levels of wealth and preparedness rather than exposure per se. In addition to human lives, globally disasters caused high economic losses, an estimated US$2,600 billion over the period 1994–2013, although this amount may be underestimated by 50 % (CRED 2015). Even if most of the economic losses are reported in higher to middle income countries, disasters are major setbacks to development efforts of lower income or small economy countries as a proportion of their gross domestic products (CRED 2015). This is especially true for countries which are struggling to improve basic amenities and economic opportunities (UNISDR 2014).

Disasters are one of the reasons why people move but usually as an additional trigger to underlying livelihoods issues. As Nepal is greatly represented in this volume, we illustrate the above point with the case of post-earthquake Nepal 2015. As Nepal is a country characterized by intense internal and international migration flows—30 % of its Gross Domestic Product (GDP) is based on remittances—remittance funds are likely to be critical for the reconstruction efforts. The affected areas have an unusually high proportion of migrant workers, roughly one in five and the Government of Nepal is inviting migrant workers to return to participate in the reconstruction and recovery process (IOM 2015). Not surprisingly, a 2009 World Bank study found that remittances tend to increase in the aftermath of a disaster (Karnik 2015). In the case of Nepal, some economists question whether the money will assist long-term projects beyond individual household reconstruction as remittances are often mainly used to support day-to-day needs. However in the absence of adequate government and international response to the earthquake aftermath, the private response through social networking, volunteerism and social consciousness is to some extent filling the public vacuum (Glencourse and Shakya 2015). Villages with good social capital, i.e. extended family members who can assist or connections to government officials are most likely to receive emergency and recovery assistance. At the same time, widespread access to social media is creating a push for greater transparency of the distribution of assistance, possibly leading to a movement towards greater government accountability (Glencourse and Shakya 2015).

Mobility is as old as human kind and corresponds to one of the most commonly used strategies for human survival: the ability to move around freely to optimize our livelihoods opportunities (Diamond 2005). However in order to explain the terms used in this book, we give some definitions. Mobility is often used as a more general term to encompass movement and flows of populations, which can include daily commutes to nearby cities or faraway places. Mobility can include migration which involves the movement or change of residence from one place to another, either temporarily, seasonally or more permanently (UNDP 2009). A displaced person becomes a refugee when he or she crosses an international border, has a legitimate fear of persecution based on race, religion, political opinion or belong to a social group and is not protected by his or her country (UNHCR 2015a). People who are forced to move due to violence, political persecution or a disaster but who does not cross an international border is considered an Internally Displaced Person (IDP) and might present specific protection needs (IFRC 2012; UNDP 2010). According to IFRC (2012), there were 72 million "forced migrants" in 2012 (although it is not clear if this figure refers to refugees and IDPs, or only IDPs) and peaks of migration in 2014 and 2015 (UNHCR 2015b). Forced migration takes many forms and is a global phenomenon, such as in Haiti post-earthquake (Chap. 10) or post Indian Ocean tsunami (Chap. 12).

The 232 million international migrants in 2013 constitute the highest reliable number on record, albeit a relatively small percentage, 3.2 % of the global population. As a consequence of the international economic crisis in 2010, there was a reverse in the migration trend of increasing numbers of migrants towards the most affluent regions towards a greater increase in migration in less wealthy countries (OECD 2013). Thus over the past decade, the number of migrants has increased more rapidly in the South than in the North, and South-South migration slightly exceeds South-North migration (OECD 2013). According to the OECD (2013), refugees, a category that does not include those moving due to environmental stresses, account for 7 % of all international migrants, or 15.7 million, a relatively small proportion of the global migrant population.

Migration is often considered to stem from various push, pull factors and multiple additional obstacles and pathways or intervening factors, which are composed of a myriad of economic, political, social, demographic and environmental drivers (Black et al. 2011; Foresight 2011; Piguet and Pecoud 2011; Chaps. 8 and 15). Yet referring to migration as mainly the outcome of push, pull and intervening factors brings a risk of oversimplification as each of the push, pull and intervening factors may apply in one circumstance but not in another. Rather, Oliver-Smith (2012) stated the need to explore migration as a process explained by how well a society has progressed in achieving mutuality between society and environment. And disasters are at the core of this mutuality between society and environment (or lack of it).

Different types of migration are discussed in this volume. Displacement/resettlement is the type of migration most often linked with disasters and often involves temporary or internal migration as part of the immediate responses of the recovery process. Jahn et al. (Chap. 10) describe the aftermath of the earthquake in Port-au-Prince, Haiti, which displaced a large share of the local population within and outside the city, but also triggered movement of people towards the capital in search of improved livelihoods. Although this type of migration was supposed to be temporary, many people actually resettled in the capital city. Other types of migration discussed include migration as an adaptive strategy to environmental stress (Chaps. 9 and 13) or as socio-ecological adaptation (Chap. 14). Migration as an adaptation strategy will be discussed in more detail below. Finally, migration as a result of conflict and violence is discussed in this book to a much lesser extent, as this type of migration is more about governance than disasters and environmental causes per se. The line between social/political and economic migration is blurry and certainly an extremely timely and important subject of debate, but one that this volume does not cover, with the exception of the chapter on "disaster diplomacy" (Chap. 12), which touches upon the relation between disasters as possible opportunities for resolving political and social conflicts. And socio-political drivers of migration are usually accompanied by indistinguishable pre-existing economic and environmental vulnerabilities. Thus, most of the cases described in this book focus on examples of intertwined environmental and economic processes, which require populations to adapt and cope, often through migration, rather than political or conflict-driven migration.

2.2 Linkages Between Climate Change and Migration

There are increasingly clear linkages between climate change and migration. This is particularly the case of small island states (Campbell and Warrick 2014; Farbotko and Lazrus 2012; Chap. 15). The causes of the mobility of their populations are certainly more complex than a single push factor and deeply rooted in societal features such as history, culture, poverty, urban slum growth, water management, unsustainable management of natural resources and poor governance, albeit compounded and multiplied by the increasingly difficult environment conditions created by global warming (Adger et al. 2013). Some chapters allude to "environmental refugees", a definition originally proposed by El-Hinnawi in 1985. However, value, coverage and use of the term are contested (Black 2001; de Sherbinin et al. 2011; Lazarus 2011; Piguet 2013; Stojanov et al. 2014; Warner 2010).

Oliver-Smith (2012) has summarized this criticism from three perspectives: theoretical, legal and political. The theoretical objection lies with the causality of factors that lead to migration already alluded to in this introduction, as population movements are always driven by social, political and ecological variables, which are all impacted by environmental processes and change (Black 2001; Castles 2002; Kibreab 1997). As described above, a "refugee" has a specific legal status as defined by the 1951 United Nations Convention Relating to the Status of Refugees, which does not qualify people displaced by environmental causes as refugees. Finally, the use of the term is criticized for provoking fear-mongering with press predictions of thousands of immigrants flocking to the North (Morrisey 2012; Castles 2002). In parallel, Hartmann (2010) asserted that the use of "climate refugees" depoliticizes the economic and political causes of migration and obfuscates the role of governance in addressing migration issues. Although less polemic, the term "environmental migration" is more commonly used and could be seen as an outcome of the lack of mutuality between nature and society, as described by Oliver-Smith (2012). However, it is also criticized based on many of the same arguments relevant for the critique of "environmental refugees", mainly based on the consideration of the multiple causality of migration (Black et al. 2011; Oliver-Smith 2012; Stojanov et al. 2014). We find the following quote appropriate in summing up the discussion about migration and climate change: "[..] the key question in relation to climate change is less whether climatic variability or climate events will drive migration, and more whether climate change will increase climatic variability or the frequency of extreme events, with the result that future flows are likely to be larger than those currently experienced" (Black et al. 2011: 435). Finally, as reflected by the diversity of examples given in this book, we need to consider the multiple scales at which we consider migration—DRR–CCA interactions, not only at the international but also at the national and local scales.

Our second emerging theme is that human mobility and migration can have both positive and negative impacts on development, risk reduction and CCA (Black et al. 2011; Foresight 2011; Piguet 2010; Kaenzig 2015; Chap. 8). As mentioned above, better employment and education in the destination place may improve development opportunities, while it could lead to a transfer of risk if people move to areas such as urban slums, amplifying exposure in the place of migration (Chaps. 5, 6 and 11).

Collins (Chap. 8) provides an in-depth overview of migration as a *positive* coping and/or adaptation strategy to environmental and economic stresses. He distinguishes between adaptation and coping strategies as a function of the type of threat, and length of time required for a population to reduce risks of various sorts. Both require making adjustments to livelihoods based on decisions and choices following an appraisal of events and possible outcomes or consequences (Burton et al. 1993; Lazarus 2011). Coping strategies may be considered a more immediate response and possibly greater adjustment to a sudden stress, such as a disaster or a sudden economic crisis, based on experience of past events, while adaptation strategies are considered small adjustments to more slow onset and new/future stresses such as climate change or an economic recession.

Temporary migration or displacement in the aftermath of a hazard event or sudden shock are very common coping strategies, and could eventually lead to more permanent migration, as in the case of Hurricane Katrina, especially when people are faced with an accumulation of impacts and few other options (Chaps. 8 and 7). Slow-onset environmental crises compounded by climate uncertainty and economic/political instability, instead, often result in longer-term adaptation strategies and more permanent migration (Chaps. 9 and 11). This type of migration is not new and can be traced back to the beginning of mankind when environmental conditions no longer support human activities. Classic examples include the "Dust Bowl migration" in the 1930s in the United States, which was largely caused by poor soil conservation practices, yet compounded by a severe drought which led to permanent outmigration, mainly to California (Mc Leman 2006:331, quoted by

Black et al. 2011:433) a plight popularized by Steinbeck's (1937) *Of Mice and Men*. In Pakistan, this point has been clearly documented: drought resulted in more permanent international migration than flooding which causes more temporary internal migration (Plamer and Smith 2014; Mueller et al. 2014).

As migrants move temporarily or permanently, they bring with them their know-how, life experiences, potential for innovation and may be more adaptive to natural hazards in the place of destination (Chap. 6). Outmigration can also have empowering effects for those who stay behind: Upreti and Shrestha (Chap. 9) describe how migration in Nepal has given greater independence to women who have been given greater responsibility for managing households as such a large number of men have migrated out for work. In addition to the large remittance inflow to the country, international and urban migration from and within Nepal has also geographically expanded and enriched social capital and networks, a key factor underpinning the current post-earthquake reconstruction process.

Jacobi et al. (Chap. 14) describe research on migration from the Bolivian high plateau (*Altiplano*) to the Amazonian lowlands, and on the pressures posed by newcomers on local indigenous groups and biodiversity. To reduce the environmental drivers of migration, a number of *Altiplano* farmers converted to organic cocoa farming as a more adapted and sustainable practice to environmental conditions. This example also highlights the potential for transfers of risks from one place to another, as a negative outcome of migration. Similar examples are given by Pfeifer et al. (Chap. 4) where long-term migration in the 1980s and 1990s from the Sahel led to settlement along the coast on dunes, former wetlands and the tripling of Dakar's population, over-pumping for drinking water and irrigation for urban agriculture, and sanitation and water quality issues. Texier and Edelblutte (Chap. 6) describe the transfer of risks from rural to urban areas: the megacities of Mumbai and Jakarta are facing increasing flooding and pollution risks due to environmental factors such as growing impervious urban surfaces and lack of waste treatment, both of which are compounded by the unmanaged inflow of migrants to slum areas. Similarly, in Nepal, farmers continue to leave the less fertile hillside terraces to move to nearby and further away valley bottoms or urban areas where they are often forced to settle illegally along river fronts, increasing their exposure to flooding and disrupting social networks (Chaps. 11 and 13). All the above examples highlight how human mobility is a key dynamic in the complex interactions between natural processes and anthropogenic actions.

3 Conclusions: Towards Sustainable Development

We live in an increasingly complex world, which academia, with our disciplinary boundaries, is not always able to capture adequately (Gall et al. 2015). The genesis of this book was an interdisciplinary workshop organized by the University of Lausanne, Switzerland in December 2013, which brought together some of the most influential researchers, policy makers and practitioners from the fields of DRR,

CCA and migration from both the South and North. Some authors are among the most well-respected in the field of DRR and migration, others are junior researchers but with rich field experiences. Although each chapter explores inter-linkages between the book's themes, they are organized around three main entry points depending on authors' main perspectives: DRR, migration and CCA. Each section features conceptual chapters, empirical studies and case studies, some with policy recommendations.

The book includes a number of chapters from Nepal as the editors' research emanated from a project focusing on Nepal and Bolivia "Sustainable Land Management In Mountain Regions Of Bolivia and Nepal in the Context Of Outmigration, Climate Change And Disaster Risk Reduction", funded by the Swiss Network for International Studies. With many of this book's authors and editors having personally experienced the Nepal 2015 earthquake, we continue to observe the post-Nepal earthquake situation as it unfolds as a timely illustration of the disaster—migration nexus. This is a country with an already fragile balance between topography and livelihood options, with some groups privy to strong social networks and others remaining highly vulnerable primarily due to inequalities wrought by the caste system. With at least one third of all Nepali households which have at least one member working and living abroad, Nepal is one of the world's most remittance-dependent countries (United Nations Nepal Information Platform 2015). Post-earthquake, the population is experiencing a level of mobility similar to that seen during the Maoist insurgency. Migration from the rural areas is likely to increase exponentially, creating new challenges but perhaps also solutions for the reconstruction process. It is also an example of "disaster diplomacy", as illustrated by Kelman (Chap. 12), where the country's new constitution was finally signed post-earthquake after a decade of political stalemate, giving promise to a more sustainable political and economic future. As governments are tasked with translating international frameworks—whether DRR or CCA-related—into action, the hope is that this volume will provide timely "food for thought" for integrating migration and sustainable development concerns into more comprehensive DRR policies and practices.

References

Adger WN, Barnett J, Brown K, Marshall N, O'Brien K (2013) Cultural dimensions of climate change impacts and adaptation. Nat Clim Chang 3(2):112–117

Black R (2001) Environmental refugees: myth or reality? New issues in refugee research. UNHCR Research Paper 34, p 20

Black R, Kniveton D, Schmidt-Verkerk K (2011) Migration and climate change: towards an integrated assessment of sensitivity. Environ Plan 43(2):431–450. doi:10.1068/a43154

Burton I, Kates RW, White GF (1993) The environment as hazard, 2nd edn. Guilford Press, New York

Campbell J, Warrick O (2014) Climate change and migration issues in the Pacific. UNESCAP, Fiji

Castles S (2002) Environmental change and forced migration: making sense of the debate. New Issues in Refugee Research. UNHCR Research Paper 70, p 16

CRED (2015) EM-DAT database. CRED and Université de Louvain, Louvain-la-Neuve

de Sherbinin A, Castro M, Gemenne F, Cernea MM, Adamo S, Fearnside P, Krieger G, Lahmani S, Oliver-Smith A, Pankhurst A, Scudder T, Singer B, Tan Y, Wannier G, Boncour P, Ehrhart G, Hugo G, Pandey B, Shi G (2011) Preparing for resettlement associated with climate change. Science 334(6055):456–457

Diamond J (2005) Collapse: how societies choose to fail or succeed. Viking, New York

El-Hinnawi E (1985) Environmental refugees. United Nations Environment Programme, Nairobi

Farbotko C, Lazrus H (2012) The first climate refugees? Contesting global narratives of climate change in Tuvalu. Glob Environ Chang 22(2):382–390

Foresight (2011) Migration and global environmental change. Future challenges and opportunities. Government Office for Science, London

Gall M, Nguyen KH, Cutter SL (2015) Integrated research on disaster risk: is it really integrated? Int J Disaster Risk Reduct 12:255–267

GDNR (2015) Everyday disasters. http://www.gndr.org/programmes/frontline-programme/everyday-disasters.html. Accessed 13 Aug 2015

Glencourse, B., S. Shakya, (2015) Shaking Up the Status Quo in Nepal, in International New York Times. June 1, 2015: New York. http://www.nytimes.com/2015/06/02/opinion/shaking-up-the-status-quo-in-nepal.html?_r=0

Hartmann B (2010) Rethinking climate refugees and climate conflict: rhetoric, reality and the politics of policy discourse. J Int Dev 22:233–46

IFRC (2012) World disasters report. 2012. http://www.ifrcmedia.org/assets/pages/wdr2012/

IOM (2015) Earthquake forces Nepal's overseas workers to make tough choices. https://www.iom.int/migrant-stories/earthquake-forces-nepals-overseas-workers-make-tough-choices. Accessed 18 May 2015

IPCC (2014a) Climate change 2014: impacts, adaptation, and vulnerability. Part A: Global and sectoral aspects. In: Field CB, Barros VR, Dokken DJ, Mach KJ, Mastrandrea MD, Bilir TE, Chatterjee M, Ebi KL, Estrada YO, Genova RC, Girma B, Kissel ES, Levy AN, MacCracken S, Mastrandrea PR, White LL (eds) Contribution of working group II to the fifth assessment report of the Intergovernmental Panel on Climate Change. Cambridge University Press, Cambridge, UK, p 1132

IPCC (2014b) Climate change 2014: impacts, adaptation, and vulnerability. Part B: Regional aspects. In: Barros VR, Field CB, Dokken DJ, Mastrandrea MD, Mach KJ, Bilir TE, Chatterjee M, Ebi KL, Estrada YO, Genova RC, Girma B, Kissel ES, Levy AN, MacCracken S, Mastrandrea PR, White LL (eds) Contribution of working group II to the fifth assessment report of the Intergovernmental Panel on Climate Change. Cambridge University Press, Cambridge, UK, p 688

Kaenzig R (2015) Can glacial retreat lead to migration? A critical discussion of the impact of glacier shrinkage upon population mobility in the Bolivian Andes. Popul Environ 36 (4):480–496

Karnik M (2015) Billions of dollars from overseas workers will be key to rebuilding Nepal's ravaged economy. Quartz, India. http://qz.com/392011/billions-of-dollars-from-overseas-workers-will-be-key-to-rebuilding-nepals-ravaged-economy/. Accessed 28 Apr 2015

Kibreab G (1997) Environmental causes and impact of refugee movements: a critique of the current debate. Disasters 21(1):20–38

Lazarus NW (2011) Coping capacities and rural livelihoods: challenges to community risk management in Southern Sri Lanka. Appl Geogr 31(1):20–34

McLeman R (2006) Migration out of 1930s rural Eastern Oklahoma: insights for climate change research. Great Plains Q 26(1):27–40

Morrisey J (2012) Rethinking the "debate on environmental refugees": From "maximalists and minimalists" to "proponents and critics". J Polit Ecol 19:36–49

Mueller V, Gray C, Kosec K (2014) Heat stress increases long-term human migration in rural Pakistan. Nat Clim Chang 4:182–185

OECD (2013) International migration outlook 2013. OECD Publishing, Paris. doi:10.1787/migr_outlook-2013-en

Oliver-Smith A (2012) Debating environmental migration: society, nature and population displacement in climate change. J Int Dev 24(8):1058–1070

Piguet E (2010) Linking climate change, environmental degradation and migration: a methodological overview. Wiley Interdiscip Rev Clim Chang 1(4):517–524

Piguet E (2013) From "primitive migration" to "climate refugees"—the curious fate of the natural environment in migration studies. Ann Assoc Am Geogr 103(1):148–162

Piguet E, Pecoud A (2011) Migration and climate change. Cambridge University Press, Cambridge

Plamer, P.I. and M.J. Smith (2014) Earth Systems: Model human adaptation to climate change, Nature 512, 365–366 (28 August 2014) doi:10.1038/512365a

Steinbeck, J. (1937) Of Mice and Men, New York, N.Y., USA, Penguin Books, 1994

Stojanov R, Kelman I, Shen S, Duží B, Upadhyay H, Vikhrov D, Lingaraj GJ, Mishra A (2014) Contextualising typologies of environmentally induced population movement. Disaster Prev Manag 23(5):508–523, http://dx.doi.org/10.1108/DPM-09-2013-0152

UNDESA (2015) Focusing on new trends in migration. http://www.un.org/en/development/desa/news/population/cpd46-trends-in-migration.html. Accessed 5 July 2015

UNDP (2009) Human Development Report 2009: Overcoming Barriers: Human Mobility and Development. Basingstoke, Palgrave, Macmillan

UNDP (2010) Mobility and Migration. Human Development Report Office. A Guidance Note for Human Development Report Teams. UNDP, New York, p 32

UNHCR (2015a) The 1951 refugee convention. http://www.unhcr.org/pages/49da0e466.html. Accessed 6 July 2015.

UNHCR (2015b) UNHCR statistical yearbook, 2013. 25 Years of global forced displacement. UNHCR, Geneva, p 13. http://www.unhcr.org/54cf99109.html

UNISDR (2014) Annual report 2014. Geneva, Switzerland: United Nations Office for Disaster Risk Reduction (UNISDR)

UNISDR (2015) Making development sustainable: the future of disaster risk management. Global assessment report on disaster risk reduction. United Nations Office for Disaster Risk Reduction (UNISDR), Geneva, Switzerland

United Nations Nepal Information Platform (2015) Migrant workers and their families. http://un.org.np/oneun/undaf/migrant_workers. Accessed 12 June 2015

United Nations News Centre (2015) Fleeing conflicts, record numbers of migrants and refugees crossed Mediterranean so far in 2015—UN. http://www.un.org/apps/news/story.asp?NewsID=51315#.VZwJ3Ea52_c

Warner K (2010) Global environmental change and migration: governance challenges. Glob Environ Chang 20(3):402–413

Chapter 2
Human Mobility in a Socio-Environmental Context: Complex Effects on Environmental Risk

Lorenzo Guadagno

1 Introduction

Mobility has been a constant feature of human communities and societies in any place and time. From the worldwide spread of the early *Homo sapiens* groups, all the way to modern migration flows in response to industrialization, urbanization and globalization processes, human mobility has shaped the world, underpinning the circulation of ideas, knowledge, and goods on short and long distances.

The magnitude of today's population flows is however unprecedented. There are in the world around one billion migrants (UNDESA 2013; UNDP 2009), and many more people move on short distances and on a temporary basis—their movement largely elusive to mobility tracking systems (Tacoli 2013). Mobility contributes to determining global and local distribution of people; location, size, density, and composition of communities; and individual access to livelihood and well-being options (Schensul and Dodman 2013). It is a key cultural, social, economic dynamic of our modern societies, both a consequence and a driving force of modernity itself (Castles and Miller 2009).

Population flows are an essential feature of the continuum of human interactions with the environment (Sanderson 2009). It is through staying in and moving into places that people access environmental resources and opportunities, and are exposed to hazardous events and processes (Wisner et al. 2012). On the other hand, population flows, and the circulation of material and immaterial resources they induce, contribute to reproducing or challenging the socio-environmental relationships that determine access to those opportunities and hazards (Castles

L. Guadagno (✉)
International Organization for Migration, 17, Route des Morillons, CH-1211 Geneva 19, Switzerland
e-mail: LGuadagno@iom.int

© Springer International Publishing Switzerland 2017

K. Sudmeier-Rieux et al. (eds.), *Identifying Emerging Issues in Disaster Risk Reduction, Migration, Climate Change and Sustainable Development*,
DOI 10.1007/978-3-319-33880-4_2

and Miller 2009). Mobility is at a time the expression and a determinant of spatial hierarchies of opportunities and risks (Skeldon 2008), including environmental ones.

Human mobility is embedded in a number of processes, including environmental change, conflicts, economic growth and crises, political and cultural evolutions taking places at all geographic scales (Black et al. 2011; Castles 2010). But it is through localized socio-environmental transformations that it is determined—and determines most of its effects on risk. Mobility transforms risk landscapes by linking places into translocal geographies (Brickell and Datta 2011). Population and resource flows decouple people's agency from localized socio-environmental conditions, creating a context for individual choices and opportunities that is rooted on and influenced by broader and more or less distant relations and dynamics (Greiner and Sakdapolrak 2012). As a consequence, social and environmental relations in the sites of origin and of destination are transformed. Mobility is therefore a key dynamic to the socio-environmental relations that produce risk.

Disaster risk reduction (DRR), the set of efforts aimed to analyze and reduce the factors that determine the human, material, economic, and environmental impacts of hazards, provides a specific theoretical and operational look on these relations, and the way mobility transforms them.[1] DRR aims to understand how socio-environmental processes produce or reduce vulnerability and risk: this requires understanding human mobility's role in shaping exposure, vulnerability, and resilience to environmental hazards. On the other hand, DRR aims to prevent these processes from resulting in disasters: this requires creating the conditions for mobility decisions to empower people to better avoid, absorb, and recover from, shocks and stresses.

While a number of theoretical perspectives on population movements provide useful elements to a risk reduction discourse, a specific, encompassing DRR perspective seems to be largely missing from the debate on migration, displacement, relocations, and other population movements. Some human mobility issues have recently been integrated in the Sendai Framework for Disaster Risk Reduction 2015–2030 (SFDRR). However, a comprehensive look on mobility as an underlying dynamic of risk is still absent from the DRR policy and operational agenda. This paper, based on the review of existing literature on migration, development and the environment, attempts to propose a theoretical framework to interpret mobility from a DRR angle. This should help identify some key questions and recommendations that could be useful to better integrate understanding and management of human mobility into DRR policy, research, and practice.

[1] For the aim of this paper this definition includes Climate Change Adaptation (understood as a subset of actions to reduce the impacts to a specific range of hazards).

2 A Livelihood Approach to Mobility and Risk

Livelihood strategies provide a useful key to interpreting the linkages between risk and mobility.[2] Livelihood choices are based on individual capacities to access human, social, physical, financial, and natural resources, but are part of complex well-being strategies defined at the level of more or less large households. The options concretely available to individuals and households are constrained by social and environmental factors: legal and political frameworks, economic dynamics, cultural specificities, and ecosystem features determine whether people are allowed or denied access to capital and opportunities, and define the boundaries of the choices they have concretely available (Sen 2000).

DRR theory (Wisner et al. 2004) looks at access to resources and livelihood strategies to determine where and how people work and live and the amount of material and immaterial resources they are able to draw upon to avoid or face a shock. Livelihood choices determine people's likelihood to be impacted by hazards, the degree of damage they are likely to sustain, and their capacity to cope with, and recover from, shocks.

The livelihood perspective is just as central for the mobility discourse. Understanding mobility choices in the context of household-level livelihood strategies helps highlight how they result from the different set of aspirations, opportunities, and constraints with which different people are faced (Mc Dowell and De Haan 1997). This perspective overcomes more deterministic approaches based on costs and benefits or on push and pull factors of movement, and reduces the need for defining categories of mobile people and population movements: "forced" and "voluntary" moving and staying are different outcomes of the same multi-causal decision-making process aimed to maintain and improve collective well-being, based on the integral consideration of existing opportunities and constraints (Black et al. 2012; de Haas 2008a, b).

The collective dimension of livelihood strategies is particularly relevant for the integration of mobility and risk reduction. Mobility choices are part of household-level decision-making processes, as they are pursued for collective projects and purposes, based on mutual obligations and commitments. Mobility allows for resource diversification, livelihood strengthening, and risk management within the household (Stark and Bloom 1985). At the same time, population movements modify the individuals' availability of resources and opportunities, producing different outcomes for different people in locations of origin, transit and destination, and beyond (Stark 1991). It is therefore difficult to single out "positive" and "negative" forms of mobility; mobility choices should rather be investigated to

[2] A livelihood comprises the capabilities, assets, and activities required for a means of living. It is sustainable if it can cope and recover from stresses and shocks and maintain capabilities and assets, while contributing net benefits to other livelihoods in the long and short term (Chambers and Conway 1992).

understand how their positive and negative effects are distributed and interplay with existing patterns of vulnerability and risk.

Two main analytical implications stem from applying a livelihood-centered approach to the study of mobility and risk: (1) the need to look at all mobility and immobility decisions and patterns for their potential of creating or reducing risk; and (2) the need to adopt a comprehensive view on risk outcomes of these decisions. These considerations inform the theoretical framework illustrated in the next section.

3 Mobility and DRR: A Theoretical Framework

Figure 2.1 outlines the progression from livelihood choices to risk dynamics, highlighting mobility patterns and related resource flows, as well as their potential outcomes that are relevant for risk reduction at different levels. The model depicts a one-on-one linkage between two socio-ecological systems. Even though in reality multiple mobility decisions are likely to create a network of population and resource flows among interconnected locales, the inclusion of more than two sites in the diagram would not add perspective to the identification of mobility-related risk outcomes.

The model highlights how, within each system, mobility choices are embedded in households' livelihood strategies. Resources, opportunities, and constraints people base their choices on depend on demographic, cultural, social, political, economic, and environmental features of the system they live in. These features, specific to any given historical and geographical context, are influenced by large-scale dynamics, such as capital accumulation, crises, structural reforms, and globalization of economies and cultures. All these dynamics interact to multiply or limit the resources and opportunities that are accessible to different households and individuals. Global environmental change, including climate change, is one among these dynamics, likely to have significant impacts on people's well-being options through its interplay with local socio-economic structures and processes.

The centrality of environmental features in mobility choices has been highlighted by the research on migration and the environment, which has mainly focused on environmental drivers of mobility and environmentally driven movements (among others: Foresight 2011; McLeman 2014; Piguet 2010). However, findings of this body of work show that environmental dynamics and economic, political, and social structures always interact to determine people's decisions. Looking at community and system-wide structures and processes is therefore necessary to understand mobility patterns and outcomes (Castles 2002; Felli and Castree 2012; Zetter and Morrissey 2014).

Research also shows that population movements are never the inevitable, nor the inevitably negative consequence of stresses and change. Before, during, and after events of any kind, mobility behaviors inevitably differ (McDowell and de Haan 1997). This is the case even in sudden-onset, catastrophic disasters, which may

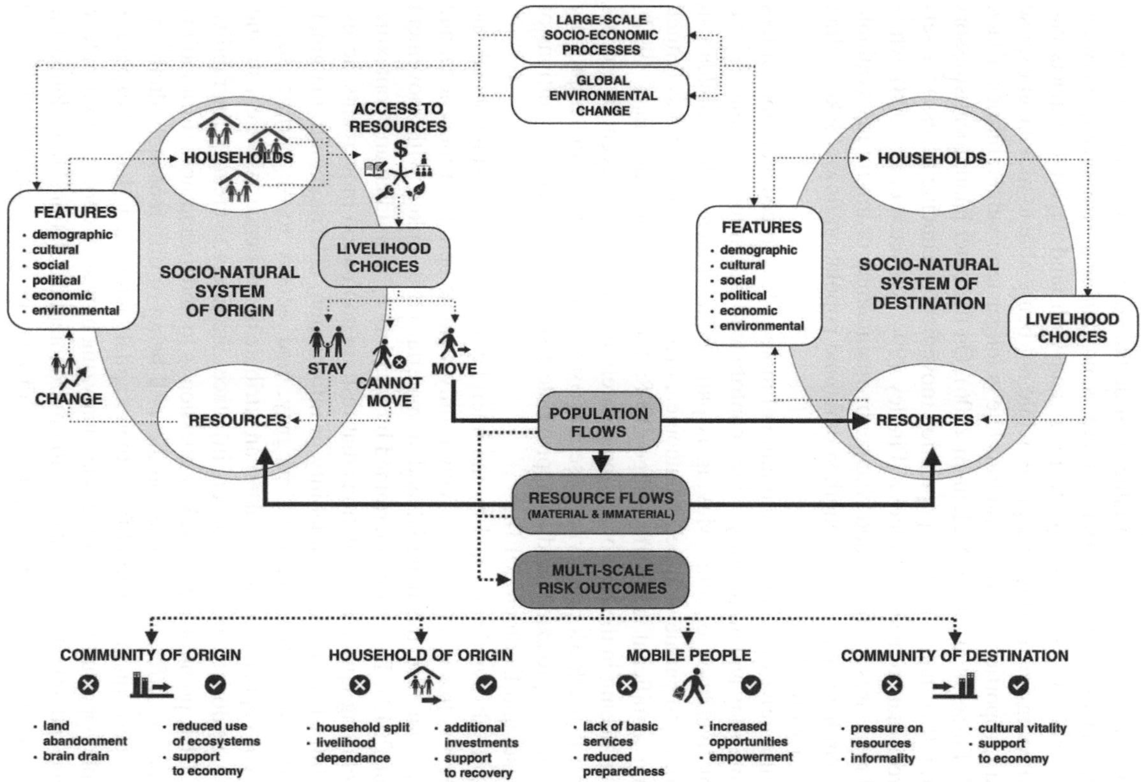

Fig. 2.1 Livelihood choices, translocality, and risks. Source: author's own elaboration. Icons: UNOCHA

produce intense shifts in local resource availability and affect in similar ways a sizable number of people, thereby inducing a massive, concentrated mobility response: some people are free to stay, some people move, some people cannot do so.

Both the need and the capacity to move of different people depend on the same social, economic, and political variables that define vulnerability and resilience to disasters (de Haas 2008a, b). Hence there is no univocal relationship between shocks and mobility. The capacity to resist and withstand negative impacts can translate in being free to stay in a place, while not being able to move can expose people to disproportionate losses and translate in longer-term vulnerabilities (Black et al. 2012; Fussel et al. 2010; McLeman 2006). On the other hand, local stressors can hinder, rather than stimulate, population movements, reducing access to assets and resources that are needed to move (Findley 1994; Kniveton et al. 2008), while successful responses to shocks and change, by maintaining or improving available resources, can help sustain or strengthen outbound population flows (Sakdapolrak et al. 2014; Tacoli 2009).

These alternatives lie along a continuum of more or less forced and more or less voluntary mobility and immobility, which contribute to determining people's in situ and distant livelihood options. Moving is part of the strategies households can pursue in order to pursue well-being outcomes, in a context of more or less limited choices and significant tradeoffs (Hugo 2008; McDowell and de Haan 1997). In fact, even fleeing in the face of an immediately life-threatening event or process is an option that allows preserving essential assets and resources and improving otherwise bleaker prospects, and an appropriate response to specific conditions of risk (Schensul and Dodman 2013).

While it is always determined by the interplay of local and distant social and ecological pressures and factors, moving is only a viable option for those local and distant who can face the material and immaterial costs linked with movement (Black et al. 2012). These are determined by a variety of individual and contextual factors, including laws and regulations that restrain population movements, accessibility of infrastructure and availability of information, discrimination, marginalization and xenophobia, physical integrity, and access to supporting social networks. The capacity to move is itself an attribute of resilience, while constraints to people's freedom to move to (or stay in) places are likely to create vulnerability.

From a DRR point of view, the key question is to understand how (im)mobility choices produce different outcomes on the well-being of different individuals and households, perpetuating or modifying their capital endowment, hazard exposure, and available opportunities. This includes analyzing how such choices contribute to reproducing and transforming the socio-environmental relations that define the systems' features, influencing local and distant people's opportunities and constraints.

The transformative effect on these system-wide features is not unique to mobility-based livelihood strategies. Mobility's distinct transformative potential lies in the establishment of relations of circulation of people and resources among systems and locales. Figure 2.1 provides a simplified view of mono-directional flows of people, which, in turn, produce bi-directional flows of material and immaterial resources. In reality, each individual movement has specific features (e.g. short or long-term, circular or permanent, from households sending away one or more members) and multiplies or reduces further mobility opportunities. However, the model highlights that movements of all kinds lead to similar risk outcomes at the different levels considered.

All population movements modify the distribution of population and the availability and access to resources and opportunities for different actors. Individual and contextual features that influence the level of freedom of people's mobility choices determine what the concrete risk outcomes of these processes will be. Research has concentrated on four main levels where consequences are likely to be produced: those moving, their household of origin, their system of origin, and their system of destination. Vulnerability and resilience can be useful heuristic tools in interpreting and comparing the complexity of these transformations.

The experience presented in the following sections shows that the outcomes at each level can be positive or negative, and that risk production and reduction can take place simultaneously within and across different levels. DRR requires therefore looking at the full spectrum of mobility patterns, as well as related flows of material and immaterial resources, across locales, to understand what is their overall effect on the production and reduction of risk for those on the move, as well as for host and home communities. Without adopting this comprehensive perspective, DRR actions, and in particular those aiming to manage human mobility in ways that minimize existing and potential vulnerabilities, can lead to redistribution, rather than reduction, of disaster risk.

4 Population Flows and Disaster Risk

4.1 Moving as a Risk Dynamic for Those Moving

As people move into a different landscape, they are confronted with different environmental features, including potentially hazardous events and processes. Studies that have looked at the changes in hazard exposure as a consequence of population movements have highlighted that flows often originate from fragile, resource-scarce ecosystems where hazards are frequent. The overall effect of movement is therefore not necessarily an increase in the number of people living in hazardous areas. However, population flows seem to be concentrating people in specific locations, some of which are becoming "hotspots" of disaster risk—in particular, fast-growing urban areas in regions highly exposed to hazards or facing significant impacts from environmental change (de Sherbinin et al. 2012; Runfola et al. 2013).

Sheer exposure to hazards, though, is insufficient to explain risk outcomes. Access to capital and opportunities are modified by mobility, and so are options and constraints that underpin people's livelihood strategies (and, as a consequence, their vulnerability and resilience). It is not the mere fact of moving that determines these outcomes; rather, it is structural features of the systems from which and to which movement takes place that determine how mobility translates into different opportunities and hazards for different people. People's origin and destination, and the legal, cultural and socio-economic barriers they encounter in moving from one to the other, matter in determining their levels to access to services and resource in receiving communities (de Haas 2008a; IOM 2013).

People generally move for the prospects of a safer, better life, and for most mobile people movement indeed results in overwhelmingly positive outcomes. By drawing on a translocal pool of resources and opportunities, including those built through previous movements, mobile people are often able to enjoy improved access to goods, services, and opportunities compared to other household members staying behind (De Moor 2011; IOM 2013; UNDP 2009). Moving can help them challenge traditional social roles and constraints and have a distinct empowering effect (de Haas 2008b), and has the potential to increase their overall satisfaction and well-being (Bartram 2013).

However, moving almost inevitably implies being disendowed of certain forms of capital (associated with, e.g. loss of local knowledge, social networks, linguistic proficiency) (Manole and Schiff 2004) and encountering a new set of boundaries (stemming, for instance, from restrictive migration regimes, discriminating policies, xenophobic stances), which can limit available well-being options (IOM 2015). As a consequence, those moving are more likely than native-born to experience insufficient access to basic services, reduced assistance by formal and informal systems and networks and lack of personal and financial security (Adams et al. 2010; Duong et al. 2011; IOM 2013; Ku and Jewers 2013; NESSE 2008; UNDP 2009). This includes increased exposure and vulnerability to natural and man-made hazards.

Limited capital endowment and increased barriers can translate in specific vulnerabilities in disaster situations, in particular through reduced awareness and preparedness due to linguistic barriers, limited trust in risk management actors and knowledge of hazard occurrence and of contingency plans and evacuation procedures, or reduced access to emergency and recovery assistance (Bolin and Stanford 1998; Perry and Mushkatel 2008; Phillips 1993; Wang et al. 2011). On the other hand, mobile people's differential capacities and resources might make them more resilient than people from their host communities: non-native groups have occasionally showed better capacities to prepare for, respond to, and recover from, disasters (Clerveaux et al. 2008; Vu et al. 2009).

Mobile people's well-being and vulnerability are rooted in continued exchange of material and immaterial resources with the household and community of origin. People at home can help manage investments, take care of dependents left behind, deal with bureaucracy or send goods and resources that support consumption and increase food security, including during crises (Frayne 2004; Long 2008; Ratha and Sirkeci 2011). Social capital, strengthened and maintained through these mutual transfers, is essential to the well-being of mobile people. However, ties to a distant household can also represent a source of obligations for distant members, limiting their opportunities and resources and potentially making them more vulnerable (Hammond 2011).

4.2 *Moving as a Risk Dynamic for Sending Households*

From the point of view of the sending household, mobility is usually a financially and psychologically costly strategy that has the potential for high well-being returns. The sending household has to invest resources to send and (especially at early stages) support one or more members to a distant location. In addition, its members have to bear the impacts linked with the absence of one or more members and with the modification of the household's structure and capacities. At the same time, if the strategy is successful, mobile people can greatly contribute to the well-being of sending households, in particular through transfers of social and financial remittances. Such transfers multiply and diversify assets and resources available to the family members, allowing for more flexibility in livelihood strategies and for improved management of risks and insecurity, and can ultimately result in enhanced resilience (Le De et al. 2013; Stark and Bloom 1985).

Financial remittances are used both to satisfy short-term needs and to support longer-term savings and investments. They are linked with increased level of consumption, lower incidence of poverty and better access to essential services (such as health and education) for the recipients (Ratha 2013; San Vincente Portes 2009; UNDP 2009; Valero-Gil 2008). Households receiving remittances tend to have more resources to invest in productive assets, better housing and means of transportation, and improved access to information and communication networks (de Haas 2006; Mohapatra et al. 2009). All these expenditures can help reduce the recipients' levels of risk, by improving their food security and health conditions, and making them more prepared and less exposed to future hazards.

Spending, both to satisfy immediate needs and to support collective socio-cultural practices, is also instrumental to improving the households' political status within the community (Dalisay 2008; Le De et al. 2014). Inflow of remittances can make households more self-reliant, contributing to empowering its members, which can make them better able to challenge their social status (Cohen 2011).

Resources transferred through remittances can be used to build up savings, which can in turn be used to cope with unexpected events, including natural hazards (Mohapatra et al. 2009; Yang and Choi 2007). Following shocks, remittances represent a flexible tool to smoothen consumptions and support reconstruction and recovery (Attzs 2008; Fagen 2006; Le De et al. 2014).

On the other hand, most households send out healthy, productive individuals (IOM 2013). As breadwinners leave, traditional livelihood patterns can be downsized or disrupted and remittance inflow can change the lifestyle of the individuals left behind in ways that reduce their self-reliance (Edward and Scott 2003; Zachariah and Rajan 2004). As a consequence, households can find themselves overly dependent on the transfer of resources from distant members—and thereby threatened by the hazards the latter are exposed to.

In addition, the individuals' psychological well-being and personal security can be negatively affected in households split as a consequence of mobility, and many of those staying behind may end up exposed to increased social risks (e.g. marginalization and violence) (Asis 2008; Dreby 2010; d'Emilio et al. 2007). However, reconfiguration of family patterns can also have beneficial effects: social networks can be enlarged (Asis 2008) and gender roles challenged, in particular as female stayers take a more prominent role in household decision-making and economic management (Deshingkar and Grimm 2005; King and Vullnetari 2006).

4.3 Mobility as a Risk Dynamic for the Home System

Impacts of population and resource flows on the system of origin are complex and affect the whole set of its intertwined human and natural components. Environmental impacts can consist of both intensification and disintensification of land use practices, and improve, reduce, or leave unchanged the levels of ecosystem health and hazard occurrence.

Reduction of population pressures can lead to the abandonment of intensive land-use practices, with positive impacts on slope stability, erosion, water availability, and sedimentation levels (Deshingkar 2012; Jokisch 2002; Schwilch et al. Chap. 11). However, reduction of available human capital can reduce the communities' capacity to maintain infrastructures (e.g. terraces, water catchment, and irrigation systems) that are essential for the functioning of traditional cultural landscapes. This can result in biodiversity loss, increased incidence of landslides, floods, fires, avalanches, soil erosion, and desertification, and ultimately reduced food, water, and livelihood security (Raj Khanal and Watanabe 2006; Rey Benayas et al. 2007).

As additional resources flow into remittance-receiving households, communities can engage in environmentally unsustainable behaviors (Gray 2009; Robson and Nayak 2010). Increased use of land stemming from development of residential housing can lead to environmental degradation and increased hazard incidence (Klaufus 2010). Resources can also be invested in conservation activities (Konseiga 2004) and intensification of agricultural practices can lead to the concentration of environmental pressures to limited areas, allowing for ecosystem recovery in areas left uncultivated (Deshingkar 2012; Vanwey et al. 2012).

Population and resource flows are also key to the perpetuation and evolution of cultural, social, and political traits of communities and societies (Asis 2008; Le De et al. 2014). People's movement can underpin exchange of knowledge and ideas, increasing available human capital and circulating technologies and skills that support hazard prevention and mitigation, sustainable resource use, improved health or livelihood strengthening and diversification (de Haas 2006; Rinke 2012; UNDP 2009). These dynamics can also help challenge class and ethnic roles, helping subaltern groups escape from traditional social constraints (Cohen 2011; de Haas 2008a, b).

The inflow and spending of remittances can support the whole economic system, while population movements can stimulate or strengthen commercial relations between home and host communities (Lucas 2005). Consequent expansion of investment and local markets has the potential to reduce overall poverty levels, including for households not receiving remittances (Cohen 2011; Jongwanich 2007).

Direct engagement of diaspora members can effectively reduce vulnerability in home communities: water management, health and education, food security, resource conservation, and post-disaster recovery have been financed through collective support mechanisms or otherwise supported by migrants and returnees (Le De et al. 2014; Orozco 2007; Scheffran et al. 2012). Public and private institutions in home countries have leveraged diasporas' investments and savings to promote economic and social development, including risk reduction (Akkoyunlu and Stern 2012; Orozco 2008). Diaspora groups in receiving countries have been involved in designing and implementing development activities in areas of origin, with the added benefit of achieving better inclusion of newcomers in host communities (Østergaard-Nielsen 2011; Sall 2005).

These transfers of resources, though, have limited potential to support system-wide development and to tackle the fundamental economic processes that produce poverty and vulnerability (de Haas 2008a, b). Remittances do not necessarily reduce income inequalities in the community of origin. In fact, they can entrench them by concentrating resources in the hands of more affluent households or leading to currency devaluation and inflation, to the detriment of non-receiving households (Mazzuccato 2008; Ratha 2013). This can result in increased marginalization of non-receiving households, and ultimately in conditions of vulnerability to future hazards (Deshingkar and Aheeyar 2006). Unbalances embedded in the sender/receiver relationship can also reproduce existing social inequalities, including those linked with gender and age (Kunz 2011).

Demographic changes induced by mobility flows, in particular through the disproportionate loss of healthy, skilled, productive individuals, can cause significant socio-economic disruption of home communities and societies. Increased dependency rates, as the proportion of old and infirm people augments, can reduce the effectiveness of public welfare systems and kin and community-level care-giving (Bernhard et al. 2009; King and Vullnetari 2006). This also has implications on the effectiveness of hazard preparedness and disaster assistance. Loss of work-force can also lead to systemic fiscal and economic consequences, and can affect availability and quality of essential services, such as health and education, as well as of hazard-resistant housing for those staying behind (Docquier et al. 2010; Tasan-Kok and Stead 2013).

4.4 Mobility as a Risk Dynamic for the Host System

Population inflows also affect the economic and social life of communities of transit and destination: mobile people increase available labor supply, stimulate local economy, and contribute to maintaining functioning fiscal and social insurance systems as well as care-giving arrangements—an effect that is especially important for countering increasing dependency rates in ageing societies (Carter 2008; Ortega and Peri 2009). Mobile people can integrate the receiving countries' skill gaps (Farrant et al. 2006; Manole and Schiff 2004), which can be especially important in the aftermath of disasters and crises, when incoming workers support relief, reconstruction, and recovery efforts (Hugo 2008). Their presence is also positively related to the receiving societies' capacity for political, technological, and cultural innovation (Ratha et al. 2010).

At the same time, demographic increase translates into additional pressures on the system of destination, as well as on all areas that supply its community with resources and services. These pressures can result in reduced access to essential services and opportunities if local markets and institutions are unable to absorb increasing demand. Sudden, massive movements (including those linked with disasters and man-made crises) are particularly challenging.

Unsustainable use of ecosystems can lead to environmental degradation, increasing hazard occurrence and affecting mobile people's and host population's food and water security (McGranahan and Tacoli 2006; UNEP 2011). Unmanaged pressures on labor and housing markets, health and education systems as well as water supply, sanitation and waste management infrastructures, can negatively affect people's safety and well-being. Increased competition for scarce resources is likely to disproportionately affect weaker individuals, most often women, youth and elderly, and unskilled workers (UNDP 2009). Unmanaged population inflows also have the potential to reduce the community's collaboration and cohesion (Freire and Xiaoye 2013) and fuel inter-communal tensions and conflicts, in particular in periods of hardship such as disasters, economic downturn, and humanitarian crises.

5 Conclusions and Recommendations

The concrete outcomes of human mobility largely depend on the features of contexts of origin and of destination, including distribution of hazards and opportunities and prevailing patterns of (and barriers to) access to resources and services, as well as on the socio-economic situation of migrants and their households. Mobility, however, transforms all these features across sites and scales. It is hence both a product and a determinant of the socio-natural context people live in, integral to human development processes that shape hazard exposure, vulnerability, and resilience. Mobility poses specific challenges and offers specific opportunities to risk reduction and should be integrated in the holistic perspective that characterizes DRR.

The risk outcomes from moving are profoundly heterogeneous for different mobile and non-mobile people in communities of origin and of destination. Positive and negative changes are inextricably intertwined: all, however, matter for risk reduction and cannot be understood in isolation. It is therefore necessary to look at the whole range of individual, household, community and system-level features, in contexts of origin and of destination that are modified by human and resource flows induced by mobility decisions, and understand how change reflects onto people's risk levels. Thinking in terms of risk and risk reduction allows to interpret this multi-sited complexity in a coherent (and potentially comparable) way: this might be the inherent value of mobility-cum-DRR research.

Human mobility is a dynamic that influences individual and collective vulnerability and capacities. As such, it should be a key concern for risk management, as well as for all policies and actions that can produce or reduce risk. Land use planning, urban and economic development, delivery of basic services and ecosystem management affect mobility decisions of more or less distant individuals: reducing risks through such activities requires factoring in their mobility consequences, as well as people's movements' impacts on risk. This requires adding one more dynamic perspective to DRR, which overcomes sites, actors and time boundaries.

On the other hand, exploring human mobility from a DRR angle shows that policies and measures that facilitate, impede, regulate or manage population movements produce different risk outcomes for different people. Such policies and measures translate into increased or reduced needs and access to capital and opportunities (and ultimately in reduced or increased risk conditions) for those moving, as well as for those living in home and host communities.

It is important to highlight that all forms of mobility (not only disaster-related or induced) matter to risk reduction: whether planned or unplanned, forced or voluntary, population movements transform the risk landscape of different individuals and communities, influencing the way they are impacted by hazards. Stakeholders, procedures, and responsibilities might differ depending on the kind of movement that concretely needs to be managed. However, preserving and improving access to assets and opportunities for those moving, while preventing negative outcomes and producing positive ones for home and host households, communities and systems, are equally the objectives of efforts to manage migration, evacuations and displacement, and planned relocations and resettlements.

These considerations raise a number of theoretical questions:

- What is the overall impact of mobility in terms of risk reduction and risk creation? How can costs and benefits be measured and compared?
- Does human mobility challenge or reproduce existing conditions of vulnerability? Is mobility a viable risk reduction strategy for the most vulnerable?

- Under what conditions is mobility's risk reduction potential best harnessed? What should be avoided in order to reduce its potential negative impacts?

  In order to answer this question, in turn, it seems necessary to:

- Adopt a translocal, multi-scale and diachronic perspective to the analysis of risk reduction and risk creation, which accounts for population movements and their in situ and distant outcomes;
- Take into account all population movements in risk reduction research—without limiting the object of study to environmentally induced or cross-border movements;
- Analyze mobility decisions under the lens of livelihood practices, and understand mobility and immobility (including in the face of shocks and stresses) in a continuum of interlinked choices that depend on people's needs and capacities, each with its own effect on well-being and vulnerability;
- Systematically disaggregate risk and disaster data by individual and household mobility patterns and status.

This understanding should help implement and evaluate risk reduction options, including those that have been highlighted in the SFDRR: the involvement of migrants in DRR efforts (both in areas of origin and of destination) and the management of environmentally induced mobility (including migration, displacement, and relocations) in ways that do not increase risk and strengthen resilience. Going beyond the SFDRR text, however, preventing and reducing risk might also require looking at the following:

- Considering the mobility consequences of policies and investments (including to manage or regulate migration, and support development and risk reduction), in order to anticipate and manage their negative impacts on risk creation;
- Making safe forms of movement more accessible and affordable;
- Improving vulnerable mobile people's access to adequate resources and opportunities in areas of transit and destination;
- Supporting host communities and institutions, and in particular their most marginalized groups, in order to reduce negative impacts of population pressures;
- Facilitating the transfer of resources stemming from movements and promoting their use in ways that reduce vulnerabilities for the broader recipient community; and
- Providing effective alternatives to mobility, when they might be needed.

Human mobility cannot be considered in isolation from existing socio-environmental structures and relations that produce vulnerability. In turn, it shapes the highly contextual and complex local conditions in which risk is created, manifested, and reduced. In an increasingly mobile world, seizing the opportunities and addressing the challenges associated with human mobility will be essential for reducing the impacts of natural hazards, as well as for promoting sustainable development.

References

Adams V, van Hattum T, English D (2010) Chronic disaster syndrome: displacement, disaster capitalism, and the eviction of the poor from New Orleans. Am Ethnol 36(4):615–636

Akkoyunlu S, Stern M (2012) An empirical analysis of diaspora bonds. Research paper 2012/3. Graduate Institute, Geneva

Asis MMB (2008) How international migration can support development: a challenge for the Philippines. In: Castles S, Delgado-Wise R (eds) Migration and development: perspectives from the South. International Organization for Migration, Geneva, p 175

Attzs M (2008) Natural disasters and remittances: exploring linkages between poverty, gender and disaster vulnerability in Caribbean SIDS. Research Paper 2008/61. United Nations University, Bonn

Bartram D (2013) Happiness and 'economic migration': a comparison of Eastern European migrants and stayers. Migrat Stud 1(2):156–175

Bernhard JK, Landolt P, Goldring L (2009) Transnationalizing families: Canadian immigration policy and the spatial fragmentation of care-giving among Latin American newcomers. Int Migr 47(2):3–31

Black R, Adger WN, Arnell NW et al (2011) The effect of environmental change on human migration. Global Environ Chang 21(S1):S3–S11

Black R, Arnell NW, Adger WN et al (2012) Migration, immobility and displacement outcomes following extreme events. Environ Sci Pol 27(S1):S32–S43

Bolin R, Stanford L (1998) The Northridge earthquake: community-based approaches to unmet recovery needs. Disasters 22(1):21–38

Brickell K, Datta A (2011) Translocal geographies: spaces, places, connections. Ashgate, Burlington

Carter SB (2008) Labor market flooding? Migrant destination and wage change during America's age of mass migration. In: DeWind J, Holdaway J (eds) Migration and development across borders: research and policy perspectives on internal and international migration. International Organization for Migration and Social Science Research Council, Geneva

Castles S (2002) Environmental change and forced migration: making sense of the debate. Working paper 70. Refugee Studies Centre, Oxford

Castles S (2010) Understanding global migration: a social transformation perspective. J Ethn Migr Stud 36(10):1565–1586

Castles S, Miller MJ (2009) The age of migration: international population movements in the modern world, 4th edn. Palgrave MacMillan, Basingstoke

Chambers R, Conway G (1992) Sustainable rural livelihoods: practical concepts for the 21st century. IDS Discussion Paper 296. Institute for Development Studies, Brighton

Clerveaux V, Spence B, Katada T (2008) Using game techniques as a strategy in promoting disaster awareness in Caribbean multicultural societies: the disaster awareness game. J Disaster Res 3(5):1–13

Cohen JH (2011) Migration, remittances, and household strategies. Annu Rev Anthropol 40:103–114

d'Emilio AL, Cordero B, Bainvel B et al (2007) The impact of international migration: children left behind in selected countries of Latin America and the Caribbean. United Nations Children's Fund, New York

Dalisay SNM (2008) Survival strategies to overcome inagosto and nordeste in two coastal communities in Batangas and Mindoro, the Philippines. Disaster Prev Manage 17(3):373–382

de Haas H (2006) Migration, remittances and regional development in Southern Morocco. Geoforum 37:565–580

de Haas H (2008a) Migration and development: a theoretical perspective. International Migration Institute, Oxford

de Haas H (2008b) North African migration systems: evolution, transformations, and development linkages. In: Castles S, Delgado-Wise R (eds) Migration and development: perspectives from the South. International Organization for Migration, Geneva, p 143
De Moor N (2011) Labour migration for vulnerable communities: a strategy to adapt to a changing environment. COMCAD Working Paper 101
de Sherbinin A, Levy M, Adamo S et al (2012) Migration and risk: net migration in marginal ecosystems and hazardous areas. Environ Res Lett 7:1–13
Deshingkar P (2012) Environmental risk, resilience and migration: implications for natural resource management and agriculture. Environ Res Lett 7:015603
Deshingkar P, Aheeyar MMM (2006) Remittances in crises: Sri Lanka after the tsunami. Overseas Development Institute, London
Deshingkar P, Grimm S (2005) International migration and development: a global perspective, vol 19, Migration research series. International Organization for Migration, Geneva
Docquier F, Marchiori L, Shen IL (2010) Brain drain in globalization: a general equilibrium analysis of the sending countries' perspective. Discussion Paper 7682. CEPR
Dreby J (2010) Divided by borders: Mexican migrants and their children. University of California Press, Berkley
Duong LB, Linh TG, Thao NTP (2011) Social protection for rural-urban migrants in Vietnam: current situation, challenges and opportunities. Institute of Development Studies, Hanoi
Edward TJ, Scott R (2003) Migration and incomes in source communities: a new economics of migration perspective from China. Econ Dev Cult Change 52:75–101
Fagen P (2006) Remittances in crises: a Haiti case study. Overseas Development Institute, London
Farrant M, MacDonald A, Sriskandarajah D (2006) Migration and development: opportunities and challenges for policymakers, vol 22, Migration research series. International Organization for Migration, Geneva
Felli R, Castree N (2012) Neoliberalising adaptation to environmental change: foresight or foreclosure? Environ Plan A 44(1):1–4
Findley SE (1994) Does drought increase migration? A study of migration from rural Mali during the 1983–1985 drought. Int Migr Rev 28:539–553
Foresight Migration and Global Environmental Change (2011) Final project report. The Government Office for Science, London
Frayne B (2004) Migration and urban survival strategies in Windhoek, Namibia. Geoforum 35 (4):489–505
Freire T, Xiaoye L (2013) How immigration reduced social capital in the US: 2005–2011. Social Science Research Network working paper. http://papers.ssrn.com/sol3/papers.cfm?abstract_id=2222650
Fussel E, Sastry N, Van Landingham M (2010) Race, socioeconomic status, and return migration to New Orleans after Hurricane Katrina. Popul Environ 31:20–42
Gray CL (2009) Rural out-migration and smallholder agriculture in the Southern Ecuadorian Andes. Popul Environ 30:193–217
Greiner C, Sakdapolrak P (2012) Rural-urban migration, agrarian change and the environment in Kenya. A critical review of the literature. Popul Environ. doi:10.1007/s11111-012-0178-0
Hammond L (2011) Obliged to give: remittances and the maintenance of transnational networks between Somalis at home and abroad. Bildhaan 10, article 11
Hugo G (2008) Migration, development and environment, vol 35, Migration research series. International Organization for Migration, Geneva
IOM (2013) World migration report 2013: migrant well-being and development. International Organization for Migration, Geneva
IOM (2015) World migration report 2015: migrants in cities. International Organization for Migration, Geneva
Jokisch BD (2002) Migration and agricultural change: the case of smallholder agriculture in highland Ecuador. Hum Ecol 30(4):523–550

Jongwanich J (2007) Workers' remittances, economic growth and poverty in developing Asia and the Pacific Countries. Working Paper 1997/1. United Nations Economic and Social Commission for Asia and the Pacific

King R, Vullnetari J (2006) Orphan pensioners and migrating grandparents: the impact of mass migration on older people in rural Albania. Ageing Soc 26(5):783–816

Klaufus C (2010) Watching the city grow: remittances and sprawl in intermediate Central American cities. Environ Urban 22(1):125–137

Kniveton D, Schmidt-Verkerk K, Smith C, Black R (2008) Climate change and migration: improving methodologies to estimate flows. International Organization for Migration, Geneva

Konseiga A (2004) Adoption of agricultural innovations in the Sahel: the role of migration in food security. Paper presented at the 2004 food and nutrition security policies regional workshop in Mali

Ku L, Jewers M (2013) Health care for immigrants: current policies and issues. Migration Policy Institute, Washington

Kunz R (2011) The political economy of global remittances: gender, governmentality and neoliberalism. Routledge, London

Le De L, Gaillard JC, Friesen W (2013) Remittances and disaster: a review. Int J Disaster Risk Reduct 4:34–43

Le De L, Gaillard JC, Friesen W, Matautia Smith F (2014) Remittances in the face of disasters: a case study of rural Samoa. Environ Dev Sustain. doi:10.1007/s10668-014-9559-0

Long N (2008) Translocal livelihoods, networks of family and community and remittances in Central Peru. In: DeWind J, Holdaway J (eds) Migration and development across borders: research and policy perspectives on internal and international migration. International Organization for Migration and Social Science Research Council, Geneva

Lucas R (2005) International migration regimes and economic development. Report for the Expert Group on Development Issues in the Swedish Ministry for Foreign Affairs. http://www.egdi.gov.se/migraiton.htm

Manole V, Schiff M (2004) Migration and diversity: human versus social capital. IZA Discussion Paper 1279. Institute for the Study of Labor, Bonn

Mazzuccato V (2008) Simultaneity and network in transational migration: lessons learned from a simultaneous matched-sample methodology. In: DeWind J, Holdaway J (eds) Migration and development across borders: research and policy perspectives on internal and international migration. International Organization for Migration and Social Science Research Council, Geneva

McDowell C, de Haan A (1997) Migration and sustainable livelihoods: a critical review of the literature. IDS working paper 65. Institute of Development Studies, Brighton

McGranahan G, Tacoli C (2006) Rural-urban migration in China: policy options for economic growth, environmental sustainability and equity. International Institute for Environment and Development, London

McLeman RA (2006) Migration out of 1930s rural Eastern Oklahoma: insights for climate change research. Great Plains Quarterly Paper 151

McLeman RA (2014) Climate and human migration. Cambridge University Press, New York

Mohapatra S, Joseph G, Ratha D (2009) Remittances and natural disasters: ex-post response and contribution to ex-ante preparedness. The World Bank, Washington

NESSE (2008) Education and the integration of migrant children. European Commission, Bruxelles

Orozco M (2007) Central American diasporas and hometown associations. Inter American Development Bank, Washington, DC

Orozco M (2008) Diasporas and development: issues and impediments. In: Brinkerhoff JM (ed) Diasporas and international development: exploring the potential. Lynne Rienner, Boulder

Ortega F, Peri G (2009) The causes and effects of international labour mobility: evidence from OECD countries 1980–2005. United Nations, New York

Østergaard-Nielsen E (2011) Codevelopment and citizenship: the nexus between policies on local migrant incorporation and migrant transnational practices in Spain. Ethnic Racial Stud 34 (1):20–39
Perry RW, Mushkatel AH (2008) Minority citizens in disasters. University of Georgia Press, Athens
Phillips BD (1993) Cultural diversity in disasters: sheltering, housing and long-term recovery. Int J Mass Emerg Disasters 11(1):99–110
Piguet E (2010) Climate and migration: a synthesis. In: Afifi T, Jäger J (eds) Environment, forced migration and social vulnerability. Springer, Berlin, p 73
Raj Khanal N, Watanabe T (2006) Abandonment of agricultural land and its consequences. Mount Res Dev 26:32–40
Ratha D (2013) The impact of remittances on economic growth and poverty reduction. Migration Policy Institute, Washington
Ratha D, Sirkeci I (2011) Remittances and the global financial crisis. Migrat Lett 7(2):125–131
Ratha D, Mohapatra S, Scheja E (2010) Impact of migration on economic and social development: a review of evidence and emerging issues. Background paper to the Global Forum on Migration and Development
Rey Benayas JM, Martins A, Nicolau JM, Schulz JJ (2007) Abandonment of agricultural land: an overview of drivers and consequences. CAB Rev 57
Rinke T (2012) Temporary and circular labor migration between Spain and Colombia. In: Gemenne F, Bruecker P, Ionesco D (eds) State of environmental migration 2011. Institute for Sustainable Development and International Relations and International Organization for Migration, Paris
Robson JP, Nayak PK (2010) Rural out-migration and resource-dependent communities in Mexico and India. Popul Environ 32:263–284
Runfola DM, Romero-Lankao P, Jiang L et al (2013) The influence of internal migration on exposure to extreme weather events in Mexico. Working paper 4, population programme. Institute of Behavioral Science, Boulder
Sakdapolrak P, Promburom P, Reif A (2014) Why successful in situ adaptation with environmental stress does not prevent people from migrating? Empirical evidence from Northern Thailand. Clim Dev 6(1):38–45
Sall B (2005) Remittances and economic initiatives in sub-Saharan Africa. In: Migration, remittances and development. OECD, Paris, p 265
San Vincente Portes L (2009) Remittances, poverty and inequality. J Econ Dev 34(1):127–140
Sanderson MR (2009) Globalization and the environment: implications for human migration. Hum Ecol Rev 16:93–102
Scheffran J, Marmer E, Sow P (2012) Migration as a contribution to resilience and innovation in climate adaptation: social networks and co-development in Northwest Africa. Appl Geogr 33:119–127
Schensul D, Dodman D (2013) Populating adaptation: incorporating population dynamics in climate change adaptation policy and practice. In: Martine G, Schensul D (eds) The demography of adaptation to climate change. United Nations International Institute for Environment and Development and El Colegio de México, New York, p 1
Sen A (2000) Development as freedom. Oxford University Press, New York
Skeldon R (2008) Linkages between internal and international migration. In: DeWind J, Holdaway J (eds) Migration and development across borders: research and policy perspectives on internal and international migration. International Organization for Migration and Social Science Research Council, Geneva
Stark O (1991) The migration of labour. Blackwell, Oxford
Stark O, Bloom DE (1985) The new economics of labour migration. Am Econ Rev 75(2):173–178
Tacoli C (2009) Crisis or adaptation? Migration and climate change in a context of high mobility. Environ Urban 21(2):513–525
Tacoli C (2013) Migration as a response to local and global transformations: a typology of mobility in the context of climate change. In: Martine G, Schensul D (eds) The demography

of adaptation to climate change. United Nations International Institute for Environment and Development and El Colegio de México, New York, p 41

Tasan-Kok T, Stead D (2013) Analysing the socio-spatial vulnerability to drivers of globalisation in Lisbon, Oporto, Istanbul, Stockholm and Rotterdam. In: Eraydin A, Tasan-Kok T (eds) Resilience thinking in urban planning. Springer, Berlin, p 71

UNDESA (2013) International migration report 2013. United Nations, New York

UNDP (2009) Human development report 2009: overcoming barriers, human mobility and development. United Nations, New York

UNEP (2011) Livelihood security: climate change, migration and conflict in the Sahel. United Nations Environment Programme, Geneva

Valero-Gil J (2008) Remittances and the household's expenditure on health. MPRA Working Paper 9572. University Library, Munich

Vanwey LK, Guedes GR, D'Antona AO (2012) Out-migration and land-use change in agricultural frontiers: insights from the Altamira settlement project. Popul Environ 34(1):44–68

Vu L, VanLandingham MJ, Do M, Bankston CL (2009) Evacuation and return of Vietnamese New Orleanians affected by Hurricane Katrina. Organ Environ 22(4):470–478

Wang MZ, Amati M, Thomalla F (2011) Understanding the vulnerability of migrants in Shanghai to typhoons. Nat Hazards. doi:10.1007/s11069-011-9902-9

Wisner B, Blaikie P, Cannon T, Davis I (2004) At risk: natural hazards, people's vulnerability and disasters, 2nd edn. Routledge, London

Wisner B, Gaillard JC, Kelman I (2012) Framing disaster: theories and stories seeking to understand hazards, vulnerability and risk. In: Wisner B, Gaillard JC, Kelman I (eds) The Routledge handbook of hazards and disaster risk reduction. Routledge, London

Yang D, Choi H (2007) Are remittances insurance? Evidence from rainfall shocks in the Philippines. World Bank Econ Rev 21(2):219–248

Zetter R, Morrissey J (2014) The environment-mobility nexus. In: Fiddian-Qasmiyeh E, Loescher G, Long K, Sigona N (eds) The Oxford handbook of refugee and forced migration studies. Refugee Studies Centre, Oxford

Zachariah KC, Rajan SI (2004) Gulf revisited Working Paper 363. Centre for Development Studies, Thiruvananthapuram

Part I
Disaster Risk Reduction

Chapter 3
Disaster Risk Reduction in the Era of "Homeland Security": The Struggle for Precautionary, Preventive, and Non-violent Approaches

Kenneth Hewitt

1 Introduction

Addressing disasters in terms of *risk reduction* has to do, mainly, with ways to avoid or prevent catastrophes and to alleviate impacts of those not prevented. A "disaster risk reduction" (DRR) agenda has become identified as, and with, in-depth analyses of conditions that put people at risk, that fail to provide known preparedness and protections, or to adequately relieve post-disaster distress. The practical and ethical bases for a DRR arise from empirical findings that *most losses in recent disasters could, in fact, have been prevented* let alone reduced, and by well-known means. Its positive goals involve, especially, long-term preventive measures addressing the conditions that can and should help people avoid dangers, promote adaptive and sustainable safety. As is also shown, where the latter is available, successful prevention is readily apparent and social action critical for it.

Typically, reports create the impression that disasters are due to unprecedented, unstoppable natural forces—in the Indian Ocean 1995, at Kobe 1995, Sichuan, China in 2008, Japan and Haiti 2010, or the Philippines in 2013. However, great as the geophysical forces were, others as great or greater are found in the historical record with, and without, calamities. More importantly, damage patterns were explained mainly by pre-existing socio-economic conditions, risky settlement, land use histories and development; only weakly if at all, by the variable intensity of geophysical stresses. Detailed assessments reveal an absence of entirely feasible warnings, or enforcement of building codes and other protective measures, which

K. Hewitt (✉)
Department of Geography and Environmental Studies, Wilfrid Laurier University, Waterloo, ON, Canada
e-mail: khewitt@wlu.ca

© Springer International Publishing Switzerland 2017
K. Sudmeier-Rieux et al. (eds.), *Identifying Emerging Issues in Disaster Risk Reduction, Migration, Climate Change and Sustainable Development*,
DOI 10.1007/978-3-319-33880-4_3

perhaps indicates a misjudgment or opportunism in post-disaster assistance (Ekine 2013; Gonzalez et al. 2007; Lewis 2008).

The underpinnings of DRR arose from critical social research in the 1970s and 1980s (Bohle 1993; Hewitt 1983; Lavell 1994; White 1974; Wijkman and Timberlake 1984). The work showed that disaster risks are rooted mainly in peoples' everyday lives, in stresses and developments that push them into more dangerous locations, increase their vulnerability, and fail to offer them resources and protections (Maskrey 1989; Mileti 1999; Ozerdem and Jacoby 2006; Wisner 1993). Spatial patterns of loss were found to follow pre-existing societal conditions (Bankoff et al. 2013; Blaikie et al. 1994; Enarson and Morrow 1998; Hewitt 1997; Steinberg 2006). In famines for example, overwhelmingly those who starve, die, or end up in refugee camps were the *already* hungry, malnourished, and otherwise deprived (Copans 1975; de Waal 1997; Sen 1981). Crucially, while environmental hazards are *indiscriminate agents*, in social terms their impacts become *discriminatory disasters*—disproportionate losses for certain types of person, sectors, or groups. They single out groups subject to precarious social histories, unsustainable relations to habitats, coercion, and weak or corrupt governance (Alexander 2010; Coelho 2007; Oliver-Smith and Hoffman 2003; Pelling 2003).

The Yokohama conference of 1995 brought a singular shift in this—a shift away from prevailing explanations based on natural hazards, or extreme, unpredictable events and away from management focused on emergency preparedness, "expert systems", and technology transfers. The latter certainly have their place, but DRR looks to other humanitarian goals and preventive forms of action, for example, as systematized internationally in the Hyogo Framework for Action (HFA). For present purposes, the social and humanitarian preoccupations are revealed in UNISDR (2005):

> Priority Action 3: Use knowledge, innovation, and education to build a culture of safety and resilience at all levels
> Priority Action 4: Reduce the underlying risk factors.

At the time of writing, intense discussion to update or replace HFA in 2015 has given greater prominence to these two priorities, partly because of their neglect in HFA's first 10 years (UNISDR 2014). The notion of "underlying risk factors" challenges the still pervasive view of disasters as due to agents like earthquake or flood; to supposedly abnormal or uncertain conditions, or to unpredictable environmental change (van Niekerk 2008).

At the broadest scale, an influential report for the DRR community found that;

> ..."while only 11 percent of the people exposed to natural hazards live in low human development countries they account for more than 53 percent of recorded deaths". Development status and disaster risks are closely linked..." (UNDP 2004:1 emphasis added)

Of course, statistical summaries are only broad indications of the exact nature of exposure, vulnerability or disadvantage; of processes and responsibility (Alexander 2011; Bankoff et al. 2015; Cutter 1993; Steinberg 2006). National wealth and higher rates of development are not necessarily associated with greater safety. For its victims, risk is always local, if under constant pressure from broader, even global, developments affecting everyday lives and capacities (Davis et al. 2004; Franklin 1998). And such dangers have a social history, not just a random, extreme event trigger (Pelling 2003). Had the 2004 Indian Ocean Tsunami occurred 50+ years earlier, the types of coastal settlement, land uses, and concentrations of population hardly existed. Most victims had come to or been born in coastal zones relatively recently, driven by livelihoods lost to modernization and urbanization, or attracted to unsafe sites by them. Disproportionate deaths among women in some areas, by caste or ethnicity in others, reflect pre-existing stresses and oppression (Hewitt 2007; Oxfam International 2005; UNIFEM 2009). Meanwhile, from Aceh and Sri Lanka to Myanmar and Somalia, people in the worst-affected coastal zones were in areas ravaged by civil war and dirty wars; by severe human rights violations and military occupation (Ferks and Klem 2011; Le Billon and Waizenegger 2007). Earthquake and tsunami impacts *revealed* their precarious state, *rather than causing it.*

In such terms, disaster risk can hardly be treated separately from migration, especially where forced, or when host countries reject, exploit, ghettoize, and demonize migrants (Loyd et al. 2012; Chaps. 1 and 6). Meanwhile, climate change may ultimately become the largest "man-made" disaster or trigger of increased and changing disaster profiles. However, the social conditions noted, for those most at risk from climate change, *already* involve the highest disaster risk. More critical here, each of these concerns face challenges, similar or overlapping, as described below in relation to DRR.

2 Contradictions and Challenges

> ... today... enlightenment turns into what it undertook to hinder; the increase of fear. The uncanny...comes to the fore again out of the means used to protect against it. (Sloterdjik 1987: 330)

As a whole, the field of disasters involves some seemingly contradictory developments. Evidence is widely cited of increasing disaster numbers and losses, and of populations affected by them (EM-DAT 2013; IMF 2012; Laframboise and Loko 2012). However, a sense of contradiction arises from the, less often mentioned, great increases in disaster-related initiatives. Funding and relevant institutions, meetings and publications, non-profit and for-profit companies, have expanded exponentially in recent decades.

One indicator is the expanded role of development banks. The largest contributor, the World Bank, sponsored 528 projects labeled DRR between 1984 and 2006. It oversaw funding for many others. The Asian Development Bank sponsored 560 DRR projects between 1987 and 2012. Other financial institutions are involved from the Eurozone, Middle East, Africa, Latin America, and Australasia. Emergency assistance from the IMF (2012) has also increased.

According to Calhoun (2004), "management of emergencies [is] a very big business... mobiliz[ing] tens of thousands of paid workers and volunteers..." Hannigan (2012: 22) finds "natural disasters" to comprise, "...a global policy field...becoming considerably more crowded and turbulent... [with] the influx of thousands of new NGOs into emergency operations..." He identifies further marked expansion—if greater confusion—as disaster management is entwined with Climate Change Adaptation. The question is why such unprecedented efforts and investment have failed to even slow the rates of increase in disaster losses?

Conventional views focus on what is lacking, in knowledge or monitoring, in transfers of available techno-science, or "under-"development. However, critical assessments suggest those governmental and corporate strategies for disaster and development that have grown are, at best, inappropriate and at worst, self-serving projects (Loyd et al. 2012). A DRR perspective offers a different view of disaster causality: that socio-economic disadvantages loom larger as in:

1. Greater influx and concentrations of vulnerable people in dangerous situations, notably rural and urban settlement in flood plains, along storm and tsunami-prone coasts, in mountainous interiors, "desertified" dry lands, and polluted wetlands (Bankoff et al. 2004; Renaud et al. 2013; UNISDR 2002);
2. Developments that increase vulnerability by gender, class, (un)employment, ethnicity, religion, sector, country, etc. (Anderson 2005; Fordham 2003; Middleton and O'Keefe 1998);
3. Multiplying risks and losses associated with social upheavals, urbanization, militarization, habitat damage, and armed conflict (Bohle 1993; Hewitt 2000, 2013; Hilhorst 2013; Schuilenburg 2012).

Davis' (2006) "planet of slums" includes a disproportionate share of disaster's victims. The overall pace of urbanization and spread of fully urbanized societies coincides with a late modern reversal that Pelling (2003) calls the "urban transition from security to risk" (see also Fernandez 1999; Mitchell 1999). However, these are all situations massively affected and maintained by the prevailing economic and political order that spur "new realities", from "resource wars" to climate change, accelerating and changing directions of global economic and strategic choices. Late modernity in general has brought massive redistribution of risks and benefits, often poorly monitored, poorly understood, or cynically accepted (Bricken and Eick 2012; Ericson and Doyle 2003).

If we live, as many believe, in Ulrich Beck's (1992) "risk society", these underpinnings of disaster can hardly be ignored; notably his echo of Sloterdjik (1987) above that, "... the production of risks is [now] the consequence of scientific and political efforts to control or minimize them..." (p. 12). Studies of the disasters at Fukushima, Christchurch, New Zealand, or "dike risk" at New Orleans, might seem to literally endorse this view. However, Beck's main thesis is that the "risk society" results from obsessive commitments to *controlling dangers or threats*, and in countries and by people who, otherwise, seem among the safest! He hardly addresses environmental disasters and may seem unhelpful where most occur except that, by default, modern managers, agencies, and science play dominant roles there too.

2.1 "Securitization" and Non-disasters

...things do not necessarily develop in step in different sectors, at a given moment, in a given society, in a given country... (Foucault 1978: 8)

The fairly negative report so far needs to be balanced and questioned by some remarkable success stories. Not unlike modern medicine, in some places modern developments bring unusual safety and much-reduced disaster risk. A certain percentage of households in modern states and enclaves, generally the wealthiest, enjoy high levels of security, services, options, and resilience. They benefit most from modern facilities and support systems. They rarely appear in the disaster lists, even when just round the corner from massive destruction. A prime example was the much larger areas of Kobe city with little or no damage or loss of life in the 1995 calamity (Hewitt 2007), not because the earthquake was weaker there, but because they benefitted from known and enforced seismic building standards and fire protection. They provide telling illustration of what *effective* DRR can do, and in striking contrast to nearby failed protections where some 6,000 died. This also becomes a standard against which to measure what went wrong there.

In most cities, the main beneficiaries of modern safety spend their days in well-protected, security-conscious work places, and homes, or what Sheptyki (1997) calls "bubbles of security". They are usually in the suburbs, sometimes inner districts or ex-urban areas still defined by "old wealth", or gated communities with private security guards. "Corridors of security" take these well-to-do people from home to downtown, to businesses and industrial parks, high-end shopping, airports, and recreation. Such contrasting "riskscapes" exist in North America and Western Europe, if more striking where they overlook or wind through the vast slums of Johannesburg, Jakarta, Mexico City, Mumbai, or Karachi.

It needs emphasis that risk reduction specifically affecting disasters—building and planning codes, evacuation and quarantine plans, public health and safety, insurance—are also available in proportion to social influence and purchasing power. In addition to better forecasting networks, flood control, fire-fighting, or emergency preparedness, community support networks are also key in the form of social capacities, sustainable livelihoods, options, and habitats. The realm of "resilience" is thus rooted, like vulnerability, in community support systems. However, a compelling observation is that well-protected people acquire safety, even against disasters, *largely outside of disaster management systems*. Their security is decided through place and type of work, income, or credit, often by inherited or class benefits, through religious institutions in many countries, and preferential access to public as well as private facilities. The other side of the "underlying risk factors" is "underlying *safety factors*". Unfortunately, to date, superior security arrangements are identified with social exclusion and problems for others, rather than as DRR (Beck 1999; Franklin 1998). However, there is a more immediate, growing impediment.

3 The Poor Relation of Development and Disaster Management

DRR is the preferred perspective here, but remains a limited and even a diminishing part of the whole disasters field. A recent report estimated global assistance for "natural disasters" between 1991 and 2010, at about US$106.7 billion (Kellett and Caravani 2013). International development aid was US$3.3 trillion;

> . . .Essentially, for every US$100 spent on development aid, just 40 cents has been invested in defending that aid from the impact of disasters. . . (ibid, p. 5)

More revealing, only 12.7 % of disaster funding went to DRR. Emergency responses received 65.5 %, and reconstruction planning 21.8 %. Poorer countries received the least for DRR, with the lowest 117 countries only 9.5 %—the reverse of HFA priorities. Meanwhile, funding was allocated in a fragmented and inequitable way. Of all projects supported, 86.5 % involved 3,188 grants of less than US$1.5 million, comprising only *5.5 % of total outlays*. A few large, expensive projects took the lion's share, mainly in middle-income countries such as China and Indonesia.

The broader picture of safety and security funding reinforces this sense of contradiction (Table 3.1). Global DRR ranks poorly compared to local support for some other risks. Above all, military budgets must be considered "the elephant in the room" here, as so many public concerns (see below).

Table 3.1 Comparing world total DRR expenditures in 2010, estimated at US$1.4 billion, with other security-related expenditures in that year

Funding area	DRR as per cent
1. US border security infrastructure	90
2. US "War on Drugs" outlays	50
3. World food aid	25
4. Security firm Booz Allen's annual income[a]	25
5. US Special Operations Command (so-called "Black Ops.")	20
6. World Peacekeeping	10
7. Canada's military budget (#14 in world)	4
8. Worldwide spending of states on homeland security services and products (Statewatch 2009, 4)	1
9. Energy *subsidies*—largely for fossil fuels[b]	0.07

[a]99 % from government contracts (Chatterjee 2013). The company employed whistleblower Edward Snowden for work at the NSA

[b]According to the IMF (2015) these energy subsidies were almost US$2 trillion in 2011, "a whopping 2½ % of global GDP that could have been used more wisely"

3.1 The Up-Staging of DRR

In wealthier, more influential states, disaster response is increasingly combined with broad management complexes led by national security concerns (Bello 2006; Tellmann 2009). Disasters are subordinate to war-preparedness, border security, disease control, protection of major infrastructure, policing of international migration, and trafficking (Loyd et al. 2012). In North America, disaster management is absorbed by "Homeland Security" whose vision is: "...to ensure a homeland that is safe, secure, and resilient against terrorism and other hazards..." (US/DHS 2012: 9). "Ensuring resilience to disasters" is *fifth* on its list of missions.

Britain's National Security Strategy (UK Government 2010) has a similar profile with terrorism first on the list concerns and "natural disasters" *last*, even while acknowledging that; "... the 2007 floods in Britain occasioned the largest ever civil emergency response since the Second World War...[and] highlighted the impact that natural disasters can have, even on fully developed, networked societies..."

According to Hayes (2010: 17)

> Within Europe too the focus has been all about counter-terrorism and border control... used as a pretext to introduce surveillance policies that would have been unthinkable in the 1990s...

The sheer scale of security apparatuses and their competing agendas have an enormous impact on DRR (Buzan et al. 1998; Hannigan 2010; Hilhorst 2013). A large military involvement is taken for granted; seemingly as an obvious use of all those personnel, vehicles, and other resources sitting around between wars, mainly serving to provide armed control and "lock down" in disaster zones. In many countries the military do have better hospitals, communications, mass feeding arrangements and emergency shelter, and typically have the best-trained and equipped personnel with critical skills such as doctors, nurses, utilities technicians, and specialists in transportation.

4 Civil Defense and "The Security Industrial Complex"

Although much expanded after the 9/11 attacks, Homeland Security is rooted in a civil defense model (Alexander 2002; Amacher 2003; Hewitt 1987). Originally focused on urban civilian populations and bombing threats in the world wars, civil defense served as a basis for disaster management through the Cold War years:

> ...In 1951, NATO established the Civil Defense Committee... It soon became apparent that the capabilities to protect our populations against the effects of war could be used to protect them against disasters. (NATO-OTAN 2001: 5)

Work that helped define what became DRR and HFA was strongly critical of the civil defense model (Piel 1962). Now, with little discussion, Homeland Security (HS) has turned the clock back, amalgamated security-related agencies, redefined others and placed them on a war footing. This has enjoyed exponential increases in funding. In 2014, the US/HS total budget was almost US$60 billion, over 5 times the annual global disasters budget, and almost 100 times global DRR. After 9 years of HS, a Washington Post investigation reported that "unprecedented spending and growth so massive that its effectiveness is impossible to determine" including:

> * Some 1,271 government organizations and 1,931 private companies work[ing] on programs related to counterterrorism, homeland security and intelligence in about 10,000 locations across the United States.
> * An estimated 854,000 people, [with] top-secret security clearances.
> * In Washington and the surrounding area, 33 building complexes for top-secret intelligence work are under construction or have been built since September 2001... occupy [ing] the equivalent of almost three Pentagons or 22 U.S. Capitol buildings...
> * Many security and intelligence agencies do the same work, creating redundancy and waste.
> * Analysts who make sense of documents and conversations obtained by foreign and domestic spying share their judgment by publishing 50,000 intelligence reports each year—a volume so large that many are routinely ignored. (Priest and Arkin 2010)

Separating the public security concerns of HS from defense spending is difficult or impossible (Higgs 2007). Indeed the reverse is happening, with military spending increasingly defined in security terms and by "new war" hazards and action more and more into non-military areas.

The overall result is termed "The Security-Industrial Complex" (SIC) (Hayes 2010; Mills 2004). Some welcome it, others are highly critical of its impact, methods, and performance (MacDonald 2006). Either way, the Federal Emergency Management Agency (FEMA) is one preexisting agency absorbed into the SIC and with overlapping responsibilities in the "prison-industrial complex", the "border-industrial complex" (Miller 2013), and the "Cyber Security Industrial Complex" (Talbot 2011).

In principle, HS could pursue DRR priorities, as FEMA had begun to do in the 1990s. Thus far, it ignores long-term risk reduction and preventive measures, or participatory and cross-cultural concerns. Crisis-mode, centralized management is preferred, along with "expert" and outsider economic strategies and technologies. An overriding commitment to militarized responses and policing strategies treats disaster areas more like war zones or crime scenes; to be aggressively contained and controlled by armed force (Bricken and Eick 2012). Moreover, contrary to a widespread impression, military and police roles in disaster are neither free nor cheap. A large fraction of assistance can be absorbed for salaries, helicopters, motor fuel, accidents, and the like. As the first decade of HFA has ended, events in Haiti illustrated these concerns in a place where DRR is needed most.

4.1 *Haiti and the SIC*

Figures in the 2010 calamity are still disputed, but death tolls ranged from not less than 100,000 to over 320,000. Some 1.5 million people lost their homes and 3 million people were adversely affected. Economic losses were estimated at US$8 billion. More than 8,000 died in the subsequent cholera epidemic.

A measure of the militarization of relief is how, within 2 weeks, the USA had deployed 18,500 troops to Haiti, and Canada 2000. Several thousand more came from the Dominican Republic, France, Brazil, Israel, Qatar among others. MINUSTAH, the UN peacekeeping force already in Haiti, was boosted to 8,300 by April. US forces took control of the international airport and later, the ports and roads. Large numbers of warships, military planes, helicopters, trucks, and other motor vehicles were deployed. In February 2010, the UN assessed "peacekeeping" funds for Haiti at US$3.8 billion. One report is of US$1.9 billion disbursed to peacekeeping forces by mid-2012 (UNOSEH 2012), certainly less than claims to that time. Especially troubling was the cholera epidemic in which some 656,000+ people were made ill and 8,100+ had died by 2013. The source was traced to the contamination of streams at a MINUSTAH camp. The UN refuses to take responsibility, but the epidemic has added enormously to the hardship, overstressed medical system, and slow pace of recovery (The Economist 2013; United Nations 2010).

The relief effort had other troubling features (Moloney 2014). According to the Special Envoy (UN/OSEH 2012), of almost US$9 billion donated through bilateral and multilateral agencies:

- Less than 0.6 % went directly to Haitian businesses and activities
- Only 2.4 % of all bi- and multi-lateral funding was in cash transfers
- Less than 10 % went to the Haitian government and *none* to its agencies
- Haitian NGOs received only 1.3 % of international appeals

The report finds such "counterproductive assistance", typical in a range of other "fragile" (poor country?) settings.

4.2 Disaster as Opportunity

> The proclamation of disasters, or the threat of disasters, open windows of opportunity for militarized response, unauthorized interventions, blank *cheques*, suspending civil rights and indeed DRR projects... (Warner 2013: 89)

The SIC involves not merely greater centralization, militarization and secrecy, but there have been major efforts to privatize and shift responsibilities to for-profit enterprises as well as NGOs (Jung 2003). They absorb ever-larger parts of humanitarian aid, spin-offs to security businesses and paid security services. "Threats" are commonly bracketed with "opportunities" by HS and its related organizations (USOHDACA 2004). President Eisenhauer's original notion of the "military industrial complex" referred to how the US military had emerged as decisive not just for geostrategy, but as an economic player and employer. Such "complexes" can become *enterprises* rather than simply services, professions, or divisions of government.

In New Orleans after "Katrina" and Haiti after the earthquake, reports show that crisis and humanitarian funding served as opportunities for gain, if not blatant exploitation, by influential players, not least in or from wealthy countries and international organizations (Hepburn and Simon 2013; Provost and Dzimwasha 2014). Large contracts, much larger than most DRR projects, were awarded without an open bidding process (UNOSEH 2012).

What has gone on in New Orleans, Port-au-Prince, Christchurch New Zealand or Fukushima, and above all the 2014 *ebola* epidemic in West Africa, appears to support the contention that we are witnessing "...bold experiments *in crisis exploitation*..." (Klein 2007: 11 *emphasis added*). And as in the same disasters, crises may be used to excuse official ignorance of popular demands or contested concerns and to hand over problems to responders and experts, chosen because they serve donor priorities. Another "opportunity" is how:

> The lawlessness and high demand for cheap labor... can create an atmosphere ripe for human trafficking. Trafficking in New Orleans and other areas devastated by Hurricanes Katrina and Rita began shortly after the storms. (Hepburn and Simon 2013:12)

The possibility is thus raised that for some, letting disasters happen is more profitable and convenient than trying to prevent them!

4.3 Mindsets

Of special concern for knowledge-workers and others who read books like this, is how organizations and professions view disaster causes. DRR is developed from, and also singled out for, a particular set of priorities as outlined earlier in the chapter. Prevailing notions in the new "security state" and SIC, are remarkably similar to those critiqued and rejected in DRR (Hewitt 1983; UNISDR 2005; van Niekerk 2008).

Today, where disaster management finds its home(land), "threats", enemies or hazards external to society loom largest, and perhaps most notably the violent ones. It is identical to the old "hazards paradigm" (Bowling and Sheptycki 2012; Hewitt 1983). The resurgence of a militarized and technocratic crisis response turns away from the pre-disaster social causes of peoples' exposure and vulnerability, and from absent protection (van Uffelen 2013). A civil defense view reinstates Gilbert's (1998) "patterns of war approach", or Buzan et al.'s (1998) "logic-of-war". Goals tend to be wrapped in very broad, abstract packaging (for example, "freedom", "development"), promoted by dominant states and institutions or directed at their constituencies and media audiences. This politics of "securitization" has been seen as "a new form of power" (Schuilenburg 2012). The main issue is how it has worked *against* the HFA principles and efforts to address "underlying risk factors" (Corry 2012). As noted, these are given even greater emphasis in the revised Sendai Framework for Disaster Risk Reduction 2015–2030 (SFDRR) (UNISDR 2014). To the time of writing, it remains to be seen whether substantial material and political support will follow.

The most difficult contradictions may arise elsewhere. On the one hand, HS and the SIC have adopted the "all-hazards", transdisciplinary, multi-agency approach that many of us see as essential to making DRR effective (Hewitt and Burton 1971). Instead, the dangers of totalizing strategies and the consequences of centralizing control over complex terrains are revealed to be in the hands of a few authoritative figures or institutions, themselves under pressure from other, narrow agendas (to the state, the alliance, the economy, the corporation, the agency). In the security landscape, they become largely unaccountable because of secrecy and the emergency powers they claim. On the other hand, HS has championed prevention at a certain level and massively invested in trying to track, preempt, and forestall the primary concern, terrorist attacks. Then again, limited success and even the expanding problems are identified with a refusal to go a step further and look at "underlying risk factors"—radicalism rooted in poverty—concentrated youth unemployment, ghettoization, crime, and abuse. In the disasters arena, a conservative mindset again prevails, happy to relegate disasters to the realms of accidents, "bad apples", luck, and Mother Nature. This is a stark contrast with DRR which addresses disaster vulnerabilities of greatest extent and concern as *socially constructed*, both in how endangerment actually occurs and in how disaster risk is viewed and can be changed.

The argument is not against emergency services or national security concerns. Disasters will continue to happen and crisis response is needed. Disaster reduction, however, must occur mainly through risk aversion before the event. Again, modern safety systems and regulations and research to better understand the planetary environment continue to be needed. New products, major infrastructure, modern life styles, and global change will continue to raise urgent, specialized safety requirements. As already shown, however, these are not the main types of risk for most who suffer in disasters and most of such losses have been proven preventable, if that were to be the priority.

5 Reviving DRR: A Preventive, Humanitarian Field

These views of modern disasters suggest that the most direct and promising inspirations for DRR are found in established *preventive* fields such as public health, fire, workplace or road accident prevention, structural or aviation safety and social safety nets. Where applied and sustained, these have greatly reduced risk and losses (Green 1997; Hacking 1990). More critically, they have succeeded to the extent that they deal with social and environmental matters that may be outside, yet strongly affect hygiene or the presence of flammable materials, driver behavior or disease vectors respectively. Risk reduction in each preventive field has required working with, and the development of improvements for, those most likely to be harmed, usually those living in disadvantaged places and groups. Almost without exception, the story of accident or urban fire prevention, workplace safety and preventive medicine, involves improving conditions for the less well off. Not all damages have been or can be prevented, but it is the attitudes and goals of preventive fields that should inspire a new vision of disaster prevention.

Technical competence and capabilities are surely important, but do not come first in successful preventive fields. Experience in public health or accident prevention shows that improvements rarely happen without the trust and participation of those most at risk. They begin with commitments not merely to serving the "public good", but those most in need of assistance and resources. Ethics come first social initiatives and movements are needed that encourage participatory responses, cooperative, and sharing models (Ericson and Doyle 2003; Strang and Braithwaite 2001; Umbreit and Arbour 2010). Cultures of precaution and prevention are needed especially in congested districts of cities and in impoverished rural communities.

DRR has to conform to humanitarian principles (Hewitt 2013). Highly relevant, from the discussion above, are certain parts of the Red Cross and Red Crescent Societies' Code of Conduct, recommended for all NGOs, especially:

"*Disaster-affected communities* have a right to expect those who seek to assist them to measure up to these standards:

- *Aid is given regardless of the race, creed or nationality of the recipients and without adverse distinctions of any kind.* Aid priorities are calculated *on the basis of need alone.*
- *Aid will not be used to further a particular political or religious standpoint.*
- We shall endeavor *not to act as instruments of government foreign policy.*
- We shall *respect culture and custom.*
- We shall *attempt to build disaster response and local capacities.*
- Ways shall be found to *involve programme beneficiaries in the management of relief aid.*
- Relief aid must strive to *reduce future vulnerabilities to disaster as well as meeting basic needs.*
- We hold ourselves *accountable to* both *those we seek to assist* and those from whom we accept resources (ICRC 1994: 3–4).

For DRR such principles must be extended to building social capacities, safety and wellbeing outside crisis zones, to ensure the non-exclusive, equitable, and fair access to security in the daily necessities of living. It opposes investment in structures and cultural norms that separate, privilege, and protect some people and regions at the expense of others, or exclude the very people who are at greatest risk.

As core values, these will inevitably set DRR apart from the current approach of HS and the SIC. Risk assessments are required that respect the connectedness of people, cultures, and habitats. In turn, this engages with broader ecological principles for living and acting *with* Nature, rather than coercing, abusing, or making war on it.

6 Concluding Remarks

This chapter is intended to introduce and critically explore DRR. It seems fair to say that it shares some common predicaments with global migration and climate change adaptation (CCA), as they relate to a "sustainability" umbrella. There is already overlap and common concerns between DRR and CCA projects, if not always in clear agreement (Hannigan 2012; IPCC 2014). The extent to which climate change is driven by fossil fuel use and habitat degradation introduces an ever-increasing social, or "man-made" responsibility for environmental disasters. Arguably, the only common DRR/CCA strategy with any long-term prospect of success would be intensive conservation programs for energy, fresh water, and most other resources, leading eventually to greater efficiency and a massive reduction in their use. To date, CCA and HS, and indeed "development" and "sustainability" at large, have parted company on these very principles, which is arguably yet another big contradiction.

Similarly, environmental disasters are not only major factors in forced migration and refugee crises but involve similar, if not identical, socially determined disadvantages, impoverishment, exposure, and vulnerabilities. The latter are ever more closely bound up with HS border strategies and disaster relief measures (Loyd et al. 2012). As with DRR, the funding, research, institutions, and actions on human migration are unprecedented, and likewise for climate change—while the scale and scope of the problems grows! In the whole area of ecosystems and sustainability, it is impossible to keep pace with the studies and meetings, the research centers, books, and journals, and the documentaries dedicated to ecosphere survival (United Nations 2011). Numerous conventions and frameworks are supposed to protect Earth's habitats. Yet, as with disasters, studies show that the biosphere is in worse shape now than in 1992, and even more so compared with 1964 when the critical initiatives of the International Biological Program began. Hardly a single major ecosystem and very few fragile species are less endangered now.

Much as described for DRR, securitization and militarization of response are evident in migration, environmental destruction, and global warming. Armed conflict, arms trade and trafficking, and coercive policies receive vastly more investment and vastly increase the range of humanitarian crises and unsustainable practices (Gutman et al. 2007; Worldwatch Institute 2014). Migration and climate change have their equivalents or adjuncts of the SIC—complexes entwining governments, knowledge, for-profit and non-profit organizations, organized crime—that flourish by causing, or refusing to address, each of these late modern crises of survival (Franklin 1998; Klein 2007).

The last hopeful, if tragic, parallel is that in all these cases too, there are well-known solutions that are readily available and affordable—*if* that were the commitment. The biggest questions, as for DRR, thus arise where improvements, let alone sustainability, are treated as incompatible with business-as-usual, or the dominant economic and geostrategic interests.

References

Alexander DE (2002) From civil defense to civil protection and back again. Disaster Prev Manag 11(3):209–213

Alexander DE (2010) The L'Aquila earthquake of 6 April 2009 and Italian government policy on disaster response. J Nat Resour Policy Res 2(4):325–342

Alexander DE (2011) Towards a practical phenomenology of architecture and natural hazards. In: Broglioli GP et al (eds) Teoria e metodi della ricerca sui paesaggi d'altura. Società Archeologica Padana, Mantova, Italy, pp 203–214

Amacher P (2003) You're on your own again. Bull At Sci 2003(59):34–43

Anderson WA (2005) Bringing children into focus on the social science disaster research agenda. Int J Mass Emerg Disasters 23(3):159–175

Bankoff G, Frerks G, Hilhorst D (eds) (2004) Mapping vulnerability: disasters, development and people. Earthscan, London

Bankoff G, Lubken U, Sand J (eds) (2013) Flammable cities: urban conflagration and the making of the modern world. University of Wisconsin Press, Madison

Bankoff G, Cannon T, Krüger F, Schipper ELF (2015) Introduction: exploring the links between cultures and disasters. In: Krüger F, Bankoff G, Cannon T, Orlowski B, Schipper ELF (eds) Cultures and disasters: understanding cultural framings in disaster risk. Routldge, London, pp 1–16

Beck U (1992) Politics of risk society. In: Franklin J (ed) The politics of risk society. Polity Press, London, pp 9–22

Beck U (1999) World risk society. Polity Press, London

Bello W (2006) The rise of the relief-and-reconstruction complex. J Int Aff 59:281–297

Blaikie P, Cannon T, Davis I, Wisner B (1994) At risk: natural hazards, people's vulnerability and disasters. Routledge, London

Bohle HG (ed) (1993) World's of pain and hunger: geographical perspectives on disaster vulnerability and food security, Freiburg studies in development geography. Verlag Breitenbach, Saarbrucken

Bowling B, Sheptycki J (2012) Policing globopolis. Soc Justice 38:184–202

Bricken K, Eick V (2012) Policing the crisis, policing in crisis. Soc Justice 38:1–2

Buzan B, Waever O, Wilde J (1998) Security; a new framework. Lynne Rieneer, London

Calhoun C (2004) A world of emergencies: fear, intervention, and the limits of cosmopolitan order. Can Rev Sociol Anthropol 41:373–395

Chatterjee P (2013) How Booz Allen made the revolving door redundant, Inter Press Service News Agency, http://www.ipsnews.net/2013/06

Coelho JPB (2007) The state, the community, and natural calamities in rural Mozambique. In: Santos B de S (ed) Another knowledge is possible: beyond northern epistemologies. Verso, London, pp 219–243

Copans J (ed) (1975) Secheresses et famines du Sahel. Maspero, Paris

Corry O (2012) Securitization and 'riskification': second-order security and the politics of climate change. Millenium 40(2):235–258

Cutter SL (1993) Living with risk: the geography of technological hazards. In: Davis I, Haghebeart B, Peppiatt D (eds) Social vulnerability and capacity analysis. Discussion paper and workshop report. ProVention Consortium, Geneva

Davis I, Haghebeart B, Peppiatt D (2004) Social vulnerability and capacity analysis. Discussion paper and workshop report. ProVention Consortium, Geneva

Davis M (2006) Planet of slums. Verso, New York

De Waal A (1997) Famine crimes: politics and the disaster relief industry in Africa. Indiana University Press, Bloomington

Ekine S (2013) From AIDS to Aid; an (un)Humanitarian story in Haiti. International reporting Project, SAIS, Johns Hopkins University, January 28.

EM-DAT (2013) The international data base of OFDA/CRED. Universite Catolique de Louvain, Brussels, www.cred.be/emdat&gt

Enarson E, Morrow BH (1998) The gendered terrain of disaster: through women's eyes. Praeger, Westport, CT
Ericson RV, Doyle A (eds) (2003) Risk and morality. University of Toronto Press, Toronto
Ferks G, Klem B (2011) Muddling the peace process. The political dynamics of tsunami aid and conflict. In: Godhand J, Korf B, Spencer J (eds) Conflict and peace-building in Sri Lanka. Caught in the peace trap? Routledge, London, pp 168–182
Fernandez MA (ed) (1999) Cities at risk: Environmental degradation, urban risks and disaster in Latin America. The Network for Social Studies on Disaster, LA RED, Lima, Peru
Fordham M (2003) Gender, development and disaster: the necessity for integration. In: Pelling M (ed) Natural disasters and development in a globalizing world. Routledge, London, pp 57–74
Foucault M (1978) Security, territory, population: lectures at the College de France, 1977–78 (trans: Burchell G). Picador, New York
Franklin J (ed) (1998) The politics of the risk society. Polity Press, London
Gilbert C (1998) Studying disasters: changes in the main conceptual tools. In: Quarantelli E (ed) What is a disaster? Perspectives on the question. Routledge, New York, pp 11–18
Gonzalez L, Romano L, Salamanca L (2007) Risks and disasters in El Salvador: economic, environmental and social aspects. In: Rose WI et al (eds) Natural hazards in El Salvador. Special Paper. Geological Society of America, Boulder, pp 461–470
UK Government (2010) A strong Britain in an age of uncertainty: the national security strategy. HMGovt. https://www.gov.uk/government/data/file/61936/national-security-strategy.pdf
Green J (1997) Risk and misfortune: the social construction of accidents. University College of London Press, London
Gutman R, Rieff D, Dworkin A (eds) (2007) Crimes of war: what the public should know, 2nd edn. WW Norton, New York
Hacking I (1990) The taming of chance. Cambridge University Press, Cambridge
Hannigan J (2010) Disasters without borders. Polity Press, Cambridge
Hannigan J (2012) Disasters without borders: the international politics of natural disasters. Polity Press, London
Hayes B (2010) The EU security-industrial complex, 3 September. http://www.tni.org
Hepburn S, Simon RJ (2013) Human trafficking around the world: hidden in plain sight. Columbia University Press, New York
Hewitt K (1983) The idea of calamity in a technocratic age. In: Hewitt K (ed) Interpretations of calamity from the viewpoint of human ecology. Allen and Unwin, London, pp 3–32
Hewitt K (1987) The social space of terror: towards a civil interpretation of total war. Environ Plan D Soc Space 5:445–474
Hewitt K (1997) Regions of risk: hazards, vulnerability and disasters. Longman/Pearson, London
Hewitt K (2000) Safe place or 'catastrophic society'? An overview of risk and disasters in Canada. Can Geogr 44(4):325–341
Hewitt K (2007) Preventable disasters: addressing social vulnerability; institutional risk and civil ethics. Geogr Rundsch 3(1):43–52
Hewitt K (2013) Disasters in 'development' contexts: contradictions and options for a preventive approach. Jàmbá 5(2):1–9
Hewitt K, Burton I (1971) The hazardousness of a place: a regional ecology of damaging events in Southern Ontario. University of Toronto Press, Research Publications, no. 6, p 120
Higgs R (2007) The trillion-dollar defense budget is already here; The independent organisation, thu. March 1 www.independent.org/newsroom/article.asp?id
Hilhorst D (ed) (2013) Disaster, conflict and society in crisis. Routledge, London
ICRC (1994) Code of conduct for the International Red Cross and Red Crescent Movement and NGOs in Disaster Relief. http://www.ifrc.org/publicat/conduct.code.asp
IMF (2012) Natural disasters hitting more people, becoming more costly. International Monetary Fund, Survey Magazine. http://www.imf.org. Accessed 10 Oct
IMF (2015) Counting the costs of energy subsidies, International Monetary Fund IMF Survey Magazine: In the News, http://www.imf.org/external/pubs/ft/survey/so/2015/NEW070215A.htm

IPCC (2014) Climate change 2014: impacts, adaptation, and vulnerability contribution to AR5, IPCC Working Group II, Intergovernmental panel on Climate Change. Cambridge University Press, Cambridge
Jung D (ed) (2003) Shadow globalization, ethnic conflicts and new wars: a political economy of intra-state war. Routledge, London
Kellett J, Caravani A (2013) Financing disaster risk reduction: a 20 year story of international aid. Global facility for Disaster Reduction and Recovery, Overseas Development Institute, London
Klein N (2007) The shock doctrine: the rise of disaster capitalism. Knopf Canada, Toronto
Laframboise N, Loko B (2012) Natural disasters: mitigating impact, managing risks. IMF Working Paper WP/12/245. International Monetary Fund, Washington, DC
Lavell A (ed) (1994) Viviendo en Riesgo: Comunidades Vulnerables y Prevención de Desastres en America Latina [Living at risk: Vulnerable communities and disaster prevention in Latin America], LA RED/FLACSO, Bogotá
Le Billon P, Waizenegger A (2007) Peace in the wake of disaster? Secessionist conflicts and the 2004 Indian Ocean tsunami. Trans Inst Br Geogr 32(3):411–427
Lewis J (2008) The worm in the bud: corruption, construction and catastrophe. In: Bosher L (ed) Hazards and the built environment: attaining built-in resilience. Routledge, London, pp 238–263
Loyd J, Mitchelson M, Burridge A (eds) (2012) Beyond walls and cages: prisons, borders and global crisis. University of Georgia Press, London
Macdonald H (2006) The security-industrial complex the wall street journal, September 7
Maskrey A (1989) Disaster mitigation; a community based approach. Oxfam, Oxford
Middleton N, O'Keefe P (1998) Disaster and development: the politics of humanitarian aid. Pluto Press, London
Mileti GS (1999) Disasters by design: a reassessment of natural hazards in the United States. Joseph Henry, Washington, DC
Miller T (2013) The border-industrial complex goes abriad. TomsDispatch, November 9th http://www.tomdispatch.com
Mills MP (2004) The security-industrial complex, http://www.forbes.com/forbes/2004/1129/044
Mitchell JK (1999) Crucibles of hazard: mega-cities and disaster in transition. UNU Press, Tokyo
Moloney A (2014) Where has all the Haiti aid money gone? U.S. to keep closer track. Thomson Reuters Foundation—Wed, 30 Jul
NATO-OTAN (2001) NATO's role in disaster assistance. Civil Emergency Planning, North Atlantic Treaty Organisation, Brussels
Oliver-Smith AS, Hoffman SM (eds) (2003) The angry earth. Routledge, New York
Oxfam International (2005) The tsunami's impact on women, Oxfam briefing note. Oxfam International, Oxford
Ozerdem A, Jacoby T (2006) Disaster management and civil society: earthquake relief in Japan, Turkey and India. I.B. Tauris, London
Pelling M (2003) The vulnerability of cities: natural disasters and social resilience. Earthscan, London
Piel G (1962) The illusion of civil defense, bulletin of the atomic scientists, 29/2, February, 2–8
Priest D, Arkin WM (2010) A hidden world, growing beyond control. Washington Post, Monday July 9th. washingtonpost.com.webarchive
Provost C, Dzimwasha T (2014) Haiti earthquake: where is US aid money going? Get the data The Guardian, 10th January. http://www.theguardian.com
Renaud FG, Sudmeier-Rieux K, Estrella M (eds) (2013) The role of ecosystems in disaster risk reduction. United Nations University Press, Tokyo
Schuilenburg M (2012) The securitization of society: on the rise of quasi-criminal law and selective exclusion. Soc Justice 38(1–2):73–89
Sen A (1981) Poverty and famines: an essay on entitlement and deprivation. Clarendon, Oxford
Sheptyki J (1997) Insecurity, risk suppression and segregation: some reflections on policing in the transnational age. Theor Criminol 1(3):303–316

Sloterdjik P (1987) Critique of cynical reason (trans: Eldred M). University of Minnesota Press, Minneapolis
Statewatch (2009) NeoConOpticon: The EU Security-Industrial Complex, Statewatch and the Transnational Institute, http://www.statewatch.org/
Steinberg T (2006) Acts of God: the unnatural history of natural disaster in America, 2nd edn. Oxford University Press, New York
Strang H, Braithwaite J (eds) (2001) Restorative justice and civil society. Cambridge University Press, Cambridge
Talbot D (2011) The cyber security industrial complex. AutomatioNation, http://automationation.wordpress.com/, December 16
Tellmann U (2009) Imagining catastrophe; scenario planning and the striving for epistemic security. Economic Sociology—the European electronic newsletter, 10/2, March, 17–21.
The Economist (2013) Cholera in Haiti The UN strain. Jul 15th 2013, New York
Umbreit MS, Arbour MP (2010) Restorative justice dialogues. Springer, New York
UNDP (2004) A global report, reducing disaster risk: a challenge for development. Bureau of Crisis Prevention and Management, United Nations Development Program, New York
UNIFEM (2009) Women building their future: gender breakthrough in post-tsunami Aceh (author: Enarson E). United Nations Development Fund for Women, Bangkok
UNISDR (2002) Living with risk: a global review of disaster reduction initiatives. International Strategy for Disaster Reduction. United Nations, Geneva
UNISDR (2005) Hyogo framework for action 2005–2015. http://www.unisdr.org
UNISDR (2014) Development of the post-2015 framework for disaster risk reduction. Zero draft submitted by the Co-Chairs of the Preparatory Committee (20 October 2014)
United Nations (2010) Final report of the Independent Panel of Experts on the Cholera Outbreak in Haiti. www.un.org/News/dh/infocus/**haiti**/UN-**cholera**-report-final.pdf
United Nations (2011) Millennium development goal 8: the global partnership for development: time to deliver. MDG Gap Task Force report, 2011, New York
UNOSEH (2012) Can more aid stay in Haiti and other fragile settings? How local investment can strengthen governments and economies. United Nations Office of the Special Envoy for Haiti, New York
USOHDACA (2004) Mission statement, Washington DC, overseas humanitarian, disaster and civic aid. http://comptroller.defense.gov/defbudget/fy2004/budget justification
US/DHS (2012) Strategic Plan, Fiscal years 2012–2016, Department of Homeland Security, (February) Washington DC. https://www.dhs.gov/xlibrary/assets/dhs-strategic-plan-fy-2012-2016
Van Niekerk D (2008) From disaster relief to disaster risk reduction: a consideration of the evolving international relief mechanism. TD 4(2):355–376
Van Uffelen J-G (2013) The de-disasterization of food crises: structural reproduction or change in policy development and response option? A case study from Ethiopia. In: Loyd J, Mitchelson M, Burridge A (eds) 2012 Beyond walls and cages: prisons, borders and global crisis. University of Georgia Press, London, p 58075
Warner J (2013) The politics of "catastrophization.". In: Hilhorst D (ed) Disaster, conflict and society in crisis. Routledge, London, pp 76–94
White GF (ed) (1974) Natural hazards: local, national, global. Oxford University Press, New York
Wijkman A, Timberlake L (1984) Natural disasters: acts of god or acts of man? Earthscan, London
Wisner B (1993) Disaster vulnerability: scale, power, and daily life. Geojournal 30(2):127–140
Worldwatch Institute (2014) State of the world 2014: governing for sustainability. Island Press, Washington, DC

Chapter 4
Water-Related Risks in the Area of Dakar, Senegal: Coastal Aquifers Exposed to Climate Change and Rapid Urban Development

Hans-Rudolf Pfeifer, Alex Amiguet, Vibeke Brandvold, Silwan Daouk, Anne Gueye-Girardet, Carmen Hitz, Mamadou Lamine Ndiaye, Seydou Niang, Tomohito Okuda, Jessica Roberts, Cyril Royez, Torsten Vennemann, and Benoît Zen-Ruffinen

1 Introduction

1.1 Water Pollution in Developing Countries

Aquifer and surface water pollution of urban coastal areas and their vulnerability due to climate change and the increase of the world's population are clearly not limited to developing countries (Barlow 2003; UNEP 2010; WWAP 2009). However, in general, the main issues are:

- Ensuring a sufficient supply of drinking and irrigation water from wells with a certain distance to the coast
- The existence of sanitation facilities with a low pollution potential

---

H.-R. Pfeifer (✉) • A. Amiguet • J. Roberts • T. Vennemann • B. Zen-Ruffinen
Institut des dynamiques de la surface terrestre, Faculté des géosciences et de l'environnement, Université de Lausanne, Géopolis, CH-1015 Lausanne, Switzerland
e-mail: hans-rudolf.pfeifer@unil.ch

V. Brandvold
Department of Geology and Mineral Resources Engineering, Norwegian University of Science and Technology, Trondheim, Norway

S. Daouk
Hôpitaux Universitaires de Genève, rue Gabrielle-Perret-Gentil 4, CH-1211 Genève 14, Geneva, Switzerland

A. Gueye-Girardet
Fedevaco, Avenue Dickens 6, CH-1006 Lausanne, Switzerland

C. Hitz
Stadt Aarau, Umweltfachstelle, Stabsstelle Stadtentwicklung, Rathausgasse 1, CH-5000 Aarau, Switzerland

© Springer International Publishing Switzerland 2017
K. Sudmeier-Rieux et al. (eds.), *Identifying Emerging Issues in Disaster Risk Reduction, Migration, Climate Change and Sustainable Development*,
DOI 10.1007/978-3-319-33880-4_4

- An industry (including waste repositories), which is often not monitored for environmental concerns
- Adequate agricultural practices that avoid excessive fertilization and use of pesticides

Developing countries, with their increasing tendency to allow for the growth of megacities along the coast, often lack the economic power to investigate these issues. Additionally, they typically lack adequate sanitation facilities and farmers are rarely informed about environmental issues and sustainable agricultural practices. The consequences include health problems and/or insufficient food supplies (Pereira et al. 2002).

This chapter is concerned with the contamination and the shallow level of groundwater in the Dakar metropolitan area and possible solutions for mitigation. Increased salt content in ground and surface waters, as typical for coastal and semiarid areas, restricts the potential use of these waters for urban and rural areas. Excessive pumping of the upper freshwater layers close to the coast often draws up higher density salt water and depletes the freshwater volumes (Barlow 2003). This is intensified by the evaporation of surface or irrigation waters. Highly saline waters are not only non-consumable, but can also inhibit plant growth by means of salinization, alkalinization, or sodication, which decreases soil fertility and can even destroy the soil structure and its natural organic matter (Table 4.1). A lack of sanitation in urban areas or intense urban agriculture can lead to the pollution of ground and surface waters by elements such as carbon, nitrogen, and phosphorus, ultimately leading to the eutrophication of surface waters and also a contamination of water by pesticides.

Many coastal cities also experience flood events during the rainy season, which are often aggravated by climate change, and cause the overflowing of rivers and rising of groundwater levels.

1.2 The Situation in Dakar

The Dakar area is characterized by two typical settings (Fig. 4.1):

M.L. Ndiaye
530 rue Rouleau #03, Saint-Hyacinthe, Québec, Canada J2S 7H6

S. Niang
Laboratoire Lateu, IFAN, Université C.A.Diop, BP 206 Dakar, Senegal

T. Okuda
International Development and Regional Planning Unit, Department of Urban Engineering, Graduate School of Engineering, University of Tokyo, 7-3-1 Hongo, Bunkyo-ku, Tokyo 113-8656, Japan

C. Royez
urbaMonde, route des Savoises 15, CH-1205 Genève, Switzerland

Table 4.1 The three principal processes linked to high salt content in soils and waters

Process	RSC	Limits for EC, pH, and SAR	Geochemical reaction/damage
Salinization	<0	$EC > 4{,}000\ \mu S/cm$ $ESP < 15\ \%$ $SAR < 12$	Buildup of high Na^+, Cl^-, and SO_4^{2-} conc. by evaporation. Can inhibit plant growth
Alcalinization	>0	$pH > 8.5$	Buildup of high conc. of HCO_3^- or CO_3^{2-} in combin. with high Na^+. Precipitation of calcite ($CaCO_3$). Loss of organic matter by dissolution
Sodication	>0	$ESP > 15\ \%$ $SAR > 12$	Replacement of Ca by Na of the exchangeable compl. on mineral surfaces. Destruction of the soil structure (dispersion), reduction of its permeability

RSC ($mmol\ L^{-1}$) $= 2\ CO_3^{2-} + HCO_3^- - 2\ (Ca^{2+} + Mg^{2+}) = Alc_{carbonates} - 2\ (Ca^{2+} + Mg^{2+})$: residual alkalinity (Chhabra 2005)
EC ($\mu S/cm$): electrical conductivity of the soil solution or a soil-water slurry
ESP (%) $= Na^+/(Na^+ + K^+ + Ca^{2+} + Mg^{2+})$ with cations in mmol L^{-1}: exchangeable sodium percentage of cation exchange capacity of the solid part of the soil
$SAR = Na^+/\sqrt{((Ca^{2+} + Mg^{2+})/2)}$ with Na^+, Ca^{2+}, Mg^{2+} in mmol L^{-1}: sodium adsorption ratio of the soil solution
$ESP = -\ 0.0126 + 0.01475\ SAR$ (Richards 1954)

(a) An urban and suburban area with some preserved lowlands with abundant freshwater lakes, where urban agriculture is still practised (Niayes of Dakar area), but where groundwater contamination and flood risk have increased, particularly during the last 15 years.
(b) A rural, natural vegetation-poor dune and savanna area NE of Dakar, including the Lake Retba area, which is subjected to seawater infiltration, strong evaporation, and rapidly increasing real estate pressure.

As is the case for most African cities, Dakar's population has increased substantially during the last 40 years, from about one million in 1976 to about four million in 2012. The main migration of the rural population to Dakar took place in the 1970s, when the whole sub-Saharan region was subject to several droughts (Balbo 2006; Beauchemin and Bocquier 2004). As in other African cities, employment possibilities in the secondary industrial sector were rare and most migrants found work in the informal urban sector, such as in small commercial enterprises, often based on a widespread social network among migrants preexisting from the rural environment. This was especially the case for the Mouride group, an Islamic brotherhood that has played an important role in the production of groundnuts during the colonial period already (Broekhuis et al. 2004). Rural–urban migration has slowed down considerably during the last 20 years and the continuing rapid population increase is mainly due to a natural growth of the urban population.

Uncontrolled settlement of the lowlands during the dry periods until the 1980s, combined with an imbalance of water supply and adequate water evacuation, strongly increases flood risk during the rainy period today (from July to October, see discussion below).

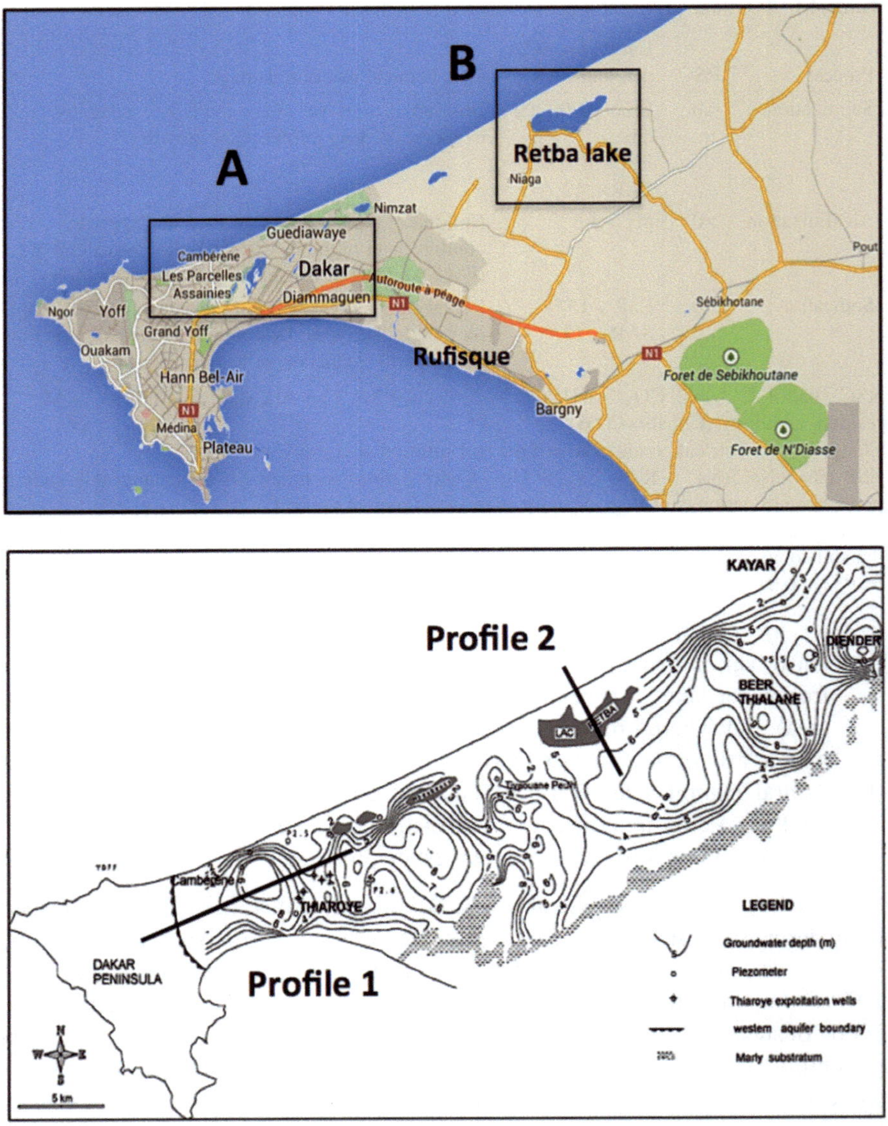

Fig. 4.1 A view of the Dakar peninsula in Senegal, Western Africa. *Top figure* (adapted from maps.google): *A*. Typical urban and suburban area with persisting urban agriculture and regular flood events in September, *B*. More or less preserved natural rural area with abundant salt lakes and important real estate pressure. *Bottom figure* (adapted from Faye et al. 2004): The study area is underlain by two shallow coastal aquifers formed of quaternary sands with a water level at a depth between 0 and 8 m: they have been used for drinking and irrigation purposes by hand-dug wells, shallow hand pump wells, and deep electric pump wells (Thiaroye between 1975 and 2005 and Beer Thialane between 1985 and 1999)

2 Methods

Between 2003 and 2012, several detailed studies on the potential impact of urban agriculture, solid waste, and irrigation practices were carried out based on detailed field studies (sampling of surface water and hand-dug wells, drilling and sampling of shallow wells, soil sampling) as a function of different locations and seasons (for example, see Daouk 2008; Gueye-Girardet 2010; Ndiaye 2009; Ndiaye et al. 2011a, b; Niang et al. 2012). In 2007, the Swiss NGO Urbamonde initiated the participatory urbanistic project urbaDTK in the often-flooded community of Djiddah Thiaroye Kao (town of Pikine). The interdisciplinary collaboration between environmental geochemists and town planners since 2011 has resulted in the studies of Amiguet (2013), Brandvold (2013), Okuda (2013), Roberts (2013) and Zen-Ruffinen (2013). These results were presented and discussed with local stakeholders, scientists of Dakar University, and engineers of several state ministries (Ndong and Royez 2012).

For water and soil, parameters such as electrical conductivity (EC, as an indirect measure of salinity), acidity (pH), and alkalinity (total amount of dissolved acid-buffering species) were measured in the field. Water was collected in plastic or glass bottles and soil in plastic bags for laboratory analyses. These included major and trace element analysis of water and soil; a stable isotope analysis of oxygen, hydrogen, carbon, and nitrogen in the water to determine the origin of the water; and dissolved ions and compounds in waters and microbiological tests for pathogenic microorganisms. In order to know more about pesticide use and epidemiological data, several inquiries were carried out (Amiguet 2013; Ndiaye 2009, 2012; Gueye-Girardet 2010). These references also provide detailed information for sampling sites and analytical details.

3 Results

3.1 Urban and Suburban Area

Most shallow freshwater aquifers have high nitrate concentrations (up to 700 mg/L, see Fig. 4.2) and are locally high in pathogenic microorganisms. Stable isotope analyses of nitrogen and oxygen can help determine possible origins of the nitrate contamination (Brandvold 2013; Gueye-Girardet 2010; see Fig. 4.3). At present, only about 70 % of the households are linked to sewers and only one wastewater treatment plant exists in Dakar, meaning that much effluent flows directly into the ocean. The remaining 30 % of households have septic tanks (D. Koné, pers. comm.). The isotope data indicates that leaking of the latter and the fact that wastewater is commonly poured on the sandy ground and then drained to the groundwater both are at the origin of a nitrate contamination. This confirms the

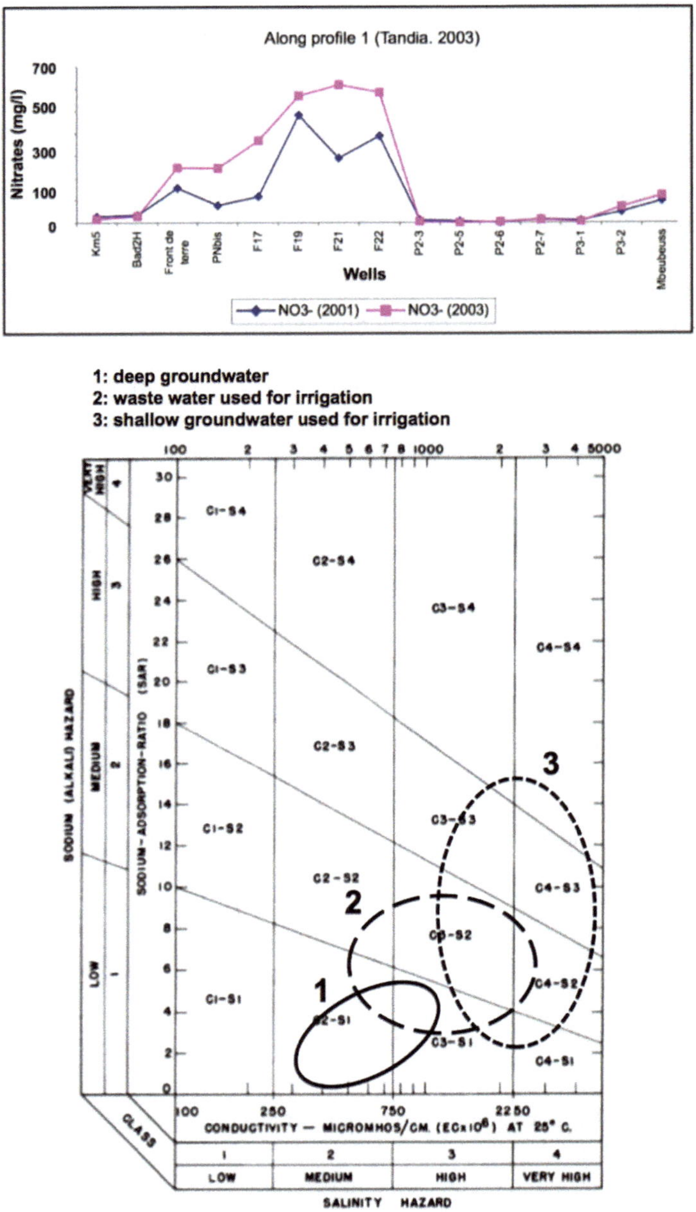

Fig. 4.2 Chemical characteristics of waters of the urban and suburban area (see A of Fig. 4.1, including the urban agricultural area of Patte d'Oie and Pikine studied in detail by Ndiaye 2009 and Gueye 2010). *Top figure*: typical nitrogen contents along profile 1 of Fig. 4.1 for 2001 and 2003, from Tandia and Demé (2003); our own results from 2004 to 2008 were rather similar. *Bottom figure*: Riverside diagram (Richards 1954) of the different water compositions used for

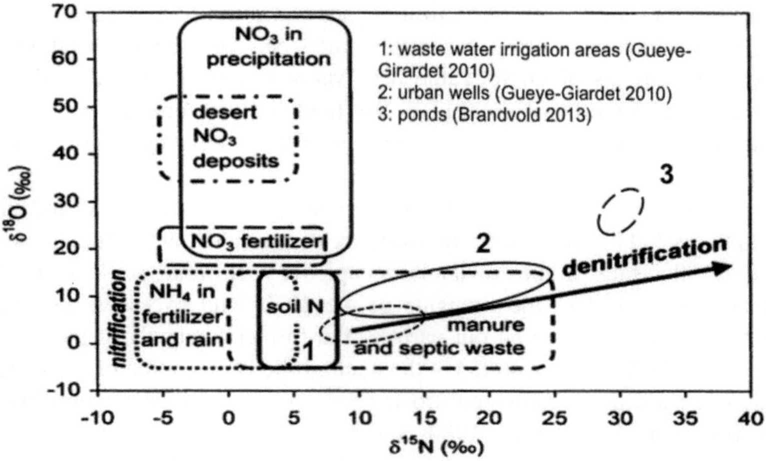

Fig. 4.3 Stable isotope composition of nitrogen and oxygen in nitrate of waters from the urban and suburban area (region A of Fig. 4.1), allowing the origin and the ongoing natural attenuation of the nitrate contamination of the groundwater by denitrification to be determined. Fields of different origins adapted from Chang et al. (2002)

hypotheses of Faye et al. (2004) and the conclusions of Ndiaye (2009), based on microbiological data.

In contrast to previous thinking, urban agriculture appears not to contribute much to this contamination. An interesting fact is that the data of Fig. 4.3 supports the existence of a natural attenuation process of this contamination via denitrification, especially in ponds, but has a negative side effect by producing a certain amount of the greenhouse gas N_2O (nitrous oxide, also known as laughing gas).

Most of the ground and surface waters have salinities (measured as EC) at the limit for drinking water, as well as for irrigation use (bottom of Fig. 4.2, EC: 3,000–5,000 μS/cm). Most drinking and irrigation water is hence provided by a 300 km long pipeline from Lake Guiers (border to Mauretania). Originally practised illegally by deviating old sewers, but today increasingly well coordinated, irrigation with partially treated wastewater has become an important practice for urban agriculture in Dakar (Faruqui et al. 2004; Gaye and Niang 2010; Gueye-Girardet 2010; Ndiaye 2009). Many farmers in suburban areas have no other choice rather than to irrigate with expensive drinking water or salty groundwater (Roberts 2013). In both cases, evaporation often leads to soil and soil waters with high

◄

Fig. 4.2 (continued) irrigation (marked with circles): 1: deep groundwater, only rarely used for irrigation, 2: waste water often used for irrigation, 3: shallow groundwater frequently used for irrigation. Their position with respect to the different quality-fields of the diagram allows judging the hazards for alkalinization (*y*-axis, sodium absorption ratio, SAR) or salinization (*x*-axis, electrical conductivity, EC) of the irrigated soil, cf. Table 4.1. Abbreviations of the different fields of the diagram: C1 S2: good water quality, C2 S2, C2 S1: medium to good quality; C1 S4, C2 S3, C3 S2, and C4 S1: medium to low water quality; C2 S4, C4 S2, and C3 S3: low water quality; C3 S4, C4 S3: very low quality; C4 S4: water not recommended for irrigation

probabilities for salinization, alkalinization, and sodication. An explanation of these terms, with which measuring parameters they can be assessed, and what the consequences are, is given in Table 4.1. Many waters have SAR values >10 and/or EC > 2,250 μS/cm plotting in the field of unacceptable waters with a pronounced sodium hazard (see the bottom of Fig. 4.2; Gueye-Girardet 2010; Roberts 2013).

The use of contaminated groundwater not only leads to health risks and changes in ecosystems, but because the groundwater is also at a shallow depth, it contributes to frequent flooding of the lowlands occupied by informal settlements. In the suburban part of Dakar, the droughts of the 1970s and excessive groundwater pumping practices have dried out parts of the former lowlands of the Niayes of Dakar (Fig. 4.4). This has resulted in the uncontrolled occupation of these areas by low-income rural immigrants and urban poor from the city center until the early 1980s. However, since the late 1980s, these informal settlements have been regularly flooded during the rainy season, usually during September (Amiguet 2013; Brandvold 2013; Cissé 2010).

Insufficient drainage canals alone do not explain these frequent floods, but the fact that some of the areas stay flooded even during the dry season suggests that there is a significant rising of the groundwater level. The groundwater level is seen to have risen for two major reasons (Ndong and Royez 2012; Okuda 2013):

1. The steady decrease in the pumping of the groundwater after 1987, because of its nitrate contamination
2. The slightly wetter climate since 1990

The steadily increasing freshwater supply from Lake Guiers and insufficient sanitation installations of the suburban part of Dakar also risk to further increase of this important imbalance of the water cycle into the future.

3.2 Rural Area Around Lake Retba

In the northeast of Dakar, the rural and natural character of the savanna with its villages, irrigated vegetable- and rain-supported agriculture is under stress, especially due to real estate speculation and the huge waste repository of Mbeubeuss-Malika (Niang et al. 2012). Two coastal aquifers extend well into this region and have served as a source for drinking and irrigation water until 1999 and should be reactivated soon (the Beer Thialane wells). At a distance of 1–2 km from the coast, the region includes younger and older dunes and shallow salt lakes of variable extension, the most important one being Lake Retba ("Lac Rose").

Recent detailed investigations have shown that at the end of the rainy season the salt concentration of the lake is up to four times that of ocean water (when surface waters are the most diluted by freshwater) (Zen-Ruffinen 2013). However, nitrates and other contaminants have low concentrations. A hydraulic connection to the ocean and the accompanying salt wedge is likely (see Fig. 4.5). Despite this,

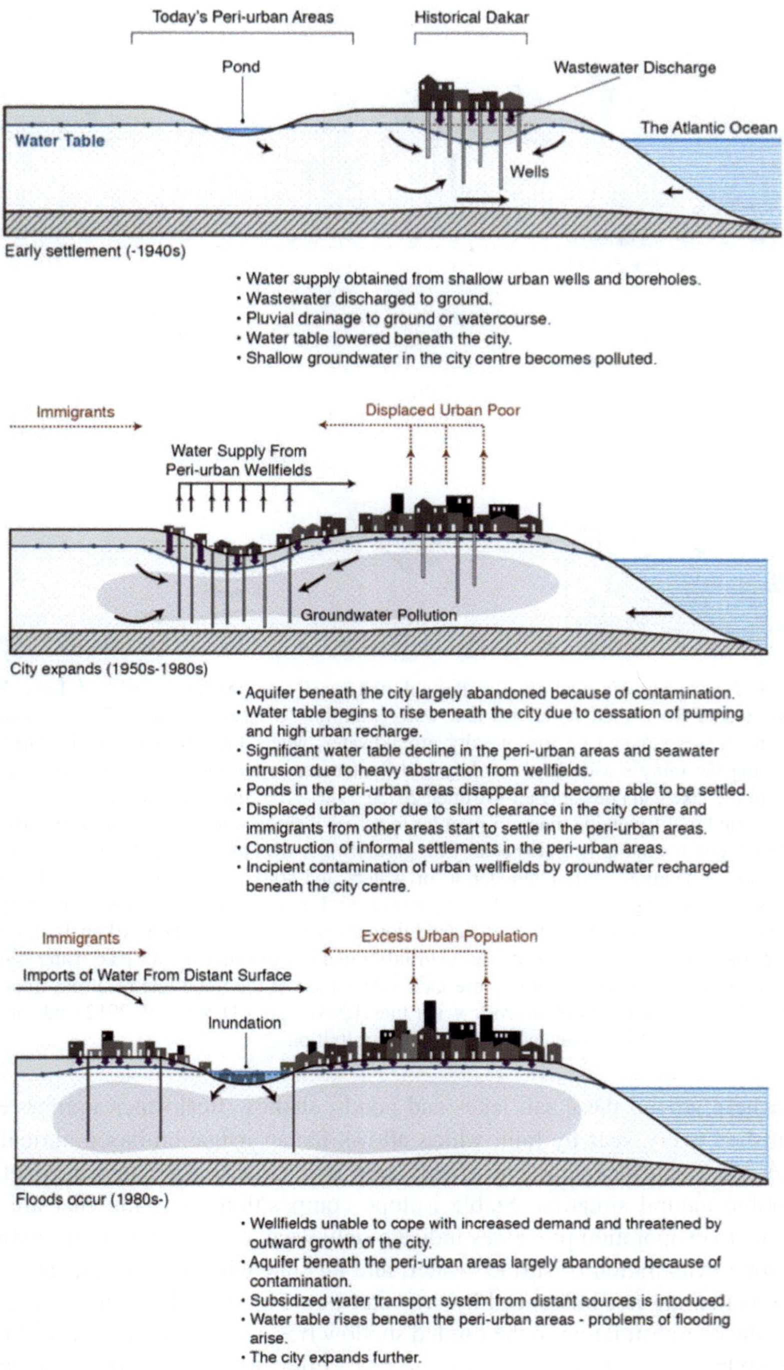

Fig. 4.4 Reconstruction of the relation between the urban development and the occurrence of floods in Dakar during the last 40 years (taken from Okuda 2013)

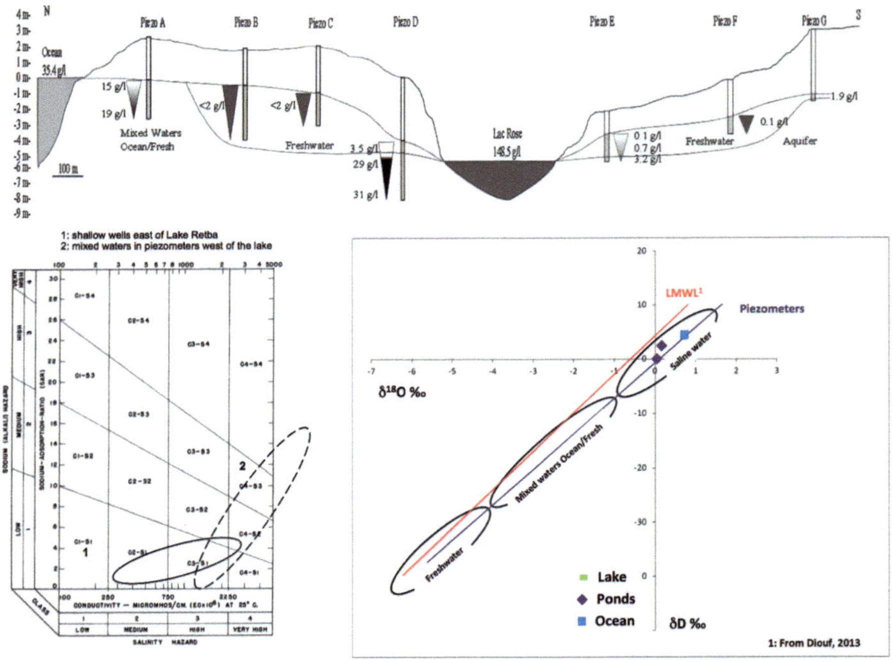

Fig. 4.5 *Top figure*: Vertical section through the rural area in the vicinity of Lake Retba, reconstructed from drilled boreholes and water sampling by Zen-Ruffinen (2013). Thanks to freshwater lenses situated on top of salty groundwater, where the lenses are replenished each year during the rainy season between July and September, irrigation-based agriculture and salt mining in the lake can coexist (Zen-Ruffinen 2013). *Lower left figure*: Chemical compositions of waters plotted in a Riverside diagram in order to assess if these waters can be used for irrigation (detailed discussion in Fig. 4.2). Especially for groundwater west of the lake, there is a high risk for salinization, sodication, and alcalinization of waters and soils. *Lower right figure*: The stable isotope values of oxygen ($\delta^{18}O$) and hydrogen ($\delta D = \delta^2 H$) of the different waters. Three types of water can be distinguished (each oval-shaped domain corresponds to about 40 analyses): freshwater (close to rain water), water mixed with infiltrated ocean water or salt lake water, surficial pond, and ocean water. The slightly more inclined position of the trend line resulting from these waters with respect to the local meteoric water line (LMWL after Diouf et al. 2012) indicates that all groundwater is subject to evaporation prior to infiltration

everywhere around these salt lakes and ponds, shallow freshwater aquifers exist, replenished every year by rain, which allows for an irrigation-based agriculture. Future real estate development of this area will have to take into account this vulnerable natural situation. Stable isotope compositions of water that are very sensitive to evaporation processes indicate only weak evaporation of groundwater but strong evaporation of surface water, and all start with typical rainwater compositions that are not influenced by seawater (see Fig. 4.5). The main risk of this area is the overabstraction of the limited shallow freshwater reserves that will result in the updraw and infiltration of saltwater as well as the salinization, sodification, and alcalinization of soils by excessive irrigation as described above (Roberts 2013; Zen-Ruffinen 2013).

Table 4.2 Summary of water related risks in the Dakar area

Type of risk	Compartment	Object	Main cause
Chemical	Groundwater	Nitrate, bacteria	High population density due to migration in the 1970s and insufficient sanitation
	Groundwater	Pesticides	Uncontrolled agricultural practices
	Groundwater	Salinization	Over pumping that induces an ocean water intrusion
	Groundwater, soils	Salinization alkalinization sodication	Increasing temperatures due to global climate change due to excessive irrigation to guarantee food security
Physical	Human settlements	Floods	Groundwater rise due to global climate change and cease of ground water pumping (due to contamination)
	Rain-supported agricultural fields	Droughts	Global climate change
Biological	Population	Infection by bacteria, viruses and helminthes	Contaminated ground- and surface waters

4 Discussion

In the Dakar area, water-related risks can be divided into three broad categories: chemical, physical, and biological and most of them can be further subdivided into two or four subcategories, respectively (see Table 4.2). Groundwater and soils are the most concerned by chemical risks, while those concerned by the physical risks are the human settlements and the land presently covered by rain-supported agriculture. Biological risks most concern the poorer populations of Dakar.

The causes that can be inferred from these different risks are a complex mixture of excessive migration and natural population growth, combined with an unsatisfactory governance of the negative effects of population increase (for example, with respect to urban sanitation and supply of clean water), inadequate agricultural practices (for example, the uncontrolled use of pesticides to help increase the crop yield), and environmental reasons that are in part induced by human actions, such as global anthropogenic climate change. In brief, the presented case study is a typical example of the complex interactions that can arise and the need to be both well monitored and subsequently managed when dealing with risk reduction, migration, climate change, as well as sustainable development, which is the general theme of this book.

5 Conclusions

In order to avoid serious problems for drinking and irrigation water in future, as well as for the health of the present ecosystem of the greater Dakar region, the already existing sewer system should be extended into low-income areas. With regard to sustainable water management and in order to reduce the general risks for the population and the ecosystem, only an interdisciplinary approach, inviting specialists to discussions about remediation plans, and including the concerned local population in such discussions, will likely be of success and reduce the risks to the local population (Ndong and Royez 2012).

Recycling as much wastewater or contaminated groundwater as possible for urban and suburban agriculture (necessary for a secure food supply for the city and to avoid long transport distances) will not only have direct economic advantages but would also help to lower the groundwater level in the urban area and therefore contribute to avoid floods. However, strict control of the water quality would also be required as well as an adequate solution to extract all types of waters during the rainy period, when agriculture does not require irrigation water.

References

Amiguet A (2013) L'eau, l'aigle et les lions: quand la population prend en main le problème des inondations de la banlieue de Dakar. Master thesis, Univ. Lausanne, 95p

Balbo M (2006) Urban growth, migration and development perspectives in Sub-Saharan Africa. Dipartimento di Pianificazione, Universita IUAV di Venezia, 7p. http://www.cespi.it/CESPI-SID/Balbo-Urban_growth.pdf

Barlow PM (2003) Ground water in freshwater-saltwater environments of the Atlantic Coast. Circular 1262, U.S. Geological Survey, Reston, Virginia, 121p

Beauchemin C, Bocquier P (2004) Migration and urbanization in francophone West Africa: a review of the recent empirical evidence. Urban Stud 41:2245–2272

Brandvold V (2013) Nitrate pollution of the Thiaroye aquifer in the suburban parts of the Dakar region, Senegal – Using chemical and stable isotope analyses to identify the sources of nitrate and the processes at the/surface water interface. Master thesis, Univ. Trondheim, 135p

Broekhuis A, De Bruijn M, De Jong A (2004) Urban-rural linkages and climate variability. In: Dietz AJ et al (eds) The impact of climate change on drylands: with focus on West-Africa. Kluwer Academic Publishers, Dordrecht, pp 301–321

Chang CY, Kendall C, Silva SR, Battaglin A, Campbell DH (2002) Nitrate stable isotopes: tools for determining nitrate sources among different land uses in the Mississippi River Basin. Can J Fish Aquat Sci 59:1874–1885

Chhabra R (2005) Classification of salt-affected soils. Arid Land Res Manag 19:61–79

Cissé O (ed) (2010) Les inondations à Dakar et banlieue: Mieux comprendre les causes pour des solutions durables. Acte du colloque 11 et 12 juin 2010, Amicale des Diplômés de l'Ecole Polytechnique de Thiès

Daouk S (2008) Caractérisation de la matière organique de différents amendements et évaluation de leurs impacts sur la matière organique d'un sol agricole de Pikine: Dakar, Sénégal. Master thesis, Univ. of Lausanne, 120p

Diouf A, Faye SC, Diehiou M, Kaba M, Faye S, Gaye B, Faye A, Englert A, Wohnlich S (2012) Combined uses of water-table fluctuation (WTF), chloride mass balance (CMB) and environmental isotopes methods to investigate groundwater recharge in the Thiaroye sandy aquifer (Dakar, Senegal). Afr J Environ Sci Tech 6:425–437

Faruqui NI, Niang S, Redwood M (2004) Untreated wastewater use in market gardens: a case study of Dakar, Senegal. In: Scott CA, Faruqi NI, Raschid-Sally L (eds) Wastewater use in irrigated agriculture. Confronting the livelihood and environmental realities. CABI Publishing, Cambridge, MA, pp 127–134

Faye SC, Faye S, Wohnlich S, Gaye CB (2004) An assessment of the risk associated with urban development in the Thiaroye area (Senegal). Environ Geol 45:312–322

Gaye M, Niang S (2010) Manuel des bonnes pratiques de l'utilisation saine des eaux usée dans l'agriculture urbaine. ENDA RUP, Dakar, 126p

Gueye-Girardet A (2010) Evaluation des pratiques d'irrigation, de fertilisation et d'application de pesticides dans l'agriculture périurbaine de Dakar, Sénégal. PhD dissertation, Faculté des Géosciences et de l'Environnement, Université de Lausanne, Suisse, 192p

Ndiaye ML (2009) Impacts sanitaires des eaux d'arrosage de l'agriculture Urbaine de Dakar (Sénégal). PhD dissertation, Université de Genève, 101p

Ndiaye ML (2012) Urban agriculture in Dakar, Senegal: health aspects related to polluted irrigation water. In: Robertson M (ed) Sustainable cities, Local solutions in the Global South. International Development Research Centre, Ottawa, pp 33–46

Ndiaye ML, Dieng Y, Niang S, Pfeifer H-R, Tonolla M, Peduzzi R (2011a) Effect of irrigation water on the incidence of Salmonella spp. on lettuces produced by urban agriculture and sold on the markets in Dakar, Senegal. Afr J Microbiol Res 5:2885–2890

Ndiaye ML, Niang S, Pfeifer H-R, Peduzzi R, Tonolla M, Dieng Y (2011b) Effect of irrigation water and processing on the microbial quality of lettuces produced and sold on markets in Dakar (Senegal). Irrigat Drain 60:509–517. doi:10.1002/ird.590

Ndong MS, Royez C (2012) Quelles synergies face aux inondations dans la Région de Dakar? Actes du séminaire de recherche-action du 16-18 nov. 2011 au Conseil Régional de Dakar. Dakar: IAGU, Urbamonde; Geneva

Niang S, Sarr B, Pfeifer H-R, Guèye-Girardet A, Gaye ML, Ndiaye ML, Dieng Y, Daouk S, Gaye CB (2012) Impact de la décharge de Mbeubeuss sur les ressources en eau et les sols. In: Cissé O, Mbeubeuss à Dakar au Sénégal (eds) Les décharges d'ordures en Afrique. Karthala and IAGU, Dakar, pp 33–87, 329p

Okuda T (2013) Comparative study of participatory development for strengthening adaptability: a case study of informal settlement upgrading in Dakar, Senegal. Master thesis, Univ. of Tokyo, Graduate school of engineering, 147p

Pereira LS, Cordery I, Iacovides I (2002) Coping with water scarcity. International hydrological program VI. Technical Documents in Hydrology, 58. UNESCO, Paris

Richards LA (1954) Diagnosis and improvement of saline and alkali Soils. Handbook 60, Salinity Laboratory. U.S. Department of Agriculture, Washington, 160p

Roberts J (2013) Assessment of elemental nutrients and of organic matter quality in cultivated lands of Rufisque-Lengdeng, Sénégal. Master thesis, Fac Geosci Env, Univ. Lausanne, 79p

Tandia A, Demé I (2003) Aquifère superficiel et pollution urbaine en afrique de l'ouest. UNEP/ UNESCO/UN-HABITAT/ECA, Rapport d'avancement du Sénégal. 39p

UNEP (2010) Afrique: Atlas de l'Eau. Division of Early Warning and Assessment (DEWA). United Nations Environment Programme (UNEP), Nairobi

WWAP (2009) The United Nations World Water Development Report 3: water in a changing world. World Water Assessment Programme. Paris, UNESCO and London, Earthscan

Zen-Ruffinen B (2013) The paralic region of the Lake Retba (Lac Rose) in Senegal: sweet and salt water layers in the groundwater. Master thesis, Fac Geosci Env, Univ. Lausanne

Chapter 5
Dike Risk: Revealing the Academic Links Between Disaster Risk Reduction, Sustainable Development, Climate Change, and Migration

Patrick Pigeon

1 Introduction: What Is "Dike Risk" and What Is the Relevance of Investigating It?

Dike risk addresses a specific and worldwide problem that disaster risk reduction (DRR) managers meet: the trend for protective measures such as dikes to unwillingly contribute to the preparation for future disasters. This has been investigated before by researchers primarily working on issues of flood risk prevention (Kates and Burton 1986; Saury-Pujol et al. 2001). In Hewitt's chapter, "Disaster risk reduction in the era of "homeland security"; the struggle for precautionary, preventive, and non-violent approaches" in this volume (Chap. 3), the issue of dike risk is mentioned as belonging to contradictions that DRR managers and researchers have to deal with. According to Hewitt, "Beck's 'risk society' itself turns upon a sense of contradiction: that increased risks are a consequence of obsessive and spectacular commitments to controlling dangers or threats, mainly in countries and by people who, otherwise, seem among the safest."

This chapter considers this issue in more detail. Dike risk occurs in cases where dikes, i.e., along rivers or seafronts, are considered as a means of reducing risk only. In these conditions, dikes encourage increased building behind them, which acts as a migration driver without necessarily taking other existing risks into account. In such cases, risk management appears contradictory: dikes supposedly try to protect; however they actually somehow contribute to the occurrence of new disasters. This becomes problematic because questioning dikes may increase an opposition to DRR policies and thus challenging their overall relevance.

P. Pigeon (✉)
Université de Savoie-CISM, Le Bourget du Lac 73376, Cedex, France
e-mail: Patrick.Pigeon@univ-savoie.fr

© Springer International Publishing Switzerland 2017

K. Sudmeier-Rieux et al. (eds.), *Identifying Emerging Issues in Disaster Risk Reduction, Migration, Climate Change and Sustainable Development*,
DOI 10.1007/978-3-319-33880-4_5

In France, dike risk has been judicially recognized since 2002 and it has been integrated into some risk prevention plans (RPP) (in French: "*plan de prévention des risques*"). In order to shed light on this contradictory perspective and how risk zoning types display it, this chapter traces back a short history of risk prevention plans in France.

Since Tazieff's law, which dates back to 1982, regulations at the national level, such as decree number 84–328 enforced on 3 May 1984, contribute to the management of risks by defining areas in which construction is or is not possible. At least two major laws include this RPP tool—*Loi Barnier* (n° 95-101, 2/2/1995) and *Loi Bachelot* (n° 2003-699, 30/7/2003)—but the basic structures of RPP are still maintained today. The *Préfet*, the French State local representative at the *département* level, enforces the risk prevention plan. This tool needs to be approved by municipalities and RPP zonings reflect trade-offs between representatives of the French state and municipalities. During the preparation phase, citizens, mainly landowners, can give advice on RPP zonings and contest them. But in the end, in cases of strong local opposition, the *Préfet* has been able to enforce contested zonation since 1995. The official statement now prohibits building in red zones while still permitting construction in blue zones. Yet landowners are encouraged to adapt their buildings to the existing risk level within the RPP zone and take building codes into account. White zones are seen not to be affected by the targeted risks. These zonings become legally binding as soon as the *préfet* takes an administrative decision ("*arrêté préfectoral*") approving the plan.

Naturally, these juridical regulations impact land-use values. Landowners and municipal authorities consider red zones a major constraint and therefore conflicts about RPP and zoning are commonplace. Measures hoping to reduce risk levels and zonings, such as dikes or other protective work, were and still are considered as a means to escape red zones and to allow for more building in flood-prone areas. During the discussion phase, areas that should have been ranked red without these protective measures could be turned into blue, but such transformations require the building of more protective measures such as dikes. This solution greatly increases the local acceptance level of RPP, especially where economic and demographic pressures challenge the relevance of their zonings.

Since the end of the twentieth century, however, some RPPs do not allow for such solutions even when there is strong opposition from local citizens or local authorities. This is exactly what creates dike risk juridically. RPP Chautagne (communes de la plaine de Chautagne 2004) and RPP Culoz clearly display a new red zone type, with a "flood-prone area in case of dike disruption, R2, to be conserved." The area takes the form of a strip on the RPP map with a width reaching 80–100 m behind the dikes, which the National Company of the Rhone River ("*Compagnie nationale du Rhône*," CNR) manages. These major dikes are maintained by CNR and are supposed to withstand at least a 100-year flood or higher; yet they prohibited the demarcation of a blue zone behind them and this seemingly contradictory decision raised strong local opposition.

On the contrary, in the municipality of Magland, flood-prone areas behind dikes are still reported as being in blue zones instead of red ones (such as at Halpades

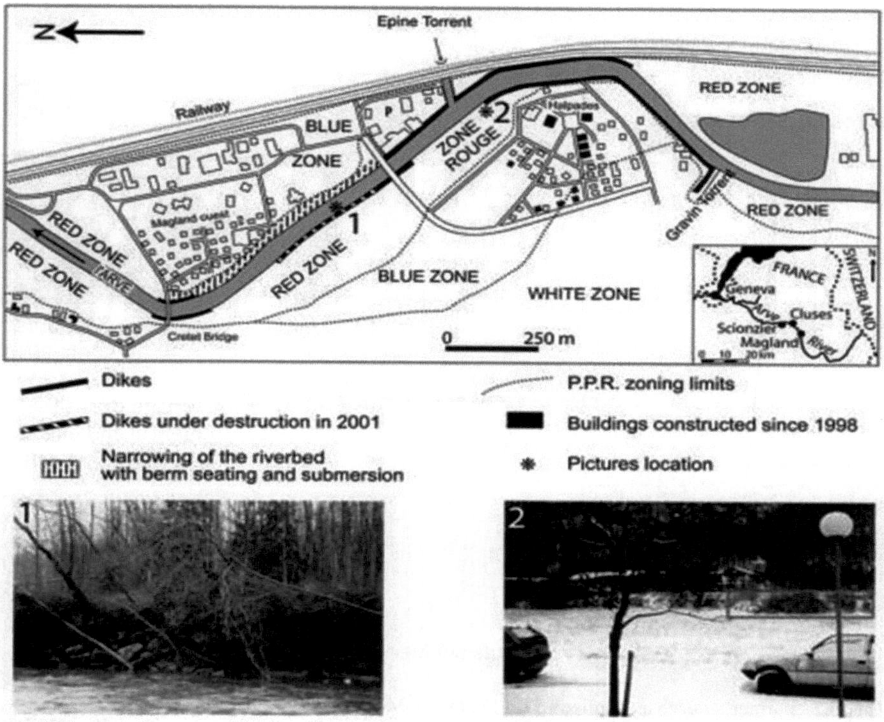

Fig. 5.1 PPR zonings contradictions in Magland (Haute-Savoie, France) display dike risk. In: Pigeon P (2013) Flood risk and watershed management conflicts in France: space for the river policies challenging resilience thinking. In: Warner JR et al. (ed.) Making Space for the River: Governance experiences with multifunctional river flood management in the US and Europe. IWA publishing, London, chap. 11; pp. 149–162, p. 155

building site, reported in Fig. 5.1). Such zoning results in significant infrastructural development, such as the newly constructed Clos-de-l'île subdivision, in the vicinity of Halpades building.

However, on the same RPP zoning map, some areas behind existing dikes have been ranked as red and this clearly shows the limitations of dike construction (such as with 1 and 2 in Fig. 5.1). RPP regulations rule that the maintenance of many of the existing dikes is not deemed necessary, considering that they could contribute to prepare a flood disaster (as with 1 in Fig. 5.1). Local landowners have hotly contested such decisions because they have reduced property values and considerably constrained land-use. Magland RPP does not mention dike risk as such, but the intent is similar. Under these conditions, the same RPP may display contradictory zonings, and the contradiction can be checked during field surveys as well.

Dike risk zoning, however, can appear even more contradictory and conflict prone when applied to already built-up areas. The 2007 RPP of La Faute-sur-Mer (Vendée) exemplifies this and observations about this can be accessed online at Pigeon (2014). The case shows how the buildings behind the dike are still seen to

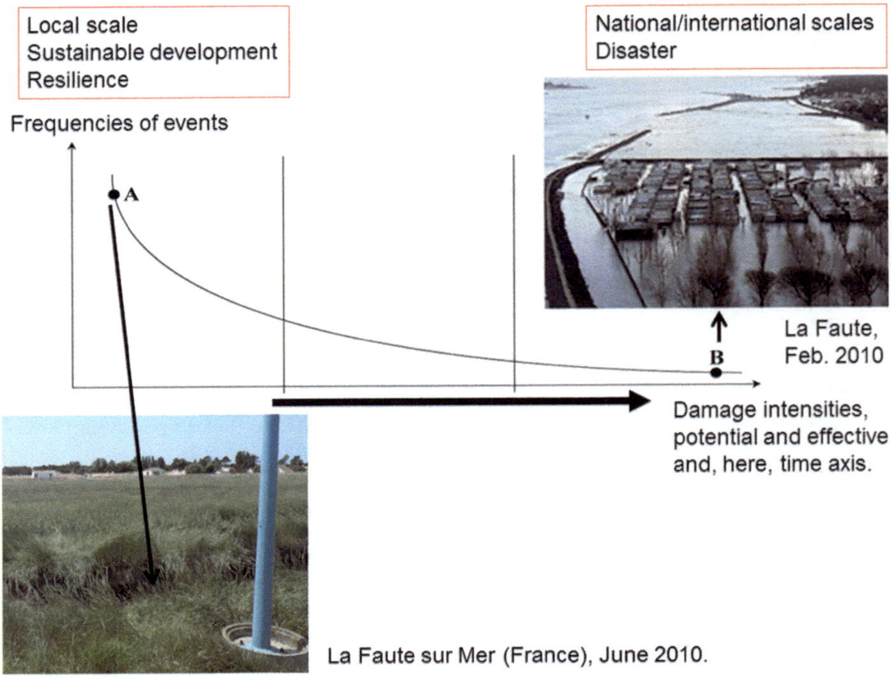

Fig. 5.2 Farmer's curve as applied to La Faute-sur-Mer disaster preparation

belong to a red zone. This zoning has been hotly challenged at the municipality level, and even more so by the landowners it has affected (Pigeon 2012a, 2014). Yet its existence has not discouraged developers from building even more in the former 2007 RPP blue zone nearby. It was here where more than 20 people drowned during the February 2010 disaster (see picture on the right, Fig. 5.2).

Dike risk is a new risk zoning that appeared officially in French risk management maps in the beginning of the twenty first century. The challenge is to understand and to better represent the contradictions this zoning displays, because it has heightened previous opposition to RPP zoning.

This new type of risk mapping shares common traits with other issues related with environmental management. They prove the possibility to prevent disasters while taking previous experiences into account (Kuhlicke 2011) and while displaying the limitations of this capability to learn from past experiences (Revault d'Allonnes 1999). Such situations may occur when discrepancies are found between what is expected from a policy or a model and what is measured during the experience period. Insurance companies meet such shortcomings today (Domenichini 2008). Poor social acceptance is commonplace in the literature on policies aiming at adapting to climate change or targeting sustainable development (Emelianoff and Theys 2001; Sheppard 2012). Understanding dike risk as a concept is therefore about demonstrating the ability to learn from the limitations of notions such as sustainable development, disaster risk reduction, climate change, and migration. There are many commonalities between all of these factors and addressing dike risk allows us to embrace a wider perspective and adopt new tools to make sense of any existing contradictions and conflicts.

2 How to Integrate Various DRR-Related Issues? Farmer's Curve and Resilience: A New Tool and Notion Consistent with SES Thinking

In DRR policy research, the Farmer's curve is often used to understand the relationship between various DRR-related issues. The curve, named after Mr. Farmer, an engineer who worked in nuclear power plants, depicts the inverse relationship between the frequency of events and the intensity of damages formerly experienced (Fig. 5.2). In other words, it belongs to the category of power laws (Bak 1999; Pigeon 2010). Power laws depict the inverse relationship between the frequency and the intensity of a given statistical population. Even more, it is linked with fractals (Binimelis Bassa 2013), and with complexity theories (Dauphiné 2003). In this sense, the Farmer's curve is consistent with the definition, Centre for Research on the Epidemiology of Disasters (CRED), gives to disasters as events of low frequency but high intensity of damages (Fig. 5.2).

The inverse relationship between event frequency and damage intensity evolves over time. Prior to diking, there is a high frequency of floods but a low-intensity damage level. The picture on the left side of Fig. 5.2 depicts such a case. It is of the estuary of Lay River in La Faute-sur-Mer, where flooding occurs on a daily basis. Diking such areas in the vicinity of a yacht harbor reduces flood frequency considerably, but contributes to the risk of a disaster. This is what happened in February 2010, as reported by the picture on the right side of Fig. 5.2. Indeed, such observations give the delusive impression that these diked-up areas have become flood risk free. Such trends become potential migration drivers towards risk-prone areas. Dike risk, however, tries to prevent diking side effects with the development of a new zoning type. Indeed, if diking is being used as a means for further construction without consideration of the associated risks, diking plays a part in preparing a low-frequency/high damage intensity level, or, in other words, a disaster.

This trend challenges the relevance of so-called natural risks or disasters (Pigeon 2005), and is in line with "the need to address the complexity of human-environment systems (also referred to as social-ecological systems)" (Renaud et al. 2013: 13). Indeed, it leads to admit links between dikes, hazards, and other elements of human settlements such as buildings, roads, and plants. Regardless of the context, all of these elements interact and even partially coevolve. After a certain time, dike risk reveals unwanted outcomes of these co-evolutions, as they contribute to prepare disasters. The fundamental question of where such unwanted outcomes manifest must still be taken into consideration, however. Under these conditions, dikes, hazards, and other components of settlement such as buildings become coevolving components of social-ecological systems (Cumming 2011; Lopez and Pigeon 2011). They can therefore be considered as the frame or basic structure of local systems, but it is crucial to remain aware of the different interacting components within these systems.

Under these conditions, it becomes possible to add a time axis to the "x" axis on the Farmer's curve (Pigeon 2010). As time progresses and if diking is being used as a means for legitimating the development of risk areas as a migration driver, Farmer's curve displays a trend towards increased disaster risk (Fig. 5.2).

Furthermore, the Farmer's curve is consistent with the notion of resilience. According to Gunderson and Holling (2002) or Walker and Salt (2006), resilience depicts the capacity of a system to maintain its basic structure during an event. Every disaster prevention policy is therefore trying to somehow increase resilience. This is also consistent with the qualitative and scalar component of the definition that CRED gives to disaster, that is, the "declaration of a state of emergency and the call for international assistance." Under such circumstances, local systems cannot maintain their basic structure, thus requiring help from other systems operating at other scales.

This is also why scalar discrepancies could be added to the Farmer's curve, for a disaster entails an event that exceeds existing local coping capacities and in turn requires external help sourced from systems operating at larger scales (Fig. 5.2). This is basically the sense that UNISDR gives to the notion of resilience (UNISDR 2013). The Farmer's curve may be used as a means of depicting the search for resilience, which is inherent to notions of sustainable development. In such a case, any measure trying to steer disaster management trends towards disaster preparation is seen as congruent with resilience searching at a variety of scales. This is also fundamental for research on climate change adaptation as not finding local acceptance of climate change adaptation policies would inevitably lead to increasing disasters in the future. It therefore stresses the necessity to develop adequate positive-feedback loops, which could somehow reduce the barriers to adaptation (Newman and Jennings 2008, p. 103). In such conditions, policies that target sustainable development and/or climate change adaptation, such as those that favor flood expansion plains and the reduction of diking in risk-prone areas, may also be found in line with resilience thinking (Mimura 2010; Pigeon 2012a, b).

Simultaneously, maintaining the resilience of a system can challenge the level of resilience in other systems. It can also demand the transfer of resources from other systems working at other scales, while co-evolving with the previous and local ones. This dynamic and multi-scalar approach of systems has been applied to case studies such as the Everglades (Walker and Salt 2006), which rely on investigations Holling (1973) previously made on forest management in Northern America. It founded the notion of social-ecological systems, which is now widely used in other research. Among the differences between these understandings of social-ecological systems, we find the reluctance to admit that humanity could function similarly to ecosystems, even if it were to be transformed by humankind. However, Berkes and Folke (2002) stress the importance of interactions between ecosystems and ecosystem management systems at various scales. Adopting such a mindset could help to reveal some of the contradictions and conflicts experienced in many management arrangements, regardless of the environmental issue taken into account, such as dike risk. This is very close to the trend that UNISDR uses to promote knowledge management systems as a tool to help find solutions to conflicts between various stakeholders at various scales within various systems (Duncan et al. 2014; Renaud et al. 2013). Whatever the interpretations, such tools and notions can only be useful in cases where they help to give us a sense of the conflicts and contradictions that each of the case studies reveal.

3 Farmer's Curve, Resilience, and SES Thinking: Methods and Applications to Various Field Cases Studies

The above presentation of what dike risk is provides us with strong rationale for investigating it, at least in France. Methods for investigating dike risk require a cross referencing of the information reported on RPP zonings with existing local field surveys. Such investigations may reveal discrepancies between what is mapped and what exists in reality. They identify DRR policy outcomes and limitations. They draw attention to zoning contradictions and to conflicts between various stakeholders affected by dike risk at various scales. In order to explain such DRR policy outcomes and limitations, it is necessary to trace the history of relationships between risk management and local settlement structuration. This is why it is crucial to consult the various archive types and cross-check them with field survey observations. Municipal archives (cadasters and Municipal council deliberations) and "*départementales*" archives allow us to trace the history of damages that are reported and the answers given, mostly in the form of corrective works such as dikes. Among other sources, they contribute considerably to investigating why and how dike risk happens. Research on hazards and their interactions with corrective works such as dikes is integrated into the assessment. This method allows us to cross reference the information, which comes from various sources, and to subsequently map the results, such as with Figs. 5.1 and 5.2.

In order to justify the relevance and operability of these methods, notions, and tools, we turn towards two main case studies. They also reveal the various roles migration plays during the processes towards disaster preparation or reduction.

The first case study is in the municipality of Magland, in the Northern French Alps. This French region experiences high demographic pressure, mostly driven by migration (Domenge 2009). Yet this is not the case with the municipality on which we focus here, at least during the periods we are considering (Léger and Gilbert 2014). This municipality displays negative net migration for recent periods, namely between 2006 and 2011. It suggests that demographic trends contribute towards greater political acceptance of dike risk zoning and of the contradictions they bear. In this case, migration was identified as a contributing solution to reducing disaster risk, at least partially.

The second case study is on the French Atlantic coast in La Faute-sur-Mer municipality. La Faute-sur-Mer municipality experienced a disaster in February 2010, which was made worse by strong local opposition to DRR policies and to dike risk (Pigeon 2012a, b). Migration played a significant role in increasing the demographic pressure in flood-prone areas. La Faute-sur-Mer is therefore a good example of the general demographic trend seen in the Vendée *département*. In France, *département* depicts a medium-size administrative subdivision, and Vendée gathers 282 municipalities. It shows how positive demographic growth in the area is primarily driven by migration. According to Bonnefoy and Gicquaud (2010), migration shapes the demographic evolution in Vendée, especially in coastal municipalities such as La Faute-sur-Mer. Positive net migration contributes to explaining the strong opposition to reducing dike risk, and can thus be seen as a contributing factor to the 2010 disaster.

The Magland municipality example demonstrates trade-offs found between a wide range of stakeholders. Such trade-offs affect municipality-wide flood-risk disaster prevention, which therefore implies relationships between dikes and corrective works, urbanization phases that are partly related with wider migration trends, and the evolution of the river's physical characteristics. Applying the methods that were presented above allows for the revealing of trends before and after the 1991 crisis (Fig. 5.3). It can therefore be seen that the decision to enforce RPP for this particular municipality may actually have led to the development of this crisis.

Before the 1991 crisis, the assumption that having more dikes justified the decisions to build more on flood-prone areas prevailed. However, it has become clear that these dikes have contributed to the transformation of physical characteristics of the mountainous river. This included the increase in the power of the river. Along diked-up parts of the river, specific power exceeds the value of 25 W per cm^2. Such an unwanted effect has been investigated by Piegay (1997) on one of river Arve tributaries, the river Giffre. It contributes to explaining why dikes and bridges are being damaged, if not being destroyed, as long as the Arve carves out its bed at the expense of surficial deposits. This trend has also been investigated and measured by Peiry et al. (1994). This is demonstrated in the two schemes coming from Piegay and Peiry, which are reported in Fig. 5.3. Until 1991, and most notably after the flood-related damages from 1968 to 1988, local authorities posited that the repairing of dikes was the best solution, and in this sense encouraged the building of more corrective works. This decision has led to the rationale for increased building behind dikes in accordance with the escalator effect that Sauri-Pujol et al. (2001) depict.

The scheme of Fig. 5.3 represents a system of the three aforementioned interacting and co-evolving components. Its evolution displays two positive-feedback loops, one in 1971 and the other in 1988. It can also be found consistent with the notion of resilience, because throughout the process, the local municipal system was able to maintain its basic structure despite any disturbances. Clearly, such was the case in 1968/1971 and 1988. However, the general trend was also towards enabling the disaster. In accordance with Gunderson and Holling (2002: 31), it proves that "resilience" not only encompasses every measure preventing a disaster but can contribute to enabling disasters as well. Should we change time scales and take into consideration the history of floods in the region (Mougin 1914), we could justify the use of the Farmer's curve here. It would depict the transition of

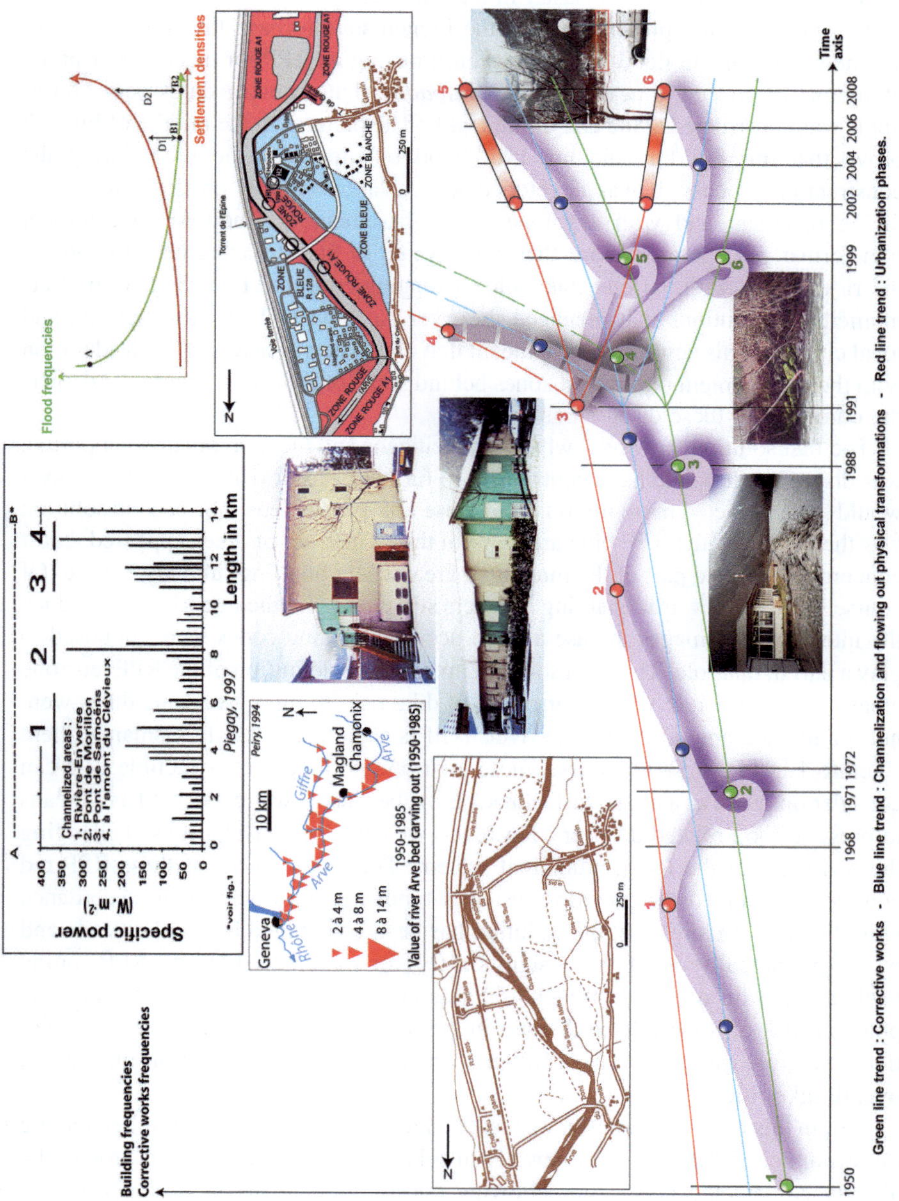

Fig. 5.3 Dike risk and resilience contradictions: socio-ecological systemic representation

these areas into places that had less flood frequency but potentially more damage intensities, for example, in the case of dike failures. On the upper right side of Fig. 5.3, we propose two separate trends for flood frequency and potential damage intensities, which display a disaster in the making.

In 1991, a local representative for the French state forbade the construction of any new buildings in flood-prone areas until having an RPP for this disaster-prone municipality. Indeed, a new building, *Halpades*, that accommodates poor households was constructed at the end of the 1980s. The pictures in the middle of Fig. 5.3 prove that this building did not take flood risk into consideration during the construction process. It was developed with a large window on the garden side facing upstream and with a garage in the basement that housed an electrical transformer post. It took more than three years to find an agreement to address the risk factors with various stakeholders ranging from the French state to inter-municipal institutions that managed the river Arve watershed, municipality, and local citizens. This new agreement meant that dike risk was taken into consideration with the enforcement of new red zones behind some of the dikes. Figure 5.3 depicts the outcomes of these discussions.

The first solution, however, which local citizens and the municipality supported, was discarded. It would have contributed to further disaster risk into the future as it would have attracted more migration to these risk-prone areas. The second solution was the zoning of the RPP blue areas. With this, a new set of dikes appeared, but it concerned a limited part of the municipal area, where buildings already existed. Of course, the trend towards enabling disasters still exists, but the intention is to reduce the intensity of damages in case a flood occurs. Building codes were supposed to play a part in that role. The third solution involves the definition of the RPP red zone areas and is consistent with the increase of dike risk. In this case, some dikes were not maintained anymore, and the river itself is seen to eventually dismantle them (Picture 1 in Fig. 5.1). Of course, these areas are no longer constructible. We can consider this zone as a negative-feedback loop, because it was conceived as a means to prevent flood-related disasters. However, we can also consider it as challenging the resilience of the local, municipal system. The enforcement of these RPP red zone areas changed the previous system's fundamental structures, in accordance with dike risk. In any case, this solution has been challenged by local citizens and by the municipality itself. Very significantly, for some stakeholders, RPP zoning itself along with dike risk recognition was considered more disturbing than the potential risk of flood-related disaster in the future. Once again, zoning contradictions therefore increased local conflicts, which inevitably revealed an increase in overall dike risk.

Yet, in this case, a consensus was reached in 1997. The trade-off implied having multi-municipal funding for a new set of dikes that were supposed to protect the already constructed areas, thus allowing construction in the new blue zones to resume. As a counterpart, new red zones and dike risk zoning would contribute to the reduction of the intensity of any future damages while simultaneously lowering flood height and speed for future floods. This expectation also concerns municipalities lying downstream. Among them, we find cities such as Cluses, Bonneville, and Annemasse (Fig. 5.1).

We suggest that this trade-off is aimed at maintaining the resilience of these systems at a regional scale and at reducing the overall damage intensities of future events. The resilience of local systems is not necessarily consistent with the resilience of other systems at regional scale, as it is here. The conflicts experienced prove the need to find a solution to various systems trying to maintain their resilience, but not evolving at the same scale. In such a case, it is also consistent with the notion of panarchy, because the solution implies a transfer of funding from the municipalities downstream. A multi-municipal institution, SM3A, working at the regional scale and in close collaboration with French state representatives, was in charge of the transfer which allowed the Magland municipality to adapt to flood risk. Indeed, the fact that the Magland municipality has recently become less attractive for in-migrants and has a higher percentage of poor households also helps to explain the trade-offs between 1991 and 1998. Construction pressure accordingly was comparatively lower here than for other municipalities downstream or upstream.

This was not the case with our second case study. The limited area and recent establishment in 1953 of La Faute-sur-Mer municipality (<7 km^2) encouraged more construction in flood-prone areas. The municipal power decided to enforce low-rise building regulations, which consequently increased even more construction pressure in relation with a positive net migration. Here again, RPP zoning and dike risk threatened the resilience of this local system. Having more red and blue zones, including any dike risk zoning, would lead to the transformation of the overall urbanization pattern of a region. Accordingly, it would require the reconsideration of the fundamental structure of entire municipal systems (Pigeon 2012a, b).

Yet, with 29 casualties in 2010, the municipal power was forced to take RPP upgraded zoning and more severe building regulation constraints into consideration. In this case again, we can easily apply the Farmer's curve. It illustrates how the zoning regulations enabled the transformation of an area which was naturally flooded daily to the construction of low-rise buildings behind dikes where 29 people drowned in 2010 (Fig. 5.2).

In turn, as a major stakeholder at the national scale, the French State tries to promote dike risk even more. Having more local, municipal systems that support the movement towards increased urbanization behind dikes implicitly increases further disaster risk. The existing institutional system linking major DRR stakeholders in France contributes to this situation. Since 1982 and the enforcement of Tazieff's law, the payment of an extra insurance premium on every contract concerning risks such as fire or car accident, that is, the so-called "cat-nat premium" is not adequate in the promotion of flood risk prevention. Over time, in view of the poor possibility to control the implementation of RPP building prescriptions, French insurers have met and are still meeting a clear, increasing trend of absolute losses. The cat-nat prime rate has increased from 2.5 % in 1982 to

12 % today. It is therefore clear that dike risk is intrinsic to efforts in sustaining the workability of existing local municipality systems.

Turning to resilience thinking, we again find conflicts between social-ecological systems at various scales, and between the various institutions that manage them (Berkes and Folke 2002: 124). Panarchy is also present here, as a means to compensate for the damages occurred by funding transfers from the state for insurance companies to operate at the national scale, in case of a major disaster. But panarchy does not mean that resilience of local systems, which contribute to the increase of disasters experienced, such as La Faute-sur-Mer, can be maintained. Now, the La Faute-sur-Mer municipality is no longer allowed to build low-rise buildings to meet growing migration pressures on the small part of its territory still open to construction. The basic structure of the municipal system thus has to be transformed. Such a transformation was driven, once again, not by the storm Xynthia in February 2010, but rather from the previous co-evolutions between institutional management systems at various scales, mainly within and between the French State and the municipalities. The impossibility to find an agreement can therefore be seen to have contributed to enabling the disaster in 2010.

4 Insights: What Can We Gain from Having a More Integrated and Systemic Approach of DRR and Environmental Related Policies?

First of all, an integrated and systemic approach contributes to the understanding of links between issues that are still considered separately today, such as migration, sustainable development, climate change, and disaster risk reduction. Contradictions and conflicts are inherent to each of these issues, and it is therefore crucial to reveal the relevance of environmental policies in spite of their limitations. The case studies of dike risk show how these contradictions and conflicts can manifest within the French context.

The trend of increasing disasters because of overreliance upon structural, corrective measures without taking their limitations into account is a global problem. Indeed, considering dikes as adequate risk prevention becomes a driver of in-migration into already risk-prone areas. Among these high-risk areas, we generally find poorer households, which thus increase the overall vulnerability of the community. Literature on climate change bears a very similar message—without an increase in adaptation acceptance, humankind is bound to experience an increase of disasters that will be fostered by gradual climate change (Mimura 2010; Newman and Jennings 2008). The Farmer's curve has the potential to describe such trends.

Second, SES understanding helps to identify the various conflicts between the political actors dealing with disaster risk reduction and climate change issues. Resilience helps us to identify the contradictions between systems and political actors managing them, at various scales (Cumming 2011). The mainstreaming of notions such as sustainable development, resilience, and disaster shows the simultaneous necessity and difficulty to find solutions to this highly conflict-prone problem—that is, unwanted trends towards the enabling of disasters—while trying to prevent them. The contradictions these notions bear may reflect the need to manage risk without having the possibility to get rid of it (Dauphiné and Provitolo 2007; Emelianoff and Theys 2001; Levy 2010; Pigeon 2012a, b). Resilience Alliance researchers have already tried to depict these contradictions through the concept of panarchy. This concept highlights the co-evolutions between systems and the political actors who manage these systems at various scales (Cumming 2011; Gunderson and Holling 2002).

This is also why today we find a trend towards using new tools such as knowledge management systems (Godfrey et al. 2010), and national observatories (as with ONRN, *Observatoire national des risques naturels*, in France, Nussbaum and Pigeon 2015). UNISDR strongly favors the implementation of such tools in accordance with international programs on disaster prevention, like KNOW-4-DRR (UNISDR 2013). Integrating environmental issues with tools as the Farmer's curve or knowledge management systems (which could be defined as managed panarchy) could therefore help to find solutions to disaster prevention and to increase the awareness of their limitations in the process.

References

Bak P (1999) Quand la nature s'organise. Flammarion, Paris

Berkes F, Folke C (2002) Back to the future: ecosystem dynamics and local knowledge. In: Gunderson L, Holling C (eds) Panarchy. Understanding transformations in human and natural systems. Washington, Island Press, pp 121–145

Binimelis Bassa MI (2013) Une nouvelle manière de voir le monde: la géométrie fractale. RBA France, Paris

Bonnefoy V, Gicquaud N (2010) Vendée: 240000 habitants supplémentaires d'ici 2040. Informations statistiques, 388. http://www.insee.fr/fr/themes/document.asp?ref_id=16922. Accessed 13 Feb 2015

Communes de la plaine de Chautagne (2004) PPR inondation, rapport de présentation. Service de la navigation Rhône-Saône, Serrières-en-Chautagne

CRED. http://www.emdat.be/criteria-and-definition. Accessed 13 Feb 2015

Cumming G (2011) Spatial resilience in social-ecological systems. Springer, Dordrecht

Dauphine A (2003) Les théories de la complexité chez les géographes. Economica-Anthropos, Paris

Dauphiné A, Provitolo D (2007) La résilience: un concept pour la gestion des risques. Annales de Géographie 654:115–125

Domenge M (2009) INSEE: Rhône-Alpes: de nombreuses arrivées de jeunes actifs et étudiants. La Lettre Analyse, 115. http://www.insee.fr/fr/themes/document.asp?ref_id=15188

Domenichini J (2008) Calcul de la sinistralité relative à des aléas "naturels" et anthropiques rares: systèmes d'information géographique (SIG), retour d'expérience et méthodes prospectives pour l'évaluation de sinistres inondation et terrorisme en réassurance. Paris, Thèse Université Paris I

Duncan C, Scherer S, Wade-Apicella S (2014) Background paper prepared for the global assessment report on disaster risk reduction 2015. HFA thematic review, relevant information on disasters is available and accessible at all levels, to all stakeholders. UNISDR, Geneva

Emelianoff C, Theys J (2001) Les contradictions de la ville durable. Le débat 113:122–135

Godfrey L, Funke N, Mbizvo C (2010) Bridging the science policy interface: a new era for south African research and the role of knowledge brokering. S Afr J Sci 106(5–6):1–8. doi:10.4102/sajs.v106i5/6.247
Gunderson L, Holling C (2002) Panarchy. Understanding transformations in human and natural systems. Island Press, Washington
Holling C (1973) Resilience and stability of ecological systems. Ann Rev Ecol Systemat 4:1–23
Kates R, Burton I (1986) Geography, resources and environment. Selected writings of Gilbert F White. Univ. Chicago Press, Chicago
Kuhlicke C (2011) The dark side of resilience. Exploring the meaning of resilience in the context of institutions and power. http://www.geographie.ens.fr/IMG/file/resilience/Kuhlicke_Dark%20side%20of%20resilience.pdf. Accessed 13 Feb 2015
Leger F, Gilbert A (2014) INSEE: Haute-Savoie: un des départements les plus dynamiques de France. La Lettre Analyses, 220. http://www.insee.fr/fr/themes/document.asp?reg_id=8&ref_id=20647. Accessed 13 Feb 2015
Lévy J (2010) Le développement urbain durable entre consensus et controverse. Info Géographique 3:39–50
Lopez J, Pigeon P (2011) Co-evolution between structural mitigation measures and urbanization in France and Colombia: a comparative analysis of disaster risk management policies based on disaster database. Habitat Int 35(4):573–581
Mimura N (2010) Scope and roles of adaptation to climate change. In: Sumi A, Fukushi K, Hirauatsu A (eds) Adaptation and mitigation strategies for climate change. Springer, Tokyo
Mougin P (1914) Les torrents de la Savoie. Soc Hist Nat Savoie, Chambéry, pp 1238
Newman P, Jennings I (2008) Cities as sustainable ecosystems. Principles and practices. Island Press, Washington
Nussbaum R, Pigeon P (2015) A national Public Private Partnership (PPP) platform for risk data sharing to stimulate DRR participative governance in France. UNISDR Scientific and Technical Advisory Group. http://www.preventionweb.net/files/workspace/7935_rnussbaumpppdrrinfrance.pdf. Accessed 17 Apr 2015
Peiry JL, Salvador PG, Nouguier F (1994) L'incision des rivières dans les Alpes du Nord: état de la question. Rev Géogr Lyon 69(1):47–56
Piegay H (1997) Principes nouveaux en matière de gestion des marges de rivières en tresses. Rev Géogr Alp 85(2):33–48
Pigeon P (2005) Géographie critique des risques. Economica, Paris
Pigeon P (2010) Catastrophes dites naturelles, risques et développement durable: Utilisations géographiques de la courbe de Farmer, VertigO - la revue électronique en sciences de l'environnement. doi:10.4000/vertigo.9491
Pigeon P (2012a) Paradoxes de l'urbanisation. L'Harmattan, Paris
Pigeon P (2012b) Apports de la résilience à la géographie des risques: l'exemple de La Faute-sur-Mer (Vendée, France), Vertig O, la revue électronique en sciences de l'environnement. doi:10.4000/vertigo.12031, 2012
Pigeon P (2014) a Notions à la une: désastre. http://geoconfluences.ens-lyon.fr/informations-scientifiques/a-la-une/notion-a-la-une/notion-a-la-une-desastre. Accessed 13 Feb 2015
Renaud F, Sudmeier-Rieux K, Estrella M (2013) The role of ecosystems in disaster risk reduction. UNU, Tokyo
Revault D'Allonnes M (1999) Le dépérissement de la politique. Flammarion, Paris
Saury-Pujol D, Roset-Pagès D, Ribas-Palom A, Pujol-Caussa P (2001) The escalator effect in flood policy: the case of Costa Brava, Catalonia, Spain. Appl Geogr 21:127–143
Sheppard S (2012) Visualizing climate change. A guide to visual communication on climate change and developing local solutions. Routledge, London
UNISDR (2013) Business and disaster risk reduction, good practices and case studies, Switzerland. http://www.unisdr.org/we/inform/publications/33428. Accessed 13 Feb 2015
Walker B, Salt D (2006) Resilience thinking. Sustaining ecosystems and people in a changing world. Island Press, Washington

Chapter 6
Jakarta: Mumbai—Two Megacities Facing Floods Engaged in a Marginalization Process of Slum Areas

Pauline Texier-Teixeira and Emilie Edelblutte

1 Introduction

Located in "developing" countries, Jakarta, the Indonesian capital, is a coastal city of more than 28.5 million inhabitants (around 10.2 million *intra-muros*). The Indian economic center, Mumbai, is the home of 12.5 million of inhabitants (see Fig. 6.1). Like numerous Asian megacities, Jakarta and Mumbai present high social and economic inequalities, especially because of high rates of rural–urban migration that forces new migrants to live in informal settlements, called *Kampungs*[1] in Jakarta. In Mumbai, more than 60 % of the population lives in informal settlements that occupy 8 % of the land area (Census 2011, see Fig. 6.1c). These populations are typically the most affected by flood-related disasters.

Exposed by their coastal location, Jakarta and Mumbai face high disaster risk. In Jakarta, the monsoon season brings intense rainfall every year (typically between October and April). Located within a deltaic plain, where 13 rivers empty into Jakarta Bay (see Fig. 6.1d), Jakarta is thus "naturally" highly prone to flooding such as those experienced in 1996, 2002, 2007 (see Fig. 6.1e), and more recently, in 2013 and 2014. At present, the city subsidence tends to worsen these floods. Water pumping activities and the weight of the city lead to natural sediment compaction, which sinks the city in some northern areas at a rate of 20 cm/year (Abidin et al. 2001). In the same way, Greater Mumbai offers a relevant context for DRR. Located in an estuary, Greater Mumbai is divided in two districts: the city of

[1] *Kampung* is the Indonesian word to designate urban villages, which compose the mosaic of Jakarta city.

P. Texier-Teixeira (✉) • E. Edelblutte
University of Lyon, CNRS, University of Lyon 3 Jean Moulin, Environnement Ville Société UMR 5600, F-69007, Lyon, France
e-mail: pauline.texier@univ-lyon3.fr

© Springer International Publishing Switzerland 2017
K. Sudmeier-Rieux et al. (eds.), *Identifying Emerging Issues in Disaster Risk Reduction, Migration, Climate Change and Sustainable Development*,
DOI 10.1007/978-3-319-33880-4_6

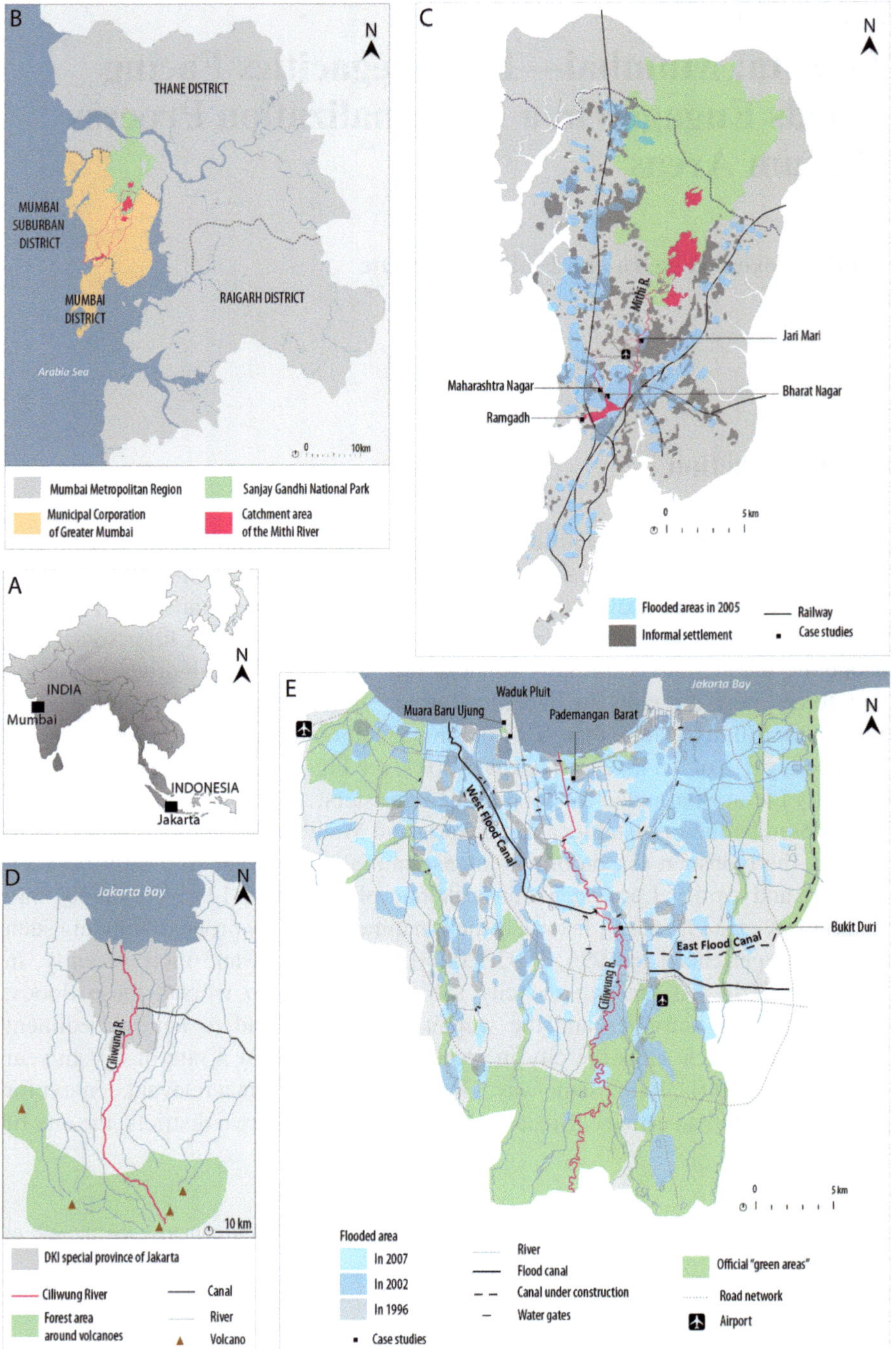

Fig. 6.1 Flood location maps of Mumbai and Jakarta (P. Texier 2009 and field work and master degree realized by E. Edelblutte, 2013)

Mumbai and the Mumbai suburban district (see Fig. 6.1b). Both these districts have been reclaimed from the Arabian Sea, linking the 11 original islands to the main land. According to a study conducted by the Organization for Economic Cooperation and Development (OECD) in 2007, Mumbai is one of the most vulnerable cities to flooding due to rising sea level (UN Habitat 2008). The city has suffered devastating floods in 2005 due to the overflow of a central urban river: the Mithi River (see Fig. 6.1c). In these two megacities, floods are also linked with bad sewage conditions, high urbanization rates, and questionable sanitation practices, which increase water stagnation and worsen injuries from floods. In 2007 in Jakarta, almost 60 % of the urban area was afflicted, where some 58–74 people died and a total of 400,000 people were affected. According to OCHA Indonesia, the 2013 flood killed 15 people and displaced 19,000, thus affecting some 250,000 people in total (Reliefweb 2013). In slums located near the main Ciliwung River, the water level reached 11.2 m from the thalweg. Floodwaters destroyed 100 houses located in poor informal settlements. In 2005,[2] Mumbai experienced heavy rainfall, with 944 mm rainfall recorded in 24 h. Besides hilly areas, forests, and lakes, some 22 % of the Greater Mumbai area was submerged by water from 26 to 27 July 2005. In some low-lying areas, the water level reached up to 6 m.

The biggest cities of India and Indonesia therefore became host to a concentration of economic, social, political, and ecological issues, but had the means and resources to cope with hazards and disasters. Both cities are the headquarters of the largest banks, financial institutions, and insurance and mutual fund companies. The two cities are thus the economic and financial centers of their respective countries.

Based on these two case studies, this comparative chapter provides an overview of the disaster risk management strategies implemented in Jakarta and Mumbai, and assesses their efficiency in reducing disaster risk in exposed slum areas. Secondly, this chapter analyzes why these strategies have failed to reduce vulnerability of communities living in slums so far, by explaining the underpinning daily social, economic, and political constraints that create this particular risky configuration. The link between disaster and migration, which is arguably the main root cause of vulnerability, is specifically discussed here. The chapter then offers some possible solutions at both an international and local scale, which propose alternatives to the current flood mitigation strategies and focus on social measures that work to de-marginalize poor vulnerable people, especially migrants, through participatory daily empowerment programs such as DRR strategies.

---

[2] No other disaster linked to floods was detected in Mumbai since then.

2 General Setting

Understanding the root causes of vulnerability is crucial within the framework of applied research, which aims to improve DRR strategies. This research focuses on vulnerability and relies on two opposed conceptual paradigms, each one explaining a different way to elaborate DRR strategies. In the first conceptual framework (often referred to as the "extreme" or "dominant" paradigm), disasters are considered mainly as resulting from natural hazards, extreme in intensity and rare in time at the human scale. In this framework, people's vulnerability is only considered within a hazard-related point of view, since response and behaviorism are exclusively considered as dependent upon the perception they have from hazards (Burton et al. 1978; Hewitt 1983; Kates 1971; White 1970). Such a paradigm premises a declaration of war on natural phenomena, and filters a strong technocratic vision of disasters into risk management policies. DRR, under this concept, only focuses on heavy structural measures (such as dyke or polder toward floods).

This hazard-related way of analyzing vulnerability was radically criticized in the 1970s (O'Keefe et al. 1976). It was only then that disasters were considered for the first time within daily patterns. Peoples' behavior is understood as being constrained by structural, social, cultural, economic, political, and non-hazard-related factors (Cannon 1994; Cuny 1983; Gaillard 2007; Hewitt 1983; O'Keefe et al. 1976; Torry 1979; Waddell 1977; Wisner et al. 2004). Vulnerability is therefore "intimately connected with the continuing process of underdevelopment recorded throughout the world," and linked with marginalization from society (O'Keefe et al. 1976:560). To understand the marginalization process experienced by vulnerable groups, it appears essential to take into account all the factors and dynamics which lead to local unsafe conditions.[3] In this framework, policies tend to be based on non-structural measures that aim to fight poverty and reduce social inequalities.

At an international scale, policies led by the United Nations Office for Disaster Reduction (UNISDR) in terms of DRR have clearly followed this new framework. The Hyogo Framework of Action 2005–2015 (HFA) set out to increase "the resilience of nations and communities to disasters" and through the establishment of National Platforms to achieve these objectives (including India and Indonesia). The Sendai Framework for Disaster Risk Reduction (SFDRR) 2015–2030 addresses the new challenges to governments and proposes four priorities for action: (1) understanding disaster risk; (2) strengthening disaster risk governance to manage disaster risk; (3) investing in disaster risk reduction for resilience (which includes a new daily vision of preparation to disasters); and (4) enhancing disaster preparedness for effective response, and to "build back better" through recovery, rehabilitation, and reconstruction. In this new framework, there is a clear prioritization of a gendered and participatory approach that facilitates strong partnerships

[3] As recommended within the PAR model from Wisner et al. (2004).

between stakeholders and whole societies. There is thus recognition of the importance to consider local knowledge as well as scientific knowledge about disasters and preparedness. The necessity to integrate DRR into sustainable development strategies in a systemic and multi-sectorial approach is also emphasized. These main points present future important challenges for both Jakarta and Mumbai when considering the specific case of their slum areas in facing floods.

3 Methods

The results analyzed in this chapter are based on two different field investigations. In Jakarta, research was carried out between 2005 and 2007 as part of a PhD (Texier 2009), with three field missions of 7 months each. Research in Mumbai was carried out in 2013 (over a period of 6 months), as part of a French research program, titled UNPEC (Urban National Parks in Emerging Countries and Cities).

Qualitative methods used were predominantly semi-structured interviews conducted with 14 institutional and non-institutional stakeholders in Jakarta and 32 in Mumbai. These were carried out to understand (1) institutional strategies and DRR-related measures and (2) stakeholders' perceptions of the root causes of disasters and societal responsibility, and what solutions they envisage for the future in slum areas. At a local scale, questionnaires were distributed in order to learn about risky behaviors (i.e., with regard to the decision-making processes of settlement in hazard-prone areas and exposition, the refusal to evacuate during flood events, the daily practices of waste disposal (such as throwing their waste into the river), water access (water pumping), risk perception, and the root causes of vulnerability from people living in informal settlements). All perception questions were left open without options provided in order not to influence the answer. A random sampling (without stratification) was applied in every hazard-designated area. One hundred and twenty surveys were conducted in Jakarta, and these were divided into four different flooded informal *kampungs* during the dry season. The *kampungs* (or towns) that we studied included: *Waduk Pluit* and *Muara Baru Ujung* near the sea of Jakarta bay, *Pademangan Barat* in the lowest part of the Ciliwung canal in North Jakarta, and *Bukit Duri* near Ciliwung River in South Jakarta (see Fig. 6.1e). In Mumbai, 60 surveys were conducted in four informal settlements along the Mithi River and its main tributary: the Vakola Nalla. The neighborhoods targeted were *Maharashtra Nagar*, *Bharat Nagar*, *Jari Mari*, and *Ramgadh* (see Fig. 6.1c). Classified as "slums" by the municipality and exposed to flooding, these settlements have homogeneous physical characteristics (for example, semi-permanent to permanent housing, access to drinking water, electricity, and sewage system). The people living in these informal districts are especially vulnerable as they are not only located in hazardous areas but also have cultural, social, and economic vulnerabilities that come predominately as a result of migration. These include high unemployment rates and poverty (their monthly incomes do not exceed US$300), low levels of education, and limited access to health care, water networks, and sewage systems. For both cases, we use descriptive statistical methods using Sphinx software to analyze the data.

Direct and participatory observations were also carried out in Mumbai and Jakarta, especially during the heavy floods that occurred in February 2007 in the district of *Bukit Duri*, in South Jakarta. An insider position for research was then adopted in order to participate in most activities in the neighborhood during the flood event, and to analyze the relationship among the community members (with close attention paid to gender and power relationships) and between community members and external stakeholders. The coping strategies that were adopted by slum dwellers marginalized from official assistance were also observed.

4 Results

1. **Institutional response from Mumbai and Jakarta municipalities to mitigate floods**

   (a) Structural measures

   The data from semi-structured interviews conducted with stakeholders combined with an analysis of literature about policies and strategies in both megacities demonstrate that the issue of flooding in Jakarta and Mumbai has been addressed through a technocratic (top-down) and hazard-focused approach. In both cities, flood control represents the largest part of the budget allowed to address flood issues. In Jakarta, the stakeholders in charge of DRR are the army, the land-use planning department, and the center of meteorology and geophysics, with a new branch of coordination through the National Disaster Mitigation Agency (BNPB). Any decisions surrounding land-use planning and management (Dinas Tata Kota and BPN) are made through the department of Public Works (PU DKI), through large-scale irrigation work, the construction of dikes, canals (e.g., the East Flood Canal, see Fig. 6.1e), and dams to channel the rivers. Still in 2015, the Jakarta administration declared an emergency response status that allowed for a regional budget of US$212 million to be spent on flood-mitigation projects that would acquiesce the land needed to build a new canal to connect East Flood Canal and West Flood Canal.[4] In Mumbai, the disaster of 2005 turned floods into the new common enemy of the city and forced DRR into the limelight as a priority for the municipality. However, to date, technical and structural solutions to contain the Mithi River have been adopted as the main strategy to mitigate flood risk, as well as restoring the river that has turned into sewage after years and years of disinterest. Such structural policies have been justified by the impending

[4] According to the Jakarta Post, 15th of February 2015.

threat of climate change that is supposedly going to increase the intensity of monsoons in Mumbai and other extreme climate events. Following this, a special agency, the MRPDA (Mithi River Protection and Development Authority), was created in order to design policies and implement them through two other agencies: the MMRDA (Mumbai Metropolitan Region Development Authority) and the MCGM (Municipal Corporation of Greater Mumbai). Based on hydrological and environmental improvement, the structural measures were supposed to address fundamental drainage issues in Mumbai.

In both cities, this technocratic "command-and-control" way of management—which can be qualified as top-down with very little collaboration with the local scale (Twigg 2004)—has led to the relocation of the city's most vulnerable people from riverbanks. In Jakarta, the government strategy was to free the riverbanks from informal settlements. In Mumbai, the Mithi Redevelopment Project aimed to rehabilitate riverbanks and support the widening of the Mithi River. Both cases involved the destruction of informal settlements and relocation (see Fig. 6.2). These evictions were often violent, generally involving police and bulldozers (Harjoko 2004; Harsono 1999). Statistics about evictions are hard to find, but between 2000

Fig. 6.2 Evictions and relocations of informal settlements in Jakarta and Mumbai

and 2005, an estimated 63,676 people were evicted and 1,592,011 people were threatened with expulsion from poor districts (Harsono 1999; Human Rights Watch 2006; Urban Poor Consortium 2005). Few accommodations were provided by the Jakarta municipality in apartments (*rumahsusun*) within the city or via transmigration to Sulawesi or Borneo (Sevin 2001), the most external provinces of Indonesia. According to the Department of Manpower and Transmigration, the Indonesian Government has, since 2006, had the objective of relocating 20,000 families per year. In Mumbai, the municipality offers accommodations to any eligible people (those who live in houses that have been built before 2000 and can prove it), but any others are denied access to any new accommodation. In Ramgadh (see Fig. 6.1c), only 20 % of evicted families were relocated; all others did not receive any compensation, which also occurred in Jakarta during the building of East Flood Canal—families often received no more than US$50 when they were evicted (Human Rights Watch 2006).

(b) Non-structural measures

Additionally, Jakarta and Mumbai have implemented non-structural measures of crisis management based on an electronic early warning system and investment in meteorological stations. These measures were based on a reactive way of management, centered on the crisis. That is, "before" is preparedness, "during" consists of providing boats and food in refugee camps, and "after" consists only of cleaning the city and repairing riverbanks and roads. Along with this, the municipalities of Jakarta and Mumbai have set up awareness campaigns on flood hazards and sanitation in informal settlements. Interviews indicate that this is because official stakeholders largely consider the poor communities of flooded informal settlements as responsible for floods. Indeed, in both cities, poor people settle on riverbanks and are said to throw their garbage into the rivers, thus increasing flooding phenomena. This practice is shown to be true because, for example, our questionnaire shows that 92 % of respondents said that they were throwing their waste into the Ciliwung River in Bukit Duri.[5] Based on such practices, it appears to Jakarta's administration that it is necessary to "educate" these people. In these cases, they are considered as ignorant about flood phenomena and undervalue the risk level that they are exposed to. The government now considers a low risk perception and low level of education as the only reason why people practice such waste disposal behaviors, which motivate the annual governmental awareness campaigns.[6] The campaigns rely upon one-way communication through

[5] The impact of this practice on flooding compared to other factors, i.e., rainwater run-off from impervious urban surfaces and land deforestation in upper catchment area, is not proved and quantified.

[6] We will see below the real reasons explaining this practice and the various solutions that can be proposed.

the erection of posters in the local municipality building (not really accessible for those who rarely go to this official building) with a lot of text that many people cannot read or understand. Indeed, people living in targeted districts are rarely involved in the conceptualization of these campaigns, especially in Jakarta, where some women's groups are now participating in door-to-door campaigns to understand people's knowledge about hygiene (especially for dengue prevention).

2. **Unadapted responses to reduce the vulnerability of people at risk**

   (a) Structural measures: a tool to serve other goals
   In both Jakarta and Mumbai, these eviction programs are presented as a tool to mitigate flooding by reducing hazard exposure and rehabilitating the riverbanks. In Jakarta, the programs are supposed to (1) increase green areas in order to prevent hazards and (2) reduce vulnerabilities by relocating slum dwellers to safer areas. These arguments are flawed since many relocation apartments are in fact built on riverbanks, and include shopping centers or luxury apartment complexes corresponding to public-private partnership projects (see Fig. 6.2). Indeed, despite the idea to preserve green areas within the city to prevent floods (which was targeted to reach up to 27.6 % of green areas in the 1965–1985 Masterplan), the allocation of green areas has not reached this, with only 6.7 % of Jakarta being green today. Pantai Indah Kapuk Residence in West Jakarta, Kelapa Gading in North, and commercial areas in the city forest of Cibubur are some examples of the shifting process of water-absorbing areas into a commercial site. From 2001 to 2008, 94 shopping centers were built (according to spatial analysis from satellite imagery; UPC 2005). These heavy concrete buildings have yet worse consequences on rainfall infiltration than previous informal housing built in wood material without concrete foundations. Moreover, the relocation of buildings presents a low quality of public services (University of Westminster and ITB 2000). In Mumbai, even if the recommendations made by engineers were well-designed and necessary to mitigate the flood risk, the MRDPA chose not to take all of these recommendations into account. The MRDPA redesigned policies and implemented them haphazardly only where informal settlements encroached the banks of the Mithi River. Additionally, there have been no discussions about the effects of the Bandra Kurla Complex (the new Business District Center), built on 220 ha of mangroves in the early 1980s that contribute to the concretization of the flood plain. Moreover, the retaining wall works as a clear division between the river (non-building land) and the land (building land). It thus serves to prevent slums from encroaching on the riverbank and also creates a clear saved space for future private development.

   There have however been some recent initiatives taken to solve the problem of flooding in Jakarta: the regreening of the upstream area of Bogor (where up to 40,000 eucalyptus trees were to be planted in 2015);

the Green Wall Project involving local families; and the two new reservoirs upstream (Megamendung and Sukamahi) built by the Public Works Department (yet one more structural measure). A program aimed at improving social security is still missing however, and while it is yet to be implemented, a project proposed by the Department of Housing in 2012 to build *rumah susun* above the river would be a huge and innovative solution since it would be the first of its kind to not imply necessary evictions (Amri and Rimadi 2012).

(b) Structural measures: unadapted to reduce vulnerability
Both the Mithi Redevelopment Project in Mumbai and the eviction and relocation project along Ciliwung River in Jakarta were based on a development-induced displacement and resettlement or DIDR (Oliver-Smith 2001), which did not appear adapted for vulnerability reduction. In Jakarta, moving from traditional *kampungs* to apartments disrupted social and cultural links by destroying a community-based way of life. Such social networks are a precious resource and play a huge role during disaster events. Inhabitants lost a real means to better cope with daily and extreme difficulties since most of them could not afford to pay rent[7] and were eventually forced to move to unsafe locations. Moreover, moving can affect their livelihoods as relocation often involves people quitting their jobs to find another one in a different location. In Mumbai, slum dwellers along the Mithi River live under constant threat of eviction. As work is still in progress in Mumbai, some informal settlements have been living under this threat since 2005. It gives them uncertain and impoverished futures, which are finally destroyed when the eviction becomes reality. Powerless, they cannot neither invest in or nor improve their housing or living conditions. In Jakarta, relocation typically results in further economic and social deprivation. Even if 69 % of the respondents state that they would like to be relocated, eviction is considered the main risk for 33 % of them as they are afraid that they would be ineligible for a new house.

(c) Non-structural strategies: based on misperceptions
Awareness campaigns as well as non-structural strategies also appear to be unadapted measures because they are conceived to educate inhabitants who live on riverbanks and thrive based on a lack of on-the-ground implementation. Results from the survey conducted in slums in Jakarta show that behaviors that are considered as risky are not linked with low perceptions of danger. On the contrary, many slum dwellers have a good knowledge of the natural and human causes of floods as almost all of them have already experienced several flood events in their lives. They have progressively built their own knowledge about floods (e.g., through precursor floods, putting their belongings in security, observing high potential water level

---

[7] A small apartment costs around 300 euros when most of the population has a monthly income of only 250 euros.

which could be reached, knowledge about risk of drowning and electrocution, and also the development of a community-based care system to cope with these disasters). Although most of them were in-migrants from rural provinces (75 % of respondents came from other provinces and 78 % grew up in rural areas), they settled in these slums more than 10 years ago, already knowing that they were settling in flood-prone areas and being fully aware of all potential consequences (e.g., material losses, drowning risks, and water-borne diseases). Waste evacuation was highlighted by 33 % of all respondents, and by 40 % of Bukit Duri respondents (Fig. 6.1e). Waste was thought by respondents to be the second cause of flooding behind rainfall from upstream (quoted by 50 % of respondents). Moreover, among the 92 % of respondents who threw their garbage into the river, 96 % felt responsible for the insalubrity of the rivers.

In Mumbai, instead of learning how to deal with floods, structural interventions (e.g., retaining walls) offer a false sense of reassurance and strengthen the misperceptions surrounding floods. They prevent the population from seeking alternatives to mitigate flood risk and enhance their sense of helplessness to flood. Based on the 2005 event, a flood is defined as a catastrophic disaster that affects the entire city (if the height of water is more than 1 m). Only 3 % of respondents defined floods as common events even though their households were commonly waterlogged. This divergence in perception was demonstrated as 30 % of the responses to the following questions—"How often do you get flooded?" and "How often do you have water in your house?"—were different from one another. Nevertheless, most slum dwellers shared that they had all experienced a flood at least once in their life (the 2005 flood) and learned from it. During heavy rainfall, 40 % of the respondents said that they stayed alert and packed their important documents and belongings. Thus, based on their experiences from the 2005 flood, they are more prepared to face and cope with future floods.

Our results from both the Jakarta and Mumbai case studies illustrate the inefficiency of institutional DRR strategies as they are based upon the dominant paradigm. As a matter of fact, they are not adapted to reduce the vulnerability of people living at risk in informal settlements. Furthermore, they do not take into account the complexity of people's vulnerability, which is embedded in daily patterns and has deep social, economic, and political causes, which the alternative radical paradigm emphasizes.

3. **Highlighting marginalization processes to understand the root causes of vulnerability and change visions of DRR**
   Slum dwellers, both from Mumbai and Jakarta, are embedded within long processes of marginalization (Texier, 2008; Texier, 2009). Their risky behaviors—which are, as seen above, settling in flood prone areas, throwing garbage within the river, and refusing to evacuate during flood events—however

can be seen as coping strategies to face daily problems (Davies 1996; Texier 2009). They are part of a system of local "arrangements and response to ensure their social and economic security," based on "ecological, economic, political and social insecurities that they face and the resources they have or can mobilize and access" to empower and improve their resilience (Nooteboom 2014: 24). Furthermore, in Jakarta, as in Mumbai, these marginalized people consider daily economic, social, and political issues as more important than flood risk, meaning that most migrants (between 40 and 95 % of respondents) choose to face flood risk and other urban disagreements rather than the poverty they had to face in their original provinces. In their opinion, floods are also not the main source of danger nor are they a prioritized concern, since 40 % of interviewees mentioned fires as the principal danger compared to the 25 % who considered floods in Jakarta. In Mumbai, 33 % of people consider evictions as a main threat. Three factors of vulnerability explain why risky behavior is practiced and why there is acceptance toward flood risk in Jakarta and Mumbai. These are detailed below.

(a) The economic factor of vulnerability
The first factor of vulnerability is linked to poverty. In spite of the possibility of being affected by floods, Indian and Indonesian migrants come first and foremost to urban centers for better employment opportunities. Sixty percent of the interviewees migrated to Jakarta under economic difficulties and could only afford to settle on unbuildable lands, which were cheaper than safer locations. In this sense, they "chose" to accept flood risk as part of their daily struggle with poverty (Pelling 1999). In Mumbai, land scarcity, real estate speculation, and a chronic shortage of housing have forced prices to soar, meaning that the poorest people must live in informal settlements. Most of these marginalized people have low incomes and high unemployment rates. They typically have informal jobs and gather all of their belongings and means of production at home and many of them work as informal sellers, owning in Jakarta their *kaki lima*, a "five-foot" handcart, which they store at home at night. It partially explains why almost 40 % of the Bukit Duri inhabitants were observed to stay at home during the 2007 flood event, even though they stated a few months before, during questionnaire campaign, that they would "put in safety their family" as a second choice behind "put their belongings in safety in high position." In this way, they are seen to prefer facing the risk of drowning or their houses collapsing, rather than leaving behind their low-valor belongings—considered unsafe without surveillance because of floods and risk of plunder—and are essential for their economic survival and recovery post-disaster.

(b) The social and societal factor of vulnerability
Social and societal factors also explain why slum dwellers are willing to stay in flood-prone districts. For example, in Jakarta, many people do not want to move from a social environment too different from their original province (i.e., when they move to their new settlement, they try to find

materials to improve their shack and sources of income so they can remain strongly linked within the community for the development of internal care, mutual assistance networks, and reciprocity bounds). During the 2007 flood event in Bukit Duri, for example, everybody within the community was observed to play a role in this micro-society, optimizing their various capabilities and resources. These were for the most part gender-specific. While men and youth organized the rescue with inflatable boats and ropes brought by a local NGO, the women took turns to manage the permanent emergency kitchens in order to feed the population for 4 days. The few houses that stayed above the water level served as a refugee camp. After the water level dropped, they organized generators and pumps to clean and secure the housing and help each other to recover from floods and limit the physical impacts more quickly. This internal social organization is thus a real livelihood strategy, which enables people to manage their own social security without a dependency upon outside stakeholders. This specific functioning explains why they hardly accept to be relocated, fearing that they might lose this function of their livelihood, even if this relocation was to improve their living conditions or be to a safer place.

In Mumbai, migrants use the slums as a transition to adapt with their arrival in the city and build a safe environment that they can live in. Slums can be a safe neighborhood for the most vulnerable groups: children, women, and elders. The first priority of respondents is the well-being of their children, and many slum dwellers rate keeping their social place within the community and in their family as crucial to migrant survival. In both cities, vulnerability and capacities to deal with risk are thus a social production (Few 2003).

(c) The political factor of vulnerability

Finally, vulnerability and the harsh hygienic conditions that are found in slum areas are rooted in the political structure. In Mumbai, people living in exposed slums along the Mithi River can be divided into three groups: (1) the Dalits, the bottom of the caste hierarchy; (2) migrants from northern states (especially Uttar Pradesh and Bihar) which bring people of all castes; and (3) Muslims, who constitute the majority of people who live along the Mithi and Vakola Nalla (Parthasarathy 2012). Municipal officials, mostly Hindus and members of the dominant Maharashtra caste, are at best indifferent to the problems of these marginalized populations, and at worst, trying to actively marginalize them. For example, migrants from northern India have been the subject of a vicious campaign referred to as "son of the soil" which accuses migrants of stealing jobs from native Maharashtrians and causing congestion, lack of housing, and infrastructure problems in Mumbai (Parthasarathy 2009). There was thus total disregard for these people during the 2005 flood, both among politicians and in the media. For several years, the poorer population of the city was considered a burden, a blot on the landscape which had to be disposed of. As victims of violence, abuse, and threats, these people were forced to live on the banks of the Mithi and Vakola Nalla, thus exposing themselves to the risk of flooding.

The case in Jakarta is similar, where the government considers in-migrant slum dwellers as “illegal.” Indeed, to move from one province to another, Indonesian citizens must ask for permission and hold a new provincial I.D. card. As the system of migration is strongly controlled to limit rural exodus, many migrants come to Jakarta without permission. Once in Jakarta, they are illegal and not allowed to obtain a Jakartanese I.D. from the local government. This illegality of demographic status has heavy consequences upon livelihood security. First, they cannot settle on buildable land or a property title, which explains why they are forced to join informal settlements that are prone to flooding (e.g., riverbanks, seafronts, along the railways, under bridges). They cannot obtain an official job without a local ID, nor can they ask for a water network connection or be covered by official waste management. Additionally, they cannot ask for the government’s monthly social support for the poor (rice and a monetary grant) or apply for care cover at the public hospital. If a local ID was to be the “free for all” in obtaining public services and the power of claims, even those who have an ID are hardly connected to the public services network since the government refuses to develop basic services in areas that they plan to clear in the middle or short term (through the evictions as explained above). The accumulation of garbage causes a spreading of vermin and disease (leptospirosis transmitted by rats, or avian flu transmitted by chickens). However, throwing waste directly into the river still appears to be the only coping strategy to mitigate daily sanitary risk.

Finally, it is the absence of rights to access public services and the subsequent lack of power to claim such services, which are linked to their fear to be evicted and not the lack of resources, in both cities that explain the prevalence of most risky behaviors. The vulnerability of these people is thus materialized by behavioral practices, which are the result of social, economic, and political processes, and converge to create a complex game of dynamic pressures—complex migrations and settlement patterns, access to land and urban public services, rapid demographic growth, and the sustenance of certain political ideologies—that themselves have deep rooted causes. This systemic and temporal process of marginalization and vulnerability has been developed and conceptualized by Wisner et al. 2004 in their famous “At Risk” book, and the PAR model (pressure-and-release). This model has been adapted for our two case studies (see Fig. 6.3).

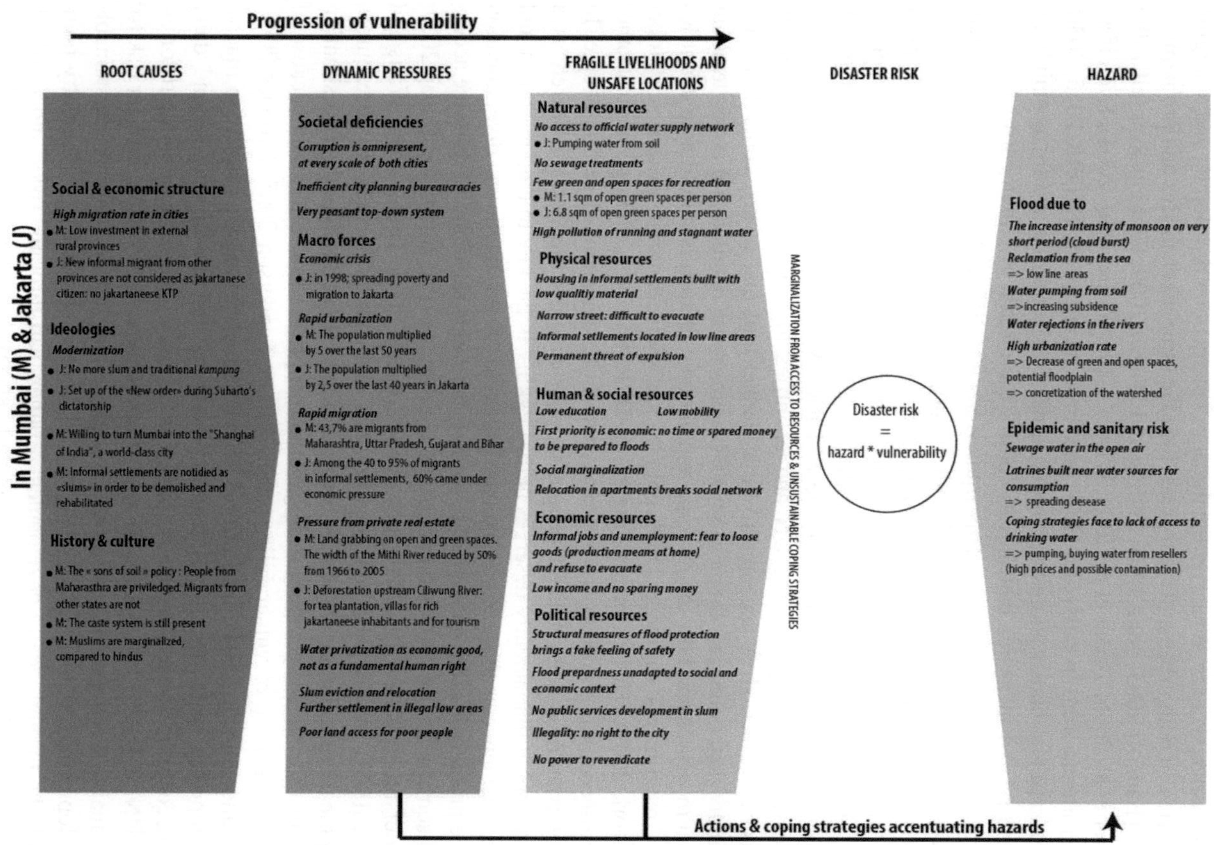

Fig. 6.3 Factors of vulnerability and marginalization process (P. Texier and E. Edelblutte 2014. Sources: Wisner et al. 2004)

5 Discussion

The results from the research conducted in Jakarta, and more recently in Mumbai, are not isolated cases. Whatever the threat is,[8] many other developing and emerging countries, in both urban and rural areas, face the same mechanisms and problems in DRR (Bankoff 2003; Nigg 1996; Pelling and Wisner 2009; Penning-Rowsell 1996; Texier et al. 2006; Texier et al. 2009; Wisner 1998; Wisner et al. 2004). Even if all governments were to ratify the HFA, and more recently, the SFDRR 2015–2030, DRR strategies hardly address the root causes of vulnerability and focus mainly on hazard-related measures that aim to control the natural environment or focus on setting awareness campaigns. This framework for DRR hides the more dominant concerns of daily hardship linked with poverty or lack of rights to access certain resources such as official IDs, water, land, and jobs. Unfortunately, this clear priority is not taken into account when local governments enact DRR strategies, thus ignoring the fourth priority of the HFA titled “Reduce the underlying risk factors,” which appears again in the SFDRR. Would it be too ambitious or an impossible challenge for the UN and countries to implement it with concrete measures? Since HFA appeared to be a blanket tool that lacked concrete advice for individual countries, the new SFDRR seems more detailed and concrete about social measures and how the regional and national platforms should help countries and governments to focus DRR policies on socioeconomic priorities (Global Network 2009). These concerns regarding DRR, however, often go against other macro-economic priorities for governments, to the detriment of certain people. In many countries, marginalized people in poor districts are not considered as a demographic that actually deserves help. Legality is the key to empowerment and better resilience as this ensures a secure access to greater livelihood diversification as the problem is often not lack of resources, but rather the claim of resource access (Benson and Twigg 2007; Chambers and Conway 1991; Gaillard and Cadag 2009; Sanderson 2008; Sen 1983, 2003; Start and Johnson 2004; Texier 2009; Watts and Bohle 1993). People who lack opportunities and rights to access these resources become more and more vulnerable because of their weak livelihoods. This situation accounts for the risky coping strategies that people are forced to adopt because of their initial state of vulnerability (Davies 1996; Sanderson 2008; Scoones 2009; Start and Johnson 2004). Institutional stakeholders tend to give priority to macro-development issues (territorial land-use planning and modernization) and daily socioeconomic threats are usually framed as the entry of development of territories. The UNISDR (2013:5) report outlines the importance of “reducing disaster risk for sustainable development.” However, it ought to aim for the contrary: “sustainable development to reduce disaster risk in a resilience approach” .... The entry point should be through “daily development,” rather than through “risks” alone, which would better address the objectives of daily survival and resilience. This becomes

[8] Other natural threats as landslides, volcanic eruptions, or even technological threats are concerned.

one of the biggest challenges for the implementation of the SFDRR. Another priority should be focused on developing a participatory and collaborative management of resources and risk. This should merge the top-down and bottom-up approaches, thus valorizing both scientific and "expert" knowledge, institutional financial and technical means, and local knowledge which would bring important resources and ideas to DRR at large.

6 Conclusions

To conclude, the two case studies on Jakarta and Mumbai show that DRR at the national level hardly address the root causes of vulnerability that are embedded within daily life. These include social, economic, and political constraints that ultimately weaken people's livelihoods, and are thus characterized by low resilience, despite there being, in many instances, strong local capacities to cope with risk and disasters. In these two cities, DRR is led by hazard-related and technocratic administrations, which naturally focus on measures aimed at controlling hazards or controlling land use, without any programs to empower the most marginalized people to valorize their existing capacities and knowledge in order to reduce their vulnerability. Migration is a root cause of vulnerability because it is linked to a lack of possibilities of empowerment and because migrants have no access or claim to basic resources required to sustain their daily needs. Legitimizing so-called illegal people (e.g., those with a local ID in Jakarta) or recognizing the lower castes as citizens in Mumbai is a key way to secure livelihoods and engage a real effort of de-marginalization. Step by step this would help to reduce poverty and provide slum dwellers with the economic opportunity and resources to move to safer places within the city without fear of being forced, injured, or displaced, and thus break all social links that ensure their social security. Engaging in such a process, however, requires a revolution of the perceptions and ideologies of institutional stakeholders who—until now—largely consider exposed marginalized people as responsible for aggravating disasters and generally act against the common interest. With a shift in focus, these institutional stakeholders can in fact contribute to the integration of DRR policy through a more participatory way of management. This way of management should focus on dimensions of daily development, which will inevitably have a relevant impact upon risk reduction in the future.

References

Abidin H, Darmawan D, Hadid A, Rajiyowiryono H, Sudiyo Y, Meilano I, Kasum AM, Kahar J, Subarya C (2001) Land subsidence of Jakarta (Indonesia) and its geodetic monitoring system. Nat Hazards 23:365–387

Amri AB, Rimadi L (2012) Ini Konsep Apartemen Apung di Atas Kali Ciliwung: Untuk itu, bahu sungai dirancang untuk kokoh terlebih dulu. Paper available online (http://metro.news.viva.co.id/news/read/369446-ini-konsep-apartemen-apung-di-atas-kali-ciliwung). Consulted on the 4th of June 2015

Bankoff G (2003) Constructing vulnerability: the historical, natural and social generation of flooding in metropolitan manila. Disasters 27(3):224–238

Benson C, Twigg J (2007) Tools for mainstreaming disaster risk reduction: guidance notes for developement organisations. Provention Consortium, Genève

Burton I, Kates RW, White GF (1978) The environment as hazard. Oxford University Press, New York

Cannon T (1994) Vulnerability analysis and the explanation of "natural" disasters. In: Varley DA (ed) Disasters, development and environment. Wiley, Chichester, pp 13–30

Census (2011) http://censusindia.gov.in/. Consulted on the 14th of May 2013

Chambers R, Conway GR (1991) Sustainable rural livelihoods: practical concepts for the 21st century. IDS Discussion paper (296)

Cuny F (1983) Disasters and developement. Oxford University Press, Oxford

Davies S (1996) Adaptable livelihoods: coping with food insecurity in the Malian Sahel. Macmillan Press, Basingstoke

Few R (2003) Flooding, vulnerability and coping strategies: local responses to a global threat. Progr Dev Stud 3(1):43–58

Gaillard JC (2007) De l'origine des catastrophes: phénomènes extrêmes ou âpreté du quotidien? Natures Sciences Sociétés 15(1):44–47

Gaillard JC, Cadag JR (2009) From marginality to further marginalization: experiences from the victimes of the july 2000 payatas transhslide in the Philippines? Jàmbá: Journal of Disaster Risk Studies 2(3): (2009), 197–215. doi: 10.4102/jamba.v2i3.27.

Global Network of Civil Society Organisations for Disaster Reduction (2009) "Clouds but little rain...", view from the Frontline, a local perspective of progress towards implementation of the Hygo Framework for Action

Harjoko TY (2004) Penggusuran or Eviction in Jakarta: Solution Lacking Resolution for Urban Kampung. 15th Biennal conference of the Asian Studies Association of Australia, p. 22

Harsono A (1999) Les spoliés de Jakarta. Le Courrier de l'Unesco, juin, pp 26–28

Hewitt K (1983) The idea of calamity in a technocratic age. In: Hewitt DK (ed) Interpretation of calamities. Allen et Unwin Inc, Boston, pp 3–32

Human Rights Watch (2006) Indonesia condemned communities: forces evictions in Jakarta, vol 18(10). HRW, NewYork, 111p

Kates RW (1971) Natural hazard in human ecological perspective: hypotheses and models. Econ Geogr 47(3):438–451

Nigg JM (1996) The social impacts of physical processes: how do we manage what we can't control? Delaware, Newark

Nooteboom G (2014) Forgotten people: poverty, risk and social security in Indonesia. The case of Madurese. Brill Academic Publishers, Leiden, 314p

O'Keefe P, Westgate K, Wisner B (1976) Taking the naturalness out of natural disasters. Nature 260(5552):566–567

Oliver-Smith A (2001, juillet) Displacement, resistance and the critique of developement: from the grass roots to the global. *Final report prepared for ESCOR R7644 and the research program on developement induced* Displacement and resèttlement, p. 123

Parthasarathy D (2009) Social and environmental insecurities in Mumbai: towards a sociological perspective on vulnerability. S Afr Rev Sociol 40(1):109–126

Parthasarathy D, Grover A (2012) Relocation and changing nature of vulnerability - a case study of Krantinagar slum, Mumbai. The Horizon – A Journal of Social Sciences 3(2):1–19

Pelling M (1999) The political ecology of flood hazard in urban Guyana. Geoforum 30:249–261

Pelling M, Wisner B (2009) Disaster risk reduction. Cases from urban Africa. Earthscan, Londres

Pennin-Rowsell E (1996) Flood-hazard response in Argentina. Geogr Rev 86(1):72–90

Reliefweb (2013) Jakarta and its greater area floods and situation report, no. 2. http://reliefweb.int/report/indonesia/floods-jakarta-and-its-greater-area-situation-report-no-02-18-January-2013
Sanderson D (2008) Villes, catastrophes naturelles et moyens de subsistance. CARE
Scoones I (2009) Liverlihoods perspectives and rural developement. J Peasant Stud 36(1):171–196
Sen A (1983) Development: which way now? Econ J 93(372):745–762
Sen A (2003) Un nouveau modèle économique, Développement, justice, liberté. Odile jacob
Sevin O (2001) Migrations, colonisation agricole et Terres Neuves en Indonésie. Collection Iles et archipels n°28
Start D, Johnson C (2004) Livelihood Options? The political economy of access, oppotunity and diversification. Overseas Developement Institute working paper (233), p 56
Texier P (2008) Floods in Jakarta: when the extreme reveals daily structural constraints and mismanagement. Disast Prev Manag 17(3):358–372
Texier P (2009) Vulnérabilité et réduction des risques liés à l'eau dans les quartiers informels de Jakarta, Indonésie – Réponses sociales, institutionnelles et non institutionnelles. Thèse de doctorat, Université Paris Diderot, 460p
Texier P, Gaillard JC, Lavigne F (2006) Les sociétés d'Asie du Sud-Est face aux menaces volcaniques: entre phénomène extrêmes et contraintes quotidiennes. In: Pascal PU (ed) Actes du colloque "Villes et volcans: relations, représentations et pratiques". Presses Universitaires Blaise Pascal, Clermont-Ferrand
Texier P, Gaillard JC, Le Masson V (2009) Les enjeux de la gestion des risques dans les quartiers urbains informels: les cas de Jakarta (Indonésie) et Angeles City (Philippines). In: Becerra DS, Peltier A (eds) Risques et environnement: recherches interdisiplinaires sur la vulnérabilité des sociétés. L'Harmattan, Paris, pp 441–455
Torry W (1979) Hazards, hazes and holes: a critique of the environment as hazard, and general reflections on disaster research. Can Geogr 23(4):368–383
Twigg J (2004) Disaster risk reduction: mitigation and preparedness in developement and emergency programming. Overseas Developement Institute Humanitarian Practice Network
UN Habitat (2008) "Coastal cities", Cities at risk from rising sea levels, pp 140–155
UNISDR (2013).GAR, Global Assessment Report o, Disaster Risk Reduction 2013, United Nations, 288p
University of Westminster and ITB (2000) Guide to good practice in core Area developemnt, DFID Research Project R 6860, Jakarta, Filed studies and Workshop, draft final report. Jakarta, p. 150
Urban Poor Consortium (2005) Eviction of Urban Poor Housing: Solution offered by Jakarta's government toward flood. Report from Dian Tri Trawaty, Networking Division of UPC, 7p
Waddell E (1977) The hazards of scientism: a review article. Hum Ecol 5(1):69–76
Watts M, Bohle H (1993) The space of vulnerability: the causal structure of hunger and famine. Progr Hum Geogr 17(1):43–67
White GF (1970) Recent development in floodplain research. Geogr Rev 60:440–443
Wisner B (1998) Marginality and vulnerability. Appl Geogr 18(1):25–33
Wisner B, Blaikie P, Cannon T, Davis I (2004) At risk, natural hazards, people's vulnerability and disasters. Routledge, New York

Chapter 7
The Necessity of Early Warning Articulated Systems (EWASs): Critical Issues Beyond Response

Irasema Alcántara-Ayala and Anthony Oliver-Smith

1 Introduction

Disaster risk reduction (DRR) and disaster risk management (DRM) are without a doubt goals that all governments urgently need to move towards. Disasters associated with tsunamis triggered by earthquakes, such as those of the Indian Ocean (2004), and on the Pacific coast of Tōhoku (2011), Japan, have represented the two sides of the same coin. On the one hand, the need for developing tsunami warning systems has indeed been a priority for the Southeastern region of Asia, but on the other hand, the evaluation of warning systems protocols for multi-risk scenarios became an urgent issue for Japanese society. Lessons from both cases clearly provided an urgent and challenging mission for the rest of the world.

Despite scientific and technological advances, current early warning systems (EWSs) cannot yet be seen as a promising answer for disaster prevention. First of all, there is a lack of effective and efficient EWSs given that risk, as a socially constructed process, possesses various spatial and temporal scales and dimensions. Secondly, in addition to individual hazard occurrences, multi-hazard forecasting has not yet been perceived as a widespread condition and therefore scenarios of hazard concatenation are usually neglected as a paradigm of integrated disaster risk research is not yet common. Thirdly, and perhaps most importantly, in addition to scientific uncertainty, social factors are indeed the greatest restraint above all.

I. Alcántara-Ayala (✉)
Departamento de Geografía Física, Instituto de Geografía, Universidad Nacional Autónoma de México (UNAM), Circuito Exterior, Ciudad Universitaria, Coyoacán, 4510 Mexico City, Mexico
e-mail: irasema@igg.unam.mx

A. Oliver-Smith
Department of Anthropology, University of Florida, Gainesville, FL, USA

© Springer International Publishing Switzerland 2017
K. Sudmeier-Rieux et al. (eds.), *Identifying Emerging Issues in Disaster Risk Reduction, Migration, Climate Change and Sustainable Development*,
DOI 10.1007/978-3-319-33880-4_7

Decision-making processes are multi-factorial dependent, and particularly in developing countries, actions are taken as a function of political momentum and individual interest. As such, lessening disaster effects has not yet been made a priority practice. As a result of the missing communication links among authorities, scientists, decision makers and populations, including their traditional knowledge, and all other involved actors in disaster risk, specifically in societies or communities characterized by high levels of vulnerability and low resilience, EWSs cannot be seen as articulated systems, but as segments of the capacity-building process needed to achieve DRR and DRM.

EWSs will fall short in true effectiveness in reducing disaster risk and impact if the science and technology do not take into account the social factors involved in decision-making processes at all levels of society. This chapter proposes that EWSs therefore must be articulated in tandem with institutions, authorities, science, decision makers and all other communities involved in DRR. EWASs can, in this sense, be defined as a coordinated structure for the implementation of specific strategies of action to achieve DRR and DRM based on the understanding of disaster risk as a socially constructed process. Disaster risk is not only influenced by the occurrence of hazards, but also possesses various spatial and temporal scales and dimensions of vulnerability that are shaped by risk drivers and causes rooted in historical processes. As such, EWASs need to consider both the single and the concatenated effects of hazards on exposed communities, capital and environmental assets.

This chapter aims to shed light on the identification of critical issues that need to be taken into account to develop and implement EWASs within a framework that goes beyond response and is based on integrated science for disaster risk and the notion of long-term communication. It is divided into five parts. An introduction to understanding disasters is given in the first section, which involves the paradigm of "unnaturalness" that seeks to demonstrate the social construction of risk and emphasize the significance of small- and medium-size disasters in building the multiple dimensions of exposure and vulnerability. Failure to recognize these factors constitutes an obstacle to achieving resilience, development, sustainability and disaster risk reduction and management.

Section 2 focuses on the general impact of disasters worldwide and the role that large-scale disasters have played in inspiring the implementation of EWSs. Particular attention is given to the Indian Ocean Tsunami (2004), Hurricane Katrina (2005) and the Tōhoku Earthquake (2011). The need for an integrated approach for the construction and implementation of sound EWASs is addressed in Sect. 3, where particular attention is given to the concepts of uncertainty, integrated science, risk communication and governance. In the final section, some concluding remarks are offered.

2 Disasters: A Prelude

2.1 *The "Unnaturalness" of Disasters: A Quick Look into Socially Constructed Processes*

Disasters associated with natural and anthropogenic hazards, climate change and morphological changes of the Earth's surface induced by both natural and human-influenced processes are indeed one of the greatest challenges for science, technology and development in the twenty-first century. As populations grow and vulnerability increases at local, regional and global scales, the human, environmental and economic consequences of disaster have increasingly become a major issue in the international agenda.

Throughout history and in stories from many lands, disasters have been interpreted as the catastrophic inner self of the planet. However, as a result of the significant shifts of scientific paradigms and a major understanding of the various spheres of interactions among nature and society, scientific thought has come to regard the unnaturalness of disasters as a more correct line of reasoning. Disasters are not natural (Maskrey 1993); they are configured by socioeconomic, historical and cultural processes linked to development and, above all, they are sculpted by the infinite dimensions of vulnerability (for example, see Blaikie et al. 1994; Burton 1997; Burton et al. 1978; Hewitt and Burton 1971; Oliver-Smith and Hoffmann 1999; Winchester 1992).

While the late 1960s saw some investigations that addressed the significance of vulnerability in the understanding of risk and disasters, it was the first decade of the twenty-first century when inquiries into disaster really stressed the social context and the different aspects of vulnerability as the key elements for risk management, integrated risk research and prevention and mitigation of disasters (for example, see Cannon 1993; Cutter 1996; Pelling 2003; Wisner et al. 2004).

2.2 *On the Significance of Small- and Medium-Scale Disasters*

Large, moderate and small-scale disasters are usually defined by quantitative factors rather than by qualitative analysis. This is therefore not surprising that the exceptional numbers of deaths, affected people and economic impact can be part of the definition of large and hence the most important disasters. However, small and moderate disasters too may have serious consequences in daily life, although their significance may not be appreciated if insufficient attention is paid to the prevailing and historically determined social and economic conditions, people's perceptions and capabilities and needs. In this way, the underestimation of their significance may impede resilience, development, sustainability and adequate disaster risk management.

Studies by La RED (Network for Social Studies on Disasters Prevention in Latin-America) have pointed out the need to evaluate the impact of small- and medium-size events, since accumulated effects are at least as important as those linked to large disasters. Indeed, both small- and large-scale disasters contribute extensively to the persistence of, or increase in, poverty (Lavell 2008; Velásquez and Rosales 1999). In this respect, for instance, research in Africa on disasters and urban development has indicated that as population and vulnerability rise in areas with an intensification of hazards, the number, territorial extent and impact of small-scale disasters rapidly increase, meaning that small events may in fact develop in time into larger events. The ability to prevent larger disasters may in this sense be enhanced through the experience gained by interventions to prevent small disasters, and this should be a major focus of risk management (Bull-Kamanga et al. 2003). Additionally, analysis in Colombia (Marulanda et al. 2010) has suggested that small events derived from socio-natural processes linked to environmental deterioration may pose a significant development problem at both the local and national scales. While this is not considered at the international level, such small events are responsible for the spatial variability and dispersion of risk.

As disasters result from unresolved problems of development (Maskrey 1989), attention must be given to any type and size of event that may impede or even counter-progress development at all scales. Indeed, disasters are generated at global and national levels, as well as at regional and local levels. Within local contexts, and even more pertinently at the household scale, the consequences for development are very important because they play a significant role in determining one's vulnerability and/or resilience.

3 Disaster Impact and Early Warning Systems

Disaster losses and mortality over the last 20 years suggest that the need for EWASs is acute. According to the United Nations Office for Disaster Reduction (UNISDR), the estimated impact of disasters worldwide has included 1.3 million deaths, 4.4 billion affected people and economic losses of US$2 trillion since 2012 (UNISDR 2012). Large and small countries, developed and developing nations, have all suffered such consequences. For example, 2.2 billion inhabitants were affected by disasters in China and the economic losses in the USA were as high as US$560 billion (without considering the impact of the storm Sandy) (UNISDR 2012). Not surprisingly, worldwide statistics show that more than 90 % of the fatalities caused by disasters associated with natural hazards occur in developing countries. The most recent data show that disasters have become more frequent over the past two decades, but the average number of affected people has decreased from 1 in 23 in 1994–2003 to 1 in 39 during 2004–2013. Population growth is partly responsible for this decline, but there is also a decrease in absolute terms. However, death rates have increased over the same period, averaging more than 99,700 deaths per year between 2004 and 2013 (EM-DAT 2015).

Before Hurricane Katrina in 2005 in New Orleans and on the Gulf Coast, disasters had been seen as inextricably linked with poverty in the developing world. It is now clear that although disasters usually hurt the poorest people most, their impact is not exclusively concentrated in the most socio-economically deprived nations (Blaikie et al. 1994; Cannon 1994). Developed and non-developed countries are vulnerable in different ways to hazards, both natural and human induced (Alcántara-Ayala 2002).

In recent years, human and economic losses due to disasters associated with the use of EWS rose dramatically as a result of the Indian Ocean Tsunami in December 2004, Hurricane Katrina in 2005, the Tōhoku Earthquake in 2011, the extensive floods in Thailand also in 2011 and Hurricane Sandy in the USA in 2012. These disasters demonstrated vulnerability in both developing and developed countries, but most notably in nations such as Japan where progress in a culture of disaster prevention and in structural and non-structural mitigation had been considered to be among the most effective in the world. What is more, this series of disastrous events brought to light a relatively new concern regarding the economic impact of high-magnitude global disasters and the resultant challenges for development.

In just a decade, the new millennium has shown that major disasters can take place in the "developed world" and consequences are beyond international and continental borders. Multiple hazards, risk and most importantly the expanding dimensions of vulnerability within the sphere of risks have proved to determine everywhere the magnitude of disastrous events. On this basis, and having no intention of undermining the significance of small- and medium-scale disasters, some of the most significant disasters that have occurred with a global impact in the past decennium are presented in this section.

3.1 The Indian Ocean Tsunami, 2004

On 26 December 2004, a tsunami originating in the Indian Ocean travelled at high speed around the globe (see Table 7.3). A 9.1-magnitude earthquake at a depth of 30 km occurred along the interplate thrust separating the oceanic India plate from the overriding Burma (Andaman) microplate that borders the larger Sunda plate (Geist et al. 2007). This produced a tsunami with heights of 30 m, run-ups of as high as 50 m a.s.l. and reaching 6 km inland (Paris et al. 2007). A 1200 km fragment of the seafloor was moved, and the energy released was comparable to that of 550 million Hiroshima atomic bomb explosions (Tsunami Global Lessons Learned Project 2009).

The aftermath involved more than 228,000 fatalities in 14 countries, including 40 nationalities in the Southeast Asia and South Asia regions, and even in regions as distant as Africa. The largest group of fatalities by percentage were women, followed by elderly and children. Human losses were highest for poor people. Economic damage accounted for US$10 billion (Tsunami Global Lessons Learned Project 2009). People displaced were estimated at $>$1.7 million. Coastal

Fig. 7.1 Gleebruk Village, district of Aceh Besar, southwest of Banda Aceh, Indonesia, before (*left*) and after (*right*) the 2004 Tsunami (Digital Globe, Imagery collected 12 April 2004)

communities, fishing, small-scale agriculture, infrastructure, private assets and tourism were considerably affected. According to estimates by the Asian Development Bank (ADB), the tsunami has resulted in two million people at risk of deeper impoverishment. Although losses were mainly of uninsured private assets, estimates of insurance pay-outs range from US$2.5 to 5 billion (Inderfurth et al. 2005).

Environmental assessments undertaken immediately after the event revealed widespread and varied damage to the natural resources, including coral reefs, mangroves, sand dunes, peat swamps and other coastal ecosystems (Fig. 7.1). Consequences for the environment derived mainly from the impact of debris, coastal erosion and depositional processes on reefs, agriculture land and rivers. This damage reflects the function of these coastal systems as the first line of defense against the tsunami. Inland waters, wetlands and agricultural land were salinized and groundwater contaminated with hazardous waste (UNEP 2006).

However, the response to the tsunami also had some positive outcomes. For example, 2 years later in Sri Lanka, the event was a catalyst for the foundation of the Ministry of National Disaster Management and Human Rights. On the same day as the tsunami, the Government of the Maldives established a National Disaster Management Centre as a coordination mechanism for disaster-related activities (Tsunami Global Lessons Learned Project 2009).

One of the most significant impacts of the devastating tsunami was the transformation of the event into a consciousness-raising mechanism for the international community; the special sessions of the 2005 second World Conference on Disaster Reduction (WCDR) in Kobe, Japan, focused attention on the disaster and on the need for EWS, public awareness and education, structured coordination, community-based approaches to DRR and multi-disciplinary work.

Fig. 7.2 Hurricane Katrina, 28 August 2005 (Source: NOAA, 2005)

3.2 Hurricane Katrina, 2005

According to the technical report by NOAA's National Climatic Data Center (Graumann et al. 2005), Hurricane Katrina (Table 7.3, Fig. 7.2) was the most costly disaster and one of the deadliest ever to strike the USA. From a meteorological perspective, the hurricane, a category 3, when it made landfall was also one of the most powerful storms to impact the coast of the USA during the past 100 years, with 127 mph winds. Effects included extensive and substantial devastation along the central Gulf Coast states, catastrophic floods in New Orleans and the displacement of more than 250,000 people. The aftermath involved over 1,800 deaths, 500,000 people affected and total damages of US$125 billion (EM-DAT database).

As a storm surge overcame the protective floodwalls of New Orleans, 80 % of the city was flooded. Buildings were considerably damaged by intense winds and torrential rains. In Louisiana, Mississippi and Alabama, more than two million people had no electricity, and more than half a million were left homeless. More than 1,200 fatalities were associated with the failure of the dikes, the storm surge, wind and rain (Congleton 2006).

Flooding in New Orleans under storm conditions, that is, a storm stronger than a category 3 hurricane, was predictable. The topography of a city located mostly below sea level, with no natural drainage, was clearly an ideal template for floods to take place. Canal and pumping networks had been implemented to control flooding, but maintenance of floodwalls and levees had not been a priority. Infrastructure capability and thresholds were thus exceeded by Katrina.

The disaster was complex. In addition to the susceptibility of the infrastructure, other factors contributed to the condition of risk. Oliver-Smith (2006) suggested that the complexity of the impact of Hurricane Katrina on New Orleans stemmed from an excessive dependence on technological protection combined with an assault on the natural defences of the environment of southern Louisiana. This gave rise to a very vulnerable city, whose population was displaced and challenged economically, socially and psychologically. Vulnerability was therefore at the root of the uneven impact of Hurricane Katrina on the inhabitants of New Orleans. Flood damage pervaded the city quarters irrespective of income, ground height and other social factors. However, pre-existing socio-economic conditions determined the ability of specific economic classes, and thus racially defined and other groups, to react to conditions during the response and recovery phases. Not surprisingly, those with the fewest resources and the least mobility suffered the most in the aftermath (Masozera et al. 2007).

The effects of Hurricane Katrina mirrored the historically constructed issues and multifaceted relationships of racism, classism, gender issues and aging (Levitt and Whitaker 2009). In spite of the accurate scenario provided by Laska (2004), this was in a nation that in the minds of many natives and outsiders, including the disaster authorities at local, state and federal levels (i.e. FEMA), had not been previously considered to be as vulnerable to such serious losses in disasters.

3.3 The Tōhoku Earthquake, 2011

On 11 March 2011, 62 municipalities in the prefectures of Aomori, Iwate, Miyagi, Fukushima, Ibaraki and Chiba in north-eastern Japan were severely damaged by the tsunami triggered by the Heisei Tōhoku 9.0 magnitude earthquake (MLIT 2011), also known as the Great East Japan Earthquake (see Table 7.3). It was a plate-boundary thrust-faulting earthquake that occurred in the subduction area with a reverse fault, a depth of 24 km, and a fault area extending 400 km and 200 km in the NS and EW directions, respectively (Headquarters for Earthquake Research Promotion 2011; Mimura et al. 2011). The Geographical Survey Institute of Japan (2011) indicated that the seismic activity was linked to a large movement of the crustal plate, including 4.0–5.0 m horizontal offshore movement and 0.4 to >1.0 m subsidence; the largest ground movement was observed at the Ojika Peninsula, Miyagi Prefecture, with 5.4 m horizontal and 1.20 m vertical movements (Mimura et al. 2011).

Tsunami waves recurred seven times during the 6 h after the earthquake, with the third being the highest. Heights varied spatially (see Table 7.1). At Ōfunato, heights up to 24 and 30 m were reported by the Port and Airport Research Institute and by a group of academic researchers from Yokohama National University and the University of Tokyo, respectively. Moreover, at Tarō, Iwate, a researcher from the University of Tokyo estimated a tsunami height of 37.9 m (Takeuchi and Chavoshian 2011).

Table 7.1 Tsunami wave height measured by offshore telemetry by the Port and Airport Research Institute

Place	Tsunami wave height (average) (m)
Port of Hachinohe	5–6
Port of Hachinohe area	8–9
Port of Kuji	8–9
Mooring GPS wave height meter at offshore of central Iwate (Miyako)	6
Port of Kamaishi	7–9
Mooring GPS wave height meter at offshore of southern Iwate (Kamaishi)	6.5
Port of Ōfunato	9.5
Run-up height, port of Ōfunato area	24
Mooring GPS wave height meter at offshore of northern Miyagi	5.6
Fishery port of Onagawa	15
Port of Ishinomaki	5
Shiogama section of Shiogama-Sendai port	4
Sendai section of Shiogama-Sendai port	8
Sendai Airport area	12

Source: Takeuchi and Chavoshian (2011)

Five aftershocks with magnitudes greater than 7.0 Mw and hundreds larger than 5.0 Mw followed the main event. The Tōhoku event was a complex multi-disaster with a chain reaction caused by seismic activity: the tsunami, floods, land subsidence, landslides and radioactive fallout from the Fukushima Daiichi Nuclear Power Station of the Tokyo Electric Power Company (TEPCO). As all power sources of TEPCO were lost, damage to the reactor cooling system and fuel disposal storages threatened inhabitants and workers in the area for several days through the diffusion of radioactive substances throughout the vicinity. Level 7 of the International Nuclear Event Scale (the same level as Chernobyl) was reached (MLIT 2011; Takeuchi and Chavoshian 2011).

Despite the existence of tsunami dikes, the surging waters overtopped tsunami thresholds and caused major damage. Structures at TEPCO designed for protection against a tsunami were 5.7 m high; however, the tsunami wave reached 15 m. Around 1,000 lives were lost because people sheltered in evacuation centres that were at an insufficient height relative to the height of the tsunami (Takeuchi and Chavoshian 2011).

The inundation extended over 561 km^2, with Ishiniomaki and Rikuzen-Takata being among the most devastated areas (Table 7.2). Seventy-six thousand houses collapsed and were washed away, with a further 244,000 partially damaged. Basic infrastructure including electricity, water supply, sewage systems and gas lines were also affected (Mimura et al. 2011). According to the National Police Agency of Japan (2015), direct damage included 4198 roads, 116 bridges, 29 railways, 45 broken dikes and 207 landslides. The debris from the destruction of buildings in Iwate, Miyagi and Fukushima were estimated at 22,600,000 tons (MLIT 2011).

Table 7.2 Examples of height and areas of inundation due to the tsunami, and the resulting human toll

Place	Inundation height (m)	Inundated area (km^2)	Area inundated (%)	Residential area inundated (%)	Dead and missing people	Rate of dead and missing people (%)
Rikuzen-Takata	15.8	13	5.6	43	2422	10.4
Kamaishi	9.3	7	1.6	22	1310	3.3
Otsuchi	12.6	4	2.0	52	1631	10.7
Ishiniomaki	15.5	73	13.1	46	5538	3.5
Onagawa	14.8	3	4.5	48	1504	15.0
Minamisanriku	15.9	10	6.1	52	1095	6.3

Source: Japan Weather Association, 2011 and Disaster Countermeasures Office, 2011, in Mimura et al. (2011)

Fig. 7.3 Devastation caused by the tsunami in the City of Ishinomaki, Miyagi, where the highest percentage of fatalities occurred (photo courtesy: Kuniyoshi Takeuchi, 2011)

The Tōhoku disaster included 19,846 fatalities, some 368,820 people affected, economic damages reaching US$210 billion (*EM-DAT*-Database) and the evacuation of 468,000 persons. Some 52,000 residents were still being sheltered 4 months after the earthquake (MLIT 2011). The highest concentration of human losses occurred in the prefectures of Miyagi (Fig. 7.3), Iwate and Fukushima. The largest numbers of deaths occurred among the elderly and women, and >90 % of deaths were due to drowning (MLIT 2011; Takeuchi and Chavoshian 2011).

The magnitude of the Tōhoku event constituted an unexpected complex multi-disaster that devastated Japan, despite being the country whose disaster preparedness and awareness had been the best in the world. Nevertheless, traditional knowledge on levels of tsunami wave height had not taken into consideration. Unquestionably, lessons learnt by Japanese society through the experience of this tragedy must contribute to a better understanding of the vulnerabilities present in supposedly "resilient" societies. The concatenated risks associated with environmental impacts will thus have to be re-evaluated.

These cases clearly demonstrated the need for EWASs. In each case, more articulated EWSs could have proved crucial in reducing the loss of life, environment and property. In addition to the scientific uncertainty, social factors restrained the delivery of crucial information. The lack of a tsunami EWS with concomitant articulations with social institutions had tragic consequences for exposed and vulnerable people in the many nations affected by the tsunami of Southeast Asia in 2004. Without question, the socially constrained distribution of information on risk and evacuation doomed hundreds of elderly and African-American citizens in the USA. In Japan, the failure to take into account traditional knowledge of tsunami risk, evidenced by the existence of nineteenth-century stone marker warning of tsunami levels, led to the exposure of people and infrastructure (Fig. 7.4 and Table 7.3).

Fig. 7.4 Traditional knowledge: Japanese stone marker for tsunami warning at Miyako City, Iwatake Prefecture, "High dwellings are the peace and harmony of our descendants and will be safe. Do not forget the disaster of the great tsunamis. Do not build any homes below this stone (1896)"

Table 7.3 Characterization of selected "large" disasters during 1900–2011

	Disaster	Indian Ocean tsunami	Hurricane Katrina	Tōhoku earthquake and tsunami
Disaster aftermath[a]	Countries affected	(a) Indonesia (b) Sri Lanka (c) India (d) Thailand	USA	Japan
	Date	26/12/2004	29/08/2005	11/03/2011
	Fatalities	(a) 165,708 (b) 35,399 (c) 16,389 (d) 8,345	1833	19,846
	Total affected people	(a) 532,898 (b) 1,019,306 (c) 654,512 (d) 67,007	500,000	368,820
	Economic damage (US$000)	(a) 4,451,600 (b) 1,316,500 (c) 1,022,800 (d) 1,000,000	125,000,000	210,000,000
Socio-economic and environmetal indicators[b]	Human Development Index (HDI) 2011	(a) 0.617 (*R*: 124) (b) 0.691 (*R*: 97) (c) 0.547 (*R*: 134) (d) 0.682 (*R*: 103)	0.910 (*R*: 4)	0.901 (*R*: 12)
	People in severe poverty (%)	(a) 7.6 (b) 0.6 (c) 28.6 (d) 0.2	NA	NA
	People living on degraded land (%)	(a) 3.1 (b) 21.1 (c) 9.6 (d) 17.0	1.1	0.3
	Quality of life (2011)	(a) 0.591 (*R*: 78) (b) 0.570 (*R*: 81) (c) 0.489 (*R*: 100) (d) 0.605 (*R*: 71)	0.806 (*R*: 31)	0.882 (*R*: 13)
	Health	(a) 0.499 (*R*: 92) (b) 0.658 (*R*: 62) (c) 0.452 (*R*: 96) (d) 0.666 (*R*: 56)	0.770 (*R*: 39)	0.907 (*R*: 3)

(continued)

Table 7.3 (continued)

	Disaster	Indian Ocean tsunami	Hurricane Katrina	Tōhoku earthquake and tsunami
	Education	(a) 0.809 (*R*: 72) (b) 0.806 (*R*: 74) (c) 0.492 (*R*: 116) (d) 0.824 (*R*: 60)	0.941 (*R*: 22)	0.946 (*R*: 14)
	Wealth	(a) 0.463 (*R*: 91) (b) 0.450 (*R*: 92) (c) 0.437 (*R*: 97) (d) 0.560 (*R*: 71)	0.818 (*R*: 20)	0.788 (*R*: 28)
	Democracy	(a) 0.696 (*R*: 58) (b) 0.456 (*R*: 85) (c) 0.734 (*R*: 54) (d) 0.417 (*R*: 90)	0.964 (*R*: 15)	0.891 (*R*: 32)
	Environment	(a) 0.344 (*R*: 113) (b) 0.653 (*R*: 51) (c) 0.413 (*R*: 104) (d) 0.633 (*R*: 59)	0.651 (*R*: 52)	0.766 (*R*: 18)

Sources: [a]EM-DAT database, [b]UNDP 2011 and Quality Life Index 2011

4 Early Warning Articulated Systems: An Integrated Approach

4.1 On the Definition of Early Warning Systems

Derived from the impact of large disasters, especially after the tsunami of Southeast Asia, particular attention has been given to the establishment and/or evaluation of EWS at international and national levels. EWS is defined as "the provision of timely and effective information, through identified institutions, that allow individuals exposed to hazards take action to avoid or reduce their risk and prepare for effective response" (UNISDR 2006). EWSs should be integrated by four major elements: risk knowledge, monitoring and predicting, dissemination of information and response (UNISDR 2006). Further recommendations from the United Nations Office for Disaster Reduction (UNISDR) include the development of a globally comprehensive EWS, rooted in existing EWSs and capacities; building national people-centred EWSs; filling the main gaps in global early warning capacities; strengthening the scientific and data foundations for early warning and developing the institutional foundations for a global EWS response (UNISDR 2006).

Based on the recommendations made by the UNISDR (2006), different warning systems have been established around the world. Despite having the same objective, namely reducing loss of lives and livelihoods, warning systems are designed and put into practice in very different ways. They vary from specific local needs to regional challenges, and in particular, they are quite frequently technical or hazard centred, leading to the question of what is actually missing from the current EWS framework.

Focusing on response rather than prevention when defining EWS perhaps inhibits the development of an integrated approach. Preparedness and risk management go beyond providing and receiving "real-time" information to take action within the traditional agenda of EWSs. Undoubtedly, warning systems aim at cautioning people, alerting and giving notifications about the occurrence of a given hazard and its likely impact so that it can be reduced. However, attention has not been given in a compulsory manner to basic questions, such as how can people be expected to know exactly what to do, when and most importantly why.

The first decade of the new millennium has demonstrated that major disasters can take place not only in the developing but also in the developed world. Indeed, consequences from disasters have no geographical borders. Within the sphere of risks, the expanding dimensions of vulnerability even of resilient societies have proved to determine the magnitude of disastrous events at all scales in space and time. If EWSs are intended to reduce or avoid human, economic and environmental losses, why are vulnerability and resilience not more commonly taken into account? Generally, the implementation of EWSs is concentrated on providing "real-time" information in order to respond, rather than enhancing the knowledge and preparedness to understand risk and thus reduce it. The answer to that question is not that simple, and indeed relates to all elements and complex interactions that need to be considered for DRR and DRM. Nonetheless, it can be pointed out that for the specific case of implementing effective and efficient EWASs, uncertainty, integrated science, communication and governance are the upmost essential ingredients.

4.2 Uncertainty

Uncertainty is a situation in which something is not known or certain. According to the Senior Seismic Hazard Analysis Committee (SSHAC), uncertainty can be divided into aleatoric uncertainty and epistemic uncertainty (SSHAC 1997). Aleatoric uncertainty, also called irreducible uncertainty, involves the intrinsic randomness of a phenomenon. It refers to the inherent uncertainty due to probabilistic variability. Epistemic uncertainty, more subjectively and professionally related, derives from the lack of knowledge of many sorts, including the inadequate understanding of underlying processes, incomplete knowledge of phenomena or a non-accurate assessment of the related characteristics. Knowledge is limited and operational modes are not known. As such, in contrast with the aleatory uncertainty, epistemic uncertainty is reducible. In this sense, producing or gathering more information would help to reduce epistemic uncertainty.

Disaster risk involves both aleatorial and epistemic uncertainties. Aleatory uncertainty is the inherent uncertainty of the hazard itself, while epistemic uncertainty is closely linked to hazard assessment in terms of modelling limitations. To this regard, the latter comprises of three components: parametric uncertainty—derived from deficient knowledge of the settings of the model's parameters, input uncertainty—resulting from incomplete knowledge of the true value of the initial state and forcing and structural uncertainty—of which, despite the knowledge of correct parameters and inputs, the model fails to represent the system (Hill et al. 2013).

Epistemic uncertainty is interrelated with the limitations of resources, information, understanding and knowledge. Therefore, it is not only related to the hazard, but above and beyond all, it is also related to the way people act, behave and take decisions. It is indeed this kind of uncertainty associated with actions and decision-making processes that are the most significant for EWAS. Epistemic uncertainty, in this sense, plays a very significant role in DRR and DRM.

4.3 Integrated Science

Scientific and technological developments have certainly led to the achievement of various aspects of social progress. However, the role of science and technology for DRR policies remains questionable, as losses and affected populations continue to rise worldwide (Cutter et al. 2015). In this regard, international efforts have been made to discuss the complexity of and achieve DRR and DRM by organizing major international events linked to global strategies, such as the World Conferences on Disaster Reduction in Yokohama (1994), Hyogo (2005) and Sendai (2015) in Japan (UNISDR 2005a, b). As an antecedent to the disaster reduction agenda, in 1992 the UN Framework Convention on Climate Change (UNFCCC) was signed, and among other commitments envisaged the social consequences of response strategies to the impacts of climate-related hazards on communities. The 2002 World Summit on Sustainable Development and the related Millennium Development Goals also considered various commitments by governments comprising "an integrated, multi-hazard, inclusive approach to address vulnerability, risk assessment and disaster management, including prevention, mitigation, preparedness, response and recovery, is an essential element of a safer world" (UNDESA 2002).

Additionally, the International Council for Science suggested the establishment of a science programme on Integrated Research for Disaster Risk (IRDR). IRDR, a global, trans-disciplinary research programme aims at addressing the major challenges of natural and human-induced environmental hazards, which entails a full integration of research expertise from all disciplines: natural, socio-economic, health and engineering sciences, in addition to the sphere of practice including policy-making, risk communication and public and political perceptions of and responses to risk. Within a framework built by positioning capacity building, case studies and demonstration projects show assessment, data management and monitoring as the three cross-cutting themes. The main objectives of the programme are (a) characterisation of hazards, vulnerability and risk; (b) understanding decision-making in complex and changing risk contexts and (c) reducing risk and curbing losses through knowledge-based actions (ICSU 2008).

Particular goals to address IRDR research objectives involve the following (IRDR 2013):

1. *Promoting integrated research*

Develop and promote integration and collaboration within the disaster risk reduction community to avoid unnecessary duplication and to maximise research outcomes.

2. *Characterizing hazards, vulnerability and risk*

Identify hazards and vulnerability leading to risks from natural hazards on global, regional and local scales; develop the capability to forecast hazard events and assess risks as well as the dynamic modelling of risk.

Address the gaps in knowledge, methodologies and types of information that prevent the effective application of science to avert disasters and reduce risk.

3. *Understanding decision-making*

Understand effective decision-making in the context of risk management—what it is and how it can be improved; identify relevant decision-making systems and their interactions; understand decision-making in the context of environmental hazards and help improve the quality of decision-making practices.

4. *Reducing risk and curbing losses*

Develop a methodology for implementing comprehensive, long-term vulnerability assessments and effective approaches to risk reduction by bringing together insights gained under Goals 2 and 3.

5. *Networking and partnership building*

Develop, strengthen and collaborate within the IRDR network at global, regional and national levels.

6. *Supporting the science and policy dialogue*

"Enhance the utilization of research findings" (IRDR 2013, pp. 7–13).

To achieve such challenging goals, IRDR is focussed on four major projects directed towards information dissemination, networking and collaboration fora: Assessment of Integrated Research on Disaster Risk (AIRDR); Disaster Loss Data (DATA); Forensic Investigations of Disasters (FORIN) and Risk Interpretation and Action (RIA). AIRDR concerns the global network of researchers involved in the first systematic and critical global assessment of integrated research on disaster risk, whereas DATA is centred in a growing network of stakeholders from different disciplines and sectors to study issues related to the collection, storage and dissemination of disaster loss data. FORIN is a methodology that investigates the underlying causes of disasters. RIA is concentrated on understanding how people, both decision-makers and ordinary citizens, perceive, interpret and make decisions, individually and collectively, regarding the risks (IRDR 2013).

4.4 Risk Communication

Communicating uncertainties and particularly communicating risk remains a major challenge. According to the US National Research Council, risk communication can be regarded as "an interactive process of exchanging of information and opinion among individuals, groups, and institutions. It involves multiple messages about the nature of risk and other messages, not strictly about risk, that express concerns, opinions, or reactions to risk messages or to legal and institutional arrangements for risk management" (National Research Council 1989).

Quite frequently the term risk communication is misinterpreted and therefore misleading. Communication within most of the common existing EWS frameworks refers to the transmission of information concerning specific circumstances or a determined event so that people can take action as a response to imminent warnings. However, the notion of risk communication needs to go beyond that apparel and rely on the understanding of how risk is a construction itself, and therefore cannot be purely based upon the certainty of existing uncertainties, in terms of both hazards and vulnerability. There is thus great value of risk perception and the influence of governance issues in the decision-making processes around risk management.

In accordance with the perspective of some researchers (Adam and Van Loon 2000; Campbell 1996; Fischhoff 1995), effective risk communication needs to consider the fact that risk perception is influenced by cultural and societal factors, and does not only focus on biophysical and technological features. Furthermore, and particularly in societies in which there is a high incidence of institutional vulnerability, the effectiveness of risk communication is linked to the social amplification of risk framework (SARF). Such a framework envisages that some aspects of the representation of hazard events interact with different processes, including psychological, social, institutional and cultural elements, and therefore the attenuation (decrease) or amplification (increase) of risk perception can affect behaviors more broadly (Kasperson et al. 1988).

People might overestimate certain risks if institutions neglect the social context of risk during decision-making processes and when delivering information to the public (Rogers et al. 2007). Moreover and contrastingly, exposed groups may also underestimate disaster risk when hazard occurrence does not fit institutional-political agendas at local or national levels, which ultimately means that there is a deficiency in information or that it is not provided accurately. A major and challenging issue is related to the fact that disaster risk is commonly underestimated according to political momenta, and current dimensions of institutional vulnerability are used to legitimise either poor decision-making or the lack of involvement and social responsibilities of authorities and stakeholders.

4.5 Disaster Governance: From Concept to Practice

To secure a safer future for the world within a disaster framework, special attention needs to be paid to disaster governance, which is defined by UNISDR as the entailment of political commitment and solid institutions that can guarantee the success of efficient and sustained DRR by the following means: improvement of DRR as a policy priority; allocation of the resources required for DRR; enforcement of DRR measures and attribution of accountability for failures and assistance for participation by civil society. These aims must be accomplished through policy and planning, legal and regulatory frameworks, resources and organization and structures (UNISDR 2004).

On the other hand, the International Risk Governance Council (IRGC) has established a risk governance framework aimed at advising risk governance institutions to recognize not only the need to consider knowledge about hazards, but also the concerns that people associate with the different causes of risks. This interdisciplinary and multi-level governance approach fully accounts for the societal context of both the risk and decision-making, and comprises of five phases: (1) - pre-assessment, (2) appraisal, (3) characterisation and evaluation, (4) management and (5) communication (IRGC 2008).

At a local scale and based on some case studies, the United Nations (2010) has acknowledged four major roles of local governments for achieving DRR that reflect the significance of disaster risk governance:

1. "To play a central role in coordinating and sustaining a multi-level, multi-stakeholder platform to promote disaster risk reduction in the region or for a specific hazard.

2. To effectively engage local communities and citizens with disaster risk reduction activities and link their concerns with government priorities.
3. To strengthen their own institutional capacities and implement practical disaster risk reduction actions by themselves.
4. To devise and implement innovative tools and techniques for disaster risk reduction, which can be replicated elsewhere or scaled up nationwide" (United Nations 2010, p. IX and X).

Quite commonly, concepts of risk governance clearly identify the totality of actors, rules, conventions, processes and mechanisms that need to be involved in data collection and analysis, communication and decision-making, in addition to the necessity of taking into account factors associated with particular contexts including institutional arrangements, political culture and risk perceptions (IRGC 2005). What is more, risk governance is also seen as the means by which "authorities, public servants, media, private sector, and civil society coordinate in communities and on regional and national levels in order to manage and reduce disaster and climate related risks" (UNDP 2013). These interactions are also expressed as the intersection between horizontal and vertical governance (Benz and Eberlein 1999; Lyall and Tait 2004). The first one consists of the relevant actors for decision-making based on a specific geographical or functional segment, whereas the latter refers to the linkages between those segments (Aven and Renn 2010).

Notwithstanding the significance of such critical, although still idealistic suggestions, disaster risk governance urgently needs to move towards application and practice.

Despite the role of international and global institutions such as the United Nations and the World Bank, disaster governance relies on state and local based actions. Although it is widely accepted that population growth, especially in cities and megacities, increases potential losses and challenges to governance, and that disaster governance is polycentric and multi-scale (Tierney 2012), DRR and DRM within sub-national and local contexts vary across countries, even in rural areas. Consequently, mechanisms, risk communication, coordination, decision-making and actions are not delineated adequately in terms of the conceptual structure of disaster risk governance. Financial, legal and organisational frameworks are politically dependant; thus transforming political borders and the interests of political parties within nations to focus on societal needs, territorial management and environmental equilibriums are priorities, but transformed into obstacles for integrated, effective and efficient action.

5 Concluding Remarks

Human and economic losses due to disaster associated with EWSs have skyrocketed significantly in recent years owing to the severe impact of the Indian Ocean Tsunami in December 2004, Hurricane Katrina in 2005, the Tōhoku

Earthquake in 2011, the extensive floods in Thailand also in 2011 and Hurricane Sandy in the USA in 2012. Such disasters have unveiled dimensions of vulnerability in all types of countries, but especially in nations such as Japan where their culture of disaster prevention and of structural and non-structural mitigation had been depicted as one of the best in the world. These series of disastrous events have also given rise to major and relatively new concern regarding the economic impact of high-magnitude global disasters and the consequent challenges for development.

The Tōhoku event for instance can be regarded as an unprecedented complex multi-disaster that devastated Japan and changed the perception of preparedness for the rest of the world. Lessons learned from this disaster include the pressing need to gather historical data and traditional knowledge on both hazards and vulnerability for EWASs. From the perspective of natural hazards, uncertainties of models and understanding of processes and mechanisms should be reconsidered, because established or expected thresholds have dramatically been exceeded. Dependence on technology has to be questioned, because disastrous events frequently threaten, damage and disrupt food supplies, water resources, electricity, transport, accessibility and communication systems. Indubitably, the Japanese experience would contribute to a better understanding of the vulnerability of resilient societies and the concatenated risks associated with environmental impacts will have to be revaluated. These signals stress the urgent need to adopt agreed and coordinated measures and policies that take a big step forward into the inclusion of the multiple dimensions of vulnerability of societies in EWASs.

The impact of small, medium and large disasters on economies and economic systems of production, and hence on the global economy, needs to be examined, and economic structures probably re-evaluated. In both developed and developing countries, societal response, organisation and recovery should be defined in terms of the dynamism of vulnerable groups that are classified by age, gender, religion, education level or psychological strengths and weaknesses for the design of EWASs. People's awareness would have to be further developed on the basis of experiences derived particularly from the latest disasters and would need to be fortified with knowledge concerning hazards and vulnerabilities. The challenge of DRR should, in this sense, probably be centered, above all, on the significance of disaster consequences as learning mechanisms for building resilience, decreasing vulnerability and achieving sustainable development.

EWASs (Fig. 7.5) should be sound processes that sustain the practices of close and continuing partnerships among communities, scientists, authorities, decision makers, stakeholders and every actor involved in the construction of risk. They should also be characterized by a responsible commitment to achieving and guaranteeing DRR and DRM in space and time. EWASs should be based on disaster risk-integrated science within a legal and ethical framework on which multi-directional and permanent risk communication plays a central role in the construction of a culture of a risk conscious society. It is not exclusively intended to serve as a coordinating system of response, but most importantly is directed towards the comprehension of disaster risk by incorporating the understanding of root causes of disasters, risk perception and the different dimensions of vulnerability,

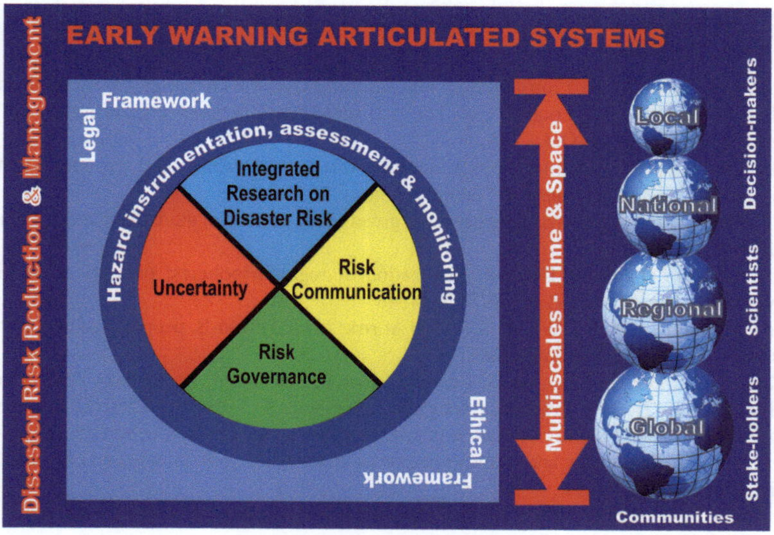

Fig. 7.5 The structure of early warning articulated systems (EWAS)

resilience and adaptation. It must also be structured as a capacity-building progression that allows people to recognize the social construction of disaster risk and its potential consequences in order to consider and assess likely disaster scenarios, risk management procedures, realistic measures, response strategies and actions, targeting preparedness, both individually and collectively, but prior to critical time frames. The failure to integrate legally enforced frameworks and ethical codes into EWASs will further exacerbate the incoherence of government policies and practices.

Integrated risk reduction, including the development of sound EWASs, demands crucial and measurable practical implementation of national and international frameworks aimed at and reinforcing the reduction of vulnerabilities and exposure, an increase in resilience and adaptation at local, regional and global scales. This is the key challenge of the present, which is derived from the lessons of the past, in light of the wisdom and desire for a safer and sustainable global future. Yet the prevailing problem of the international and national agendas is the lack of political will and commitment to reducing the vulnerability mosaic at all scales along with fostering resilience and adaptation in societies at risk.

Acknowledgements The authors thank the support kindly provided by CONACyT through the research project 156242. Prof. Kuniyoshi Takeuchi (ICHARM, Japan) kindly provided photographs of the Tōhoku Earthquake. Disaster data were kindly provided by and downloaded from the ODFA/CRED International Disaster Database Centre for Research into the Epidemiology of Disasters (CRED), Catholic University of Louvain.

References

Adam B, Van Loon J (2000) Introduction: repositioning risk; the challenge for social theory. In: Adam B, Beck U, Van Loon J (eds) The risk society and beyond: critical issues for social theory. Sage, Thousand Oaks, CA

Alcántara-Ayala I (2002) Geomorphology, natural hazards, vulnerability and prevention of natural disasters in developing countries. Geomorphology 47:107–124

Aven T, Renn O (2010) Risk management and governance concepts, guidelines and applications. Springer, Heidelberg

Benz A, Eberlein B (1999) The Europeanization of regional policies: patterns of multi-level governance. J Eur Public Policy 6(2):329–348

Blaikie P, Cannon T, Davis I, Wisner B (1994) At risk: natural hazards, people's vulnerability, and disasters. Routledge, London, p 284

Bull-Kamanga L, Diagne K, Lavell A, Leon E, Lerise F, MacGregor H, Maskrey A, Meshack M, Pelling M, Reid H, Satterthwaite D, Songsore J, Westgate K, Yitambe A (2003) From everyday hazards to disasters: the accumulation of risk in urban areas. Environ Urban 15(1):193–204

Burton I (1997) Vulnerability and adaptive responses in the context of climate and climate change. Clim Change 36:185–196

Burton I, Kates RW, White GF (1978) The environment as hazard. Oxford University Press, New York

Campbell KK (1996) The rhetorical act, 2nd edn. Wadsworth, Belmont, CA

Cannon T (1993) A hazard need not a disaster make: vulnerability and the causes of 'natural' disasters. In: Merriman PA, Browitt CWA (eds) Natural disasters: protecting vulnerable communities. Thomas Telford, London, pp 92–105

Cannon T (1994) Vulnerability analysis and the explanation of natural disasters. In: Varley A (ed) Disasters development and environment. John Wiley, Chichester, pp 13–30

Congleton RD (2006) The story of Katrina: New Orleans and the political economy of catastrophe, Public Choice, 127:5–30

Cutter SL (1996) Vulnerability to environmental hazards, progress in human geography. London 20(4):529–539

Cutter SL, Ismail-Zadeh A, Alcántara-Ayala I, Altan O, Baker DN, Briceño S, Gupta H, Holloway A, Johnston D, McBean GA, Ogawa Y, Paton D, Porio E, Silbereisen RK, Takeuchi K, Valsecchi GB, Vogel C, Wu G (2015) Global risks: pool knowledge to stem losses from disasters. Nature 522:277–279

EM-DAT 2015. http://emdat.be/human_cost_natdis

Fischhoff B (1995) Risk perception and communication unplugged: twenty years of process. Risk Anal 15(2):137–145

Geist EL, Titov VV, Arcas D, Pollitz FF, Bilek SL (2007) Implications of the 26 December Sumatra-Andaman earthquake on tsunami forecast and assessment models for great subduction-zone earthquakes. Bull Seismol Soc Am 97:S249–S270, http://walrus.wr.usgs.gov/reports/reprints/Geist_BSSA_97.pdf

Geographical Survey Institute (2011) About crustal movement in March, 2011. http://www.gsi.go.jp/WNEW/PRESS-RELEASE/2011-goudou0408.htm (in Japanese)

Graumann A, Houston T, Lawrimore J, Levinson D, Lott N, McCown S, Stephens S, Wuertez D (2005) Hurricane Katrina, a climatological perspective. Technical Report 2005-01. Asheville, N.C., 28801-5696, NOAA/NESDIS, p 27

Headquarters for Earthquake Research Promotion (2011) Evaluation of the 2011 off the Pacific Coast of Tōhoku Earthquake. http://www.jishin.go.jp/main/chousa/11mar_sanriku-oki2/index.htm (in Japanese)

Hewitt K, Burton I (1971) The hazardousness of a place: a regional ecology of damaging events. Department of Geography Research publication 6, University of Toronto, Toronto, Canada

Hill LJ, Sparks RSJ, Rougier JC (2013) Risk assessment and uncertainty in natural hazards. In: Rougier J, Sparks S, Hill LJ (eds) Risk and uncertainty assessment for natural hazards. Cambridge University Press, Cambridge, pp 1–18

ICSU (2008) A science plan for integrated research on disaster risk: addressing the challenge of natural and human-induced environmental hazards. International Council for Science, Paris

Inderfurth KF, Fabrycky D, Cohen S (2005) The 2004 Tsunami: six month report. The SigurCenter Asia Papers, 23, p 27. http://www2.gwu.edu/~elliott/assets/docs/research/tsunamiinderfurth.pdf

IRDR (2013) Annual report for the integrated research on disaster risk (IRDR). http://www.irdrinternational.org/wp-content/uploads/2014/08/2013-IRDR-Annual-Report_web-ver.-7MB.pdf

IRGC (2005) IRGC White Paper No. 1. Risk governance—towards an integrative approach. IRGC, Geneva

IRGC (2008) An introduction to the IRGC Risk Governance Framework. IRGC, Geneva, Switzerland

Kasperson RE, Renn O, Slovic P, Brown HS, Emel J, Goble R, Kasperson JX, Ratick S (1988) The social amplification of risk: a conceptual framework. Risk Anal 8:177–187

Laska S (2004) What if hurricane Ivan had not missed New Orleans? Natural Hazards Observer, November 2004, XXIX. http://www.colorado.edu/hazards/o/archives/2004/nov04/nov04c.htm

Lavell A (2008) Community and local level disaster risk management: considerations as regards relations with poverty alleviation. A Contribution to the 2009 ISDR Global Assessment Report on Disaster Risk Reduction. http://www.desenredando.org/public/varios/2008/2008_ISDR_GAR_Lavell_CBRMPoverty.pdf

Levitt JI, Whitaker MC (2009) Introduction: "Truth crushed to earth will rise again": Katrina and its aftermath. In: Levitt JI (ed) Hurricane Katrina, America's unnatural disaster. University of Nebraska Press, Lincoln, NE, pp 1–21

Lyall C, Tait J (2004) Shifting policy debates and the implications for governance. In: Lyall C, Tait J (eds.) New modes of governance: developing an integrated policy approach to science, technology, risk and the environment. pp 3–17

Marulanda MC, Cardona OD, Barbat AH (2010) Revealing the socioeconomic impact of small disasters in Colombia using the DesInventar database. Disasters 34:552–570

Maskrey A (1989) Disaster mitigation: a community based approach. Oxfam, London

Maskrey A (compiler) (1993) Los desastres no son naturales. LA RED, p 166 (in Spanish)

Masozera M, Bailey M, Kerchner C (2007) Distribution of impacts of natural disasters across income groups: a case study of New Orleans. Ecol Econ 63:299–306

Mimura N, Yasuhara K, Kawagone S, Yokoki H, Kazama S (2011) Damage from the Great East Japan earthquake and tsunami—a quick report. Mitigation and Adaptation Strategies for Global Change, pp 1–16

MLIT (2011) White paper on land, infrastructure, transport, and tourism in Japan, 2011. Ministry of Land, Infrastructure, Transport, and Tourism. http://www.mlit.go.jp/report/press/toshi07_hh_000056.html

National Police Agency of Japan (2015) Damage situation and police countermeasures associated with 2011 Tōhoku district-off the Pacific Ocean Earthquake, February 10, 2015. Emergency Disaster Countermeasures Headquarters. http://www.npa.go.jp/archive/keibi/biki/higaijokyo_e.pdf. Accessed 13 Feb 2015

National Research Council (1989) Improving risk communication. National Academy Press, Washington, DC, p 21

Oliver-Smith A (2006) Disasters and forced migration in the 21st century. http://understandingkatrina.ssrc.org/Oliver-Smith/. Accessed 15 June 2008

Oliver-Smith A, Hoffmann SM (eds) (1999) The angry earth: disaster in anthropological perspective. Routledge, New York

Paris R, Lavigne F, Wassimer P, Sartohadi J (2007) Coastal sedimentation associated with the December 26, 2004 tsunami in LhokNga, west Banda Aceh (Sumatra, Indonesia). Mar Geol 238(1–4):93–106

Pelling M (2003) The vulnerability of cities: natural disasters and social resilience. Earthscan, London, p 212

Rogers MB, Amlot R, Rubin GJ, Wessely S, Krieger K (2007) Mediating the social and psychological impacts of terrorist attacks: the role of risk perception and risk communication. Int Rev Psychiatry 19(3):279–288

SSHAC (1997) Recommendations for probabilistic seismic hazard analysis: guidance on uncertainty and use of experts (US Nuclear Regulatory Commission Report NUREG/CR-6372). Lawrence Livermore National Lab, Livermore, CA

Takeuchi K, Chavoshian A (2011) Quick report of Great Eastern Japan earthquake and tsunami (As of 22 April 2011). International Centre for Water Hazard and Risk Management, ICHARM-UNESCO, Tsukuba 305-8516, Japan. http://www.iwra.org/doc/Tsunami_report_Kuniyoshi_Takeuchi_ICHARM.pdf

Tierney K (2012) Disaster governance: social, political, and economic dimensions. Annu Rev Environ Resour 37:341–363, http://dx.doi.org/10.1146/annurev-environ-020911-095618

Tsunami Global Lessons Learned Project (2009) Tsunami legacy. Innovations, breakthroughs and changes. p 105

UNDESA (2002) Johannesburg plan of implementation of the world summit on sustainable development. Division for Sustainable Development, UN Dept. of Economic and Social Affairs, New York, www.un.org/esa/sustdev/documents/WSSD_POI_PD/English/POIToc.htm

UNDP (2013) Issue brief: disaster risk governance. Bureau for Crisis Prevention and Recovery, UNDP. http://www.undp.org/content/dam/undp/library/crisis%20prevention/disaster/Issue_brief_disaster_risk_reduction_governance_11012013.pdf

UNEP (2006) After the tsunami, rapid environmental assessment. http://www.unep.org/tsunami/reports/TSUNAMI_report_complete.pdf

UNISDR (2004) Disaster risk reduction, governance and development. Africa Educational Series, Vol 2, Issue 4, December 2004, p 56. http://www.unisdr.org/files/8546_governacedevelopment1.pdf

UNISDR (2005a) Hyogo declaration. World conference on disaster reduction, 18–22 January 2005, Kobe, Hyogo, Japan. www.unisdr.org/wcdr/intergover/official-doc/L-docs/Hyogo-declaration-english.pdf

UNISDR (2005b) Hyogo framework for action 2005–2015: building the resilience of nations and communities to disasters. World Conference on Disaster Reduction, 18–22 January 2005, Kobe, Hyogo, Japan. www.unisdr.org/eng/hfa/

UNISDR (2006) Developing early warning systems: a checklist. Report from the Third International Conference on Early Warning, Geneva

UNISDR (2012) Impacts of disasters since the 1992 Rio de Janeiro earth summit impacts of disasters since the 1992 Rio de Janeiro Earth Summit. http://www.unisdr.org/files/27162_infographic.pdf

United Nations (2010) Local governments and disaster risk reduction: good practices and lessons learned. UNISDR, Geneva

Velásquez A, Rosales C (1999) Escudriñando en los desastres a todas las escalas. Red de Estudios Sociales en Prevención de Desastres en América Latina—LA RED, p 123 (in Spanish)

Winchester P (1992) Power, choice and vulnerability: a case study in disaster management in South India. James & James, London, p 225

Wisner B, Blaikie P, Cannon T, Davis I (2004) At risk: natural hazards, people's vulnerability and disasters. Routledge, New York, p 471

Part II
Migration

Chapter 8
Applications of Disaster Risk Reduction Principles and Operational Mechanisms to Migration in Contexts of Instability

Andrew E. Collins

Key points of this chapter are that:

- Disaster risk reduction applications span underlying principles of prevention, mitigation, and response including through operational mechanisms of early warning, risk management, communication, appropriate response standards, and longer term recovery.
- Uncertainty in contexts of instability and its potential impacts condition the application of DRR to migration.
- Localized community approaches are central to understanding reactions that realize the efficacy of DRR in migration.
- Rights, well-being aspirations, and sustainable development are core principles for interpreting the application of DRR to unstable migration settings.

1 Introduction

Disaster risk reduction (DRR) principles and associated operational mechanisms are applicable to migration in contexts of instability.[1] The review from which this chapter has been developed was first commissioned to the author for a Foresight project on "Migration and Global Environmental Change" (Foresight 2011).

This chapter is a modified version of Collins' (2013) "Applications of the Disaster Risk Reduction Approach to Migration Influenced by Environmental Change" *Environmental Science and Policy*, 27s, s112–s125. The updated version is reprinted by permission of Elsevier.

[1] Instability for the case of this chapter refers to conflicts or environmental disasters, the combination of the two, and where these may be slow or rapid onset.

A.E. Collins (✉)
Department of Geography/Disaster and Development Network (DDN),
Faculty of Engineering and Environment, Northumbria University,
B201 Ellison Building, Newcastle upon Tyne NE1 8ST, UK
e-mail: andrew.collins@northumbria.ac.uk

© Springer International Publishing Switzerland 2017

K. Sudmeier-Rieux et al. (eds.), *Identifying Emerging Issues in Disaster Risk Reduction, Migration, Climate Change and Sustainable Development*,
DOI 10.1007/978-3-319-33880-4_8

The term "migrant" is used in this work for those who have moved within a country or crossed an international border. Further, "displacement" is considered to be a particular form of migration where movement is forced against the will of the individual (UNDP 2009). The chapter is derived for the processes of migration and displacement rather than long-term recovery issues.

Mechanisms for reacting to migration associated with environmental shocks depend on the nature of hazards and vulnerabilities experienced over the short or long term. For this chapter emphasis has been put on migration and displacement contexts where people's movements may be reactive, involuntary, and non-administered. However, much of the chapter is also applicable to in situ displacement, where people affected by environmental shocks or other forms of instability may undergo major disruption, either choosing to stay put or unable to move away.

There is no assumption here that policy directions aimed at reducing disaster impacts with migrant or displaced people in mind can be a straightforward analysis. Furthermore, linking DRR and migration in contexts of instability has already attracted relatively wide-scale interest, including that of the International Organization of Migration (IOM 2009, 2010). This chapter however offers a consideration of how ascendant principles and operating mechanisms relatively common to DRR might be interpreted for the case of migration.

Four prominent operating mechanisms of DRR are early warning and engagement, reactions to risk, disaster communication, and humanitarian standards in disaster response (Collins 2009a, b). While the chapter does not intend to cover all principles and mechanisms of DRR that may be applicable to migration, it is argued that these readily show some of the specific limitations and opportunities there are. In particular, DRR and migration-related decisions are conditioned by reactions to event uncertainty and their potential impacts. Secondly, localized community approaches in DRR are central to understanding the efficacy of the DRR approach to migration in instability, which in turn draws on wider development principles of rights, responsibilities, and sustainability.

1.1 Reacting to Environmental Shocks and Stressors

Reactions to environmental shocks and stressors include self-help activities that individuals, households, or communities use to protect themselves when facing adversity. For example, in Bangladesh, rural communities in the exposed coastal zone apply techniques of mutual assistance before, during, and after the onset of cyclones, including those requiring temporary altering local religious and cultural norms (Alam and Collins 2010). It had also been noted by Zaman (1993) that 88 % of people displaced by cyclone and flood from river deltas in Bangladesh migrated within 2 miles of displacement location, therefore emphasizing localized coping. It can be argued that coping is also a form of adaptation in terms of change in a system of mutual assistance, albeit driven by necessity.

Reactions to extreme event environmental shock include what people do prior to the "shock," during the event, and post-event. Migration over a short or long distance, or not at all, is one of the fundamental choices to be made. With adequate early warning, and when people believe the message (Erwin-Atwood and Major 1998), a majority will take action to protect themselves and their assets, though not necessarily completely move away from danger. As experienced for cyclones in both Bangladesh and the USA, the choice of reaction to remove from home may be mediated by the relative risk of loss of possessions, incapacity, or the need to detach. In other instances, warnings are not heeded on the basis that they were incorrect previously. In Bangladesh, not all people move to cyclone shelters on hearing of incoming cyclones because the shelter might be considered inadequate in terms of serving basic needs. People question how they can leave behind a much-loved cow, or ask themselves what would God's will be regarding this oncoming threat. Meanwhile, in pastoralist areas of Africa (Angola, Kenya, Tanzania, and others) pastoralist communities have carried out their migrations in response to environmental change across seasons as part of normal living throughout the millennia. While a process of early warning can apply in these examples, a distinction therefore needs to be made between reactions where infrastructure is predicted to remain intact and may be returned to in the short term, and those where more complete destruction of a place or instability occurs.

The propensity to adapt, be resilient, and to preserve or pursue livelihoods and well-being lies at the heart of understanding migration in instability and mechanisms for reacting to ascendant risks. The International Organization of Migration has developed an outline framework for approaching this topic but actual applications of DRR to migration in contexts of instability are potentially more complex. Those reacting to migration from instability include people who are directly exposed to it, people on-looking, and those designated to respond as part of emergency response systems. Formalized response cultures vary, from "blue light" services operational in developed regions to international humanitarian relief that gets applied to emergency migrations anywhere. The military is a prevalent component of emergency responses to environmental shocks.

Population displacement through droughts, earthquakes, tsunamis, floods, volcanoes, cyclones, hurricanes, tidal waves, landslides, and other environmental disasters usually stimulates both national and international level humanitarian reactions. Accompanying emergency management roles involve evacuation, at times forced, of people exposed to hydrological, climatic, and tectonic hazards, either short or long term, to areas where there is apparent safety. However, first migration responses from environmental shocks or conflicts are usually a more spontaneous self-preservation action of individuals, families, and communities in the absence of official assistance or alternative options to stay. They are actions that prevent a further deterioration of conditions through exposure to a known hazard and as such this is a form of risk management or adaptation, not just coping.

Slower moving environmental events and the inseparable "other drivers of migration" (Foresight 2011; Black et al. 2011a), or integrated disaster and development issues (Collins 2009a), such as a failing harvest, encroaching violence, or economic destitution, put people on the move, though with subjective differentiations. However, the consideration of instability as shocks and trends that induce migration and displacement is not solely an academic milieu, but influences the very nature of humanitarian reactions to conflict and environmental events. Misrepresentation equals misreaction, complicating the application of DRR to migration. Mindful of these points of departure and complexities, the next section explores further pertinent aspects of a DRR approach that effect its application to migration in contexts of instability.

2 DRR Principles and Migration in Contexts of Instability

Risk reduction includes prevention, mitigation, and response as underlying principles corresponding to pre-, during, and post-event actions that may be more or less formally managed processes. Where a formally managed process it may be more appropriate to use the term disaster risk management (DRM). Though there are ongoing nuances in the application of this terminology depending on location and context DRR can be considered broader, potentially more political, and draws its main principle from the longstanding health care analogy of prevention as better than cure. However, where they are long term, disaster management phases of prevention, mitigation, and response can occur simultaneously (Frerks et al. 1995). Furthermore, the author's work with DRR projects, international forums set up to progress a DRR approach, and a further review of literature for this study, including the points made in the introduction to this chapter, suggest the need for DRR practice to include the following: (1) integrated, multi-sectored, and interdisciplinary perspectives; (2) recognition of civil societal engagement beyond formalized emergency management; (3) longer term reactions beyond an immediate relief phase; and (4) improved prevention through political investment in local level risk reduction (Collins 2013; Collins et al. 2015a). These characteristics of DRR, referred to elsewhere by the author, here are considered to be in effect principles and challenges for its successful operation and application in relation to contexts of environmental migration. They are principles that inform the operational mechanisms of early warning, risk assessment (including resilience and capacity assessment), communication processes, and adhering to response standards, amongst others.

2.1 *Integrated, Multi-Sectored, and Interdisciplinary Perspectives*

For integrated emergency plans government departments with responsibilities in planning, infrastructure, finance, health, welfare, education, environment, and trade are amongst those included alongside or within disaster management coordinating bodies. However, emergency management often struggles to deliver coordination prior to or during major environmental events, such as flash floods or earthquakes, being also unable to identify the displaced until varied time frames after initial impact. This can be observed for almost all recent disaster events involving mass population displacements, such as earthquakes in China in 2009 and Haiti in 2010, floods in Pakistan in 2010 or Thailand in 2011, and tsunami/earthquake and nuclear incident in Japan in 2011. A salient question here is therefore how integrated emergency management can better respond to highly variable displacements and instabilities. Limitations in cross-sector emergency reactions where displacements are involved may be due to lines of communication not being well integrated between departments in advance and being under-resourced or due to damage from the event itself.

Beyond political infrastructure, multi-sectoralism also encourages interdisciplinary research, learning, and capacity building. The academic sector recognizes the demand since multi-causal environmental shocks and disasters draw from different knowledge environments (Alexander 1997). This is reflected across paradigms of disaster studies, from a focus on natural hazards (Smith and Petley 2009) to the construction of disaster causality (Bankhoff 2001; Lupton 1999). It extends also from an emphasis on the production of human vulnerability (Bankhoff et al. 2004; Wisner et al. 2004) to an emphasis on resilience (Manyena 2006; Paton and Johnston 2006). Through these aspects, the study of disaster, development, and associated displacement combines multiple streams of inquiry for integrated environmental, social, and economic analyses, amongst others (Collins 2009a). This includes notions of those enduring an environmental shock or conflict as survivors rather than victims, resilient rather than vulnerable, and desiring of well-being rather than just to cope. This, however, depends on the duration of the migration. Beyond its immediate dimensions of people moving from one location to the other, migration becomes a mechanism or adaptive strategy (Black et al. 2011b) through which these characteristics and aspirations are expressed over time.

Linked to development interventions, disaster studies also demand highly integrated analyses (Collins 2009a). Meanwhile, integrated models of displacement, such as that of Cernea's (1997) "impoverishment, risk and livelihood reconstruction" (IRLR) model, share similar explanations of human vulnerability to those used in disaster vulnerability analyses. A further example is provided by McLeman (2011) for which a model of progression from vulnerability to population decline to settlement abandonment is applied in contexts of environmental disasters at Montserrat, Pakistan, and other locations. There is immediate resonance here in that vulnerability assessment has also informed Integrated Emergency Management (IEM) (McLoughlin 1985), and Integrated Disaster Risk Management (IDRM) (Okada 2004). In summary, ongoing consideration of those displaced physically and mentally by conflict or an environmental shock, whether remaining in situ or relocating to be potentially hosted by others, requires an integrated view of adaptive migration effects, although not all aspects of change through migration are adaptive.

2.2 Civil Societal Engagement Beyond Formalized Emergency Management

The link between current challenges in DRR practice and the application of derived principles in reaction to migration in contexts of instability goes well beyond formalized emergency management systems. It was pointed out in the introduction to this chapter that formal responses are rarely present when people are in the act of initial flight from harm. Furthermore, civil societal reactions through mutual assistance in disasters have been under-represented or overlooked in analyses of response mechanisms. The gap here is widened in that emergency management responses get evaluated through the analyses of participating sectors based on the learning cycles of their official responders. This tends to be characterized by closed-door "lessons learnt" sessions that may take place post-event to improve formalized reactions next event, without participation of displaced survivors or those remaining at risk.

Rapid response that is not knowledgeable of situations confronted is less effective than response involving prior experience and assessment of the situation (Price 2006). However, by definition a major disaster and the resultant population displacement it causes push the boundaries of human capacity beyond its regular limits. For the case of migrations from instability, first response can be measured in terms of those who help others relocate, those who stay behind, and those who facilitate resettlement in hosting areas. In this sense both migrant and host may be a first responder. Citizen or otherwise identified independent first responders (Palen et al. 2007) represented the main reaction force when there were no emergency units available during the early stages of the earthquakes at Sichuan, China, and Haiti; flooding of Pakistan; and the Japan earthquake, tsunami, and nuclear meltdown. However, questions arise as to why, when faced with significantly similar threats, some chose to flee while others do not. Beyond physical ability, factors such as information availability, and the presence of official emergency agents, socio-economic, cultural, and psychological reasons are relevant (Collins 2015b; Krüger et al. 2015).

Early response, though not necessarily a first reaction, may include the setting up of reception areas for displaced people close to the point of environmental shock, as witnessed in Japan in 2011 and previously in New Orleans in relation to Hurricane Katrina, though in Japan a 20-km exclusion zone was set up around the Fukushima nuclear plant. Response then includes people making a decision to migrate, being formally evacuated or relocated further afield. This selective behavior indicated a complex of processes faced by a significant percentage of New Orleans residents in moving to other parts of the USA. In rapid-onset emergencies, those in charge may call upon official citizen groups, such as the US Citizen Corps, while unofficial citizen responders can be considered a hindrance to rescue services.

For external responses to emergency displacements, difficulties have been associated with unwanted forms of relief that crop up post-disaster, exemplified by the case of relief to pastoralists (Kilby 1993), and of food and medicine during the great flood of Mozambique (Christie and Hanlon 2001). Impacts of inappropriate medical aid in emergency relief included the treatment of thousands of refugees from Rwanda residing at Goma displacement area, resulting in mass death (Goma Epidemiology Group 1995). Inappropriate external interventions may contrast at times with unrecognized, though appropriate, indigenous reactions to environmental shocks. The merits of indigenous practices in disaster risk reduction are acknowledged (Kelman et al. 2012), including in relation to cyclones (Alam and Collins 2010), droughts (Richards 1985), disease epidemic, and so forth. There is also a call for increased recognition of the role of self-help approaches for disaster reduction more widely, referred to as self-care in the context of health disasters (Edgeworth 2010; Edgeworth and Collins 2006). Relocation, migration, or displacement from hazard is not necessarily a pre-requisite to the demand or effectiveness of these approaches. However, they are referred to here as aspects needing care both within and without migration in contexts of instability.

Displacement through instability may mean loss of livelihoods. Whereas formalized emergency assistance may allow for its latter restoration, such as applied through the UNDP-facilitated flood resettlement initiatives in Mozambique (Collins and Artur 2012), alternative reactions of local people are to try not to relocate. For example, not migrating away from an environmental hazard can be for want of hanging on to the potential of rebuilding a livelihood within the impact zone and to protect ongoing assets held there despite ongoing risks. Entrapment (Foresight 2011) is associated with poverty. The act of staying behind to protect livelihoods and capital assets has been exemplified by those sitting it out with weapons in the upper rooms of their houses in New Orleans, people perched with their livestock in river islands of Pakistan, and farmers continuing to produce unsellable food in irradiated zones of northeast Japan. In summary, this section has emphasized that understanding the application of DRR principles to migration in contexts of instability reaches far beyond formalized emergency response processes.

2.3 *Longer Term Reactions*

A limitation of post-disaster recovery has been that emergency relief is short term and that interests in crises tend to wane once cameras and political opportunism move on. Meanwhile, preparedness in contexts where hazards are not yet experienced is generally neglected until such time as visible threats are recognized or pressure mounts through human casualties. Reactions tend to be short term with limited longer term strategy. A response to this from the DRR paradigm has been to promote longer term preparedness. This would mean potential environmental migrants, displaced people, and those in hosting areas being aware of forthcoming scenarios in advance. This can then assist early preventative actions that offset the impact of conflict and environmental shocks. Ultimately, intergenerational justice, core to sustainable development, becomes part of future protection and implementation of a precautionary principle (Harremoës et al. 2002). Longer term reactions are also important to those injured or traumatized by crises, people for whom impacts can be less visible in the short term and requiring psychosocial rehabilitation (NCPTSD 2005; NICE 2005). However, many assumptions concerning these impacts, particularly amongst displaced groups, have failed to understand human coping beyond medical diagnoses (often Western) of mental health (Ingleby 2005).

Slow-onset environmental events such as drought leading to displacement and forced migration, famine, infectious disease epidemics, economic collapse, societal breakdown, or conflict require long-term commitments to change. Emergency services generally remain inadequate in providing sufficiently long-term responses where mass casualties are less concentrated or less predictable. This has led to one of the driving aspects of a long-established call to link disaster relief with longer term development actions (UNDP 2004). The two-way linkage between disaster and development is behind the drive to mainstream disaster reduction into development (DFID 2005, 2006), get development out of disaster, address development-induced disaster, and other variants of this equation (Collins 2009a). Though an oversimplification in operational terms, it has become conceptually reasonable in a world of global and local environmental change to equate disaster reduction directly to sustainable development, as two sides of the same issue. This has eventually become more recognized and mainstreamed through the Sendai Framework for DRR (SFDRR) (2015–2030) (UNISDR 2015) that goes further in integrating disaster and development than the previous Hyogo Framework for Action (HFA) (UNISDR 2005). Migration is often at the core of this equation as the mobility of people, either self-instigated or externally managed, from a place, time, or lack of protection to environmental threats can defuse immediate risks. However, where resettlement fails to provide livelihoods and security, perpetuated conditions of displacement further extend the effects of environmental shocks. DRR and its application to understanding and reacting to migration require a longer time frame developmental approach.

2.4 Political Investment in Local Level Risk Reduction

The logic of DRR in avoiding crises, reducing environmental impacts, or responding appropriately to emergencies remains with vast gaps in investment and accessibility to those most at risk. However, a shift to inclusion of local level capacity in HFA and SFDRR emphasized risk management and resilience building. Achieving this in relation to migration may be a long way off, though there are some good examples close to this agenda, such as through the work of Cyclone Preparedness Committees in Bangladesh (IFRC 2010), Practical Action (2010), and other NGO and local organizational efforts that recognize population displacement from disaster outcomes, but less obviously tackle questions of DRR that apply to migration.

Local level risk reduction was analyzed through action research approaches for risk and resilience committees facilitated by the Disaster and Development Centre

(DDC), currently Disaster and Development Network (DDN), in Mozambique and Nepal (Collins 2009b) including investment from the Government of Mozambique, UK Department for International Development (DFID), World Health Organization, and UNICEF. While these did not specifically target forced migrants the communities within which they have been active lie in post-conflict areas where population displacement has been commonplace. What more progressive examples of these approaches provide is a dialogue and actions at the local level that to varying degrees are devised and owned by those who will implement them, namely everyday occupants of the area at risk.

Using this approach, a reaction to flee, migrate, stay, return, and rebuild is based on closely sensed and locally governed information and strategies. For example, urban risk committees in Mozambique experimented with risk reduction approaches to build community resilience to pervasive infectious disease risk, particularly diarrheal diseases. Community committee activities included the use of visual representation of what local people considered being a risk in their neighborhood, working on the basis that both a perceived or real risk was relevant to confronting threats to community health (Williams et al. 2010). Risk committees engage local authorities once their own assessment of a risk has been confirmed and own the process of implementing their solution to the confirmed risks (Collins 2009b).

Further committees of this type in Nepal uncovered a range of factors in group organization that influence the local risk governance process, with higher success rates being found, but causes unproven, for committees more embedded in state government than functioning entirely independently (Jones et al. 2013). A core feature of the risk and resilience committee approach relevant to evacuation and migration preparedness is that they can be applied with common procedures in preparation for multiple types of oncoming disaster risk. A second key ingredient is that they can be active during times of relative normality with ongoing community-strengthening activities, while also being ready to transform into response units with the onset of instability. However, despite increased recognition of principles of DRR, their applications remain underutilized for migration contexts.

3 DRR Operational Mechanisms for Migration in Contexts of Instability

The summary analysis in Table 8.1 outlines selected operational mechanisms for migration into a disaster reduction frame for principles of prevention, mitigation, and response. The selection of these mechanisms is not intended as comprehensive but indicative of this field for ongoing development.

Table 8.1 Examples of DRR operational mechanisms for migration in contexts of instability

Preventing and preparing for migration	Managing and mitigating migration		Addressing the effects of migration
1. Early-warning engagement	2. Reactions to risks	3. Disaster communication	4. Response standards
Awareness building to strengthen motivation and capacity to negotiate migration from instability	Interpretation and representation of migrant, non-migrant, and host community reactions to risks	Integrated communication to mediate reactions to instability for those who migrate, those who stay, and those in hosting areas	Minimum and extended standards for migration in contexts of instability
Participation in early action through potential migrant and host capacity to address uncertainty	Resourcing of potential migrants		Locally grounded management of emergency relief and recovery for the most vulnerable in areas of fleeing, migration routes, and destinations
Knowledge of when and how best to evacuate or stay, relative to changes in risk, where necessary with improved risk assessment and early warning technologies	Understanding of culturally derived reactions under varying social economic and environmental contexts	Locally, grounded information between those experiencing, those observing, and those reporting reactions to migration in instability	
	Ongoing risk assessment and management	Access and use of communication technology	Longer term recovery and rehabilitation
	Making global assistance pledges accountable	Communicating past lessons for modified decision making	
	Evaluation of risk governance contexts of migrants, non-migrants, and hosting communities	Use of the media, other communication channels, and migration education	Migration as development-oriented adaptation through upholding of migrant and host rights
			Honored pledges for reconstruction in areas of instability

Source: Author

3.1 *Early-Warning Engagement*

In the case of migration as an adaptive strategy for coping with instability, early-warning engagement translates to early warning of optimal opportunities to relocate. Though variants of the disaster management "cycle" have been contested as limited in their resemblance of actual process (Frerks et al. 1995), the need to prepare for decision making before, during, and after a disaster event is not in question. As prevention, early warning should include early action (IFRC 2009) involving a contingency plan. This has already been part developed for mass migration emergencies. However, early warning does not guarantee early action due to a complex of interpretational and communication limitations (note Sect. 3.3 of this chapter).

Ultimately, preparedness challenges human organization, vision, and commitment as it involves working in preparation for "uncertain" or unproven events. Resources to invest in preparedness are often difficult to source as the evidence base is incomplete, leading to potential loss of motivation for action amongst at-risk groups. Politically it is difficult to invest in unseen risks in comparison to more obvious and well-reported scenes that follow an environmental shock involving mass emergency migrations of people. The unpredictability of migration decisions complicates prediction capacity further. The implication has been that the field of prevention and risk reduction, though receiving interest and pledges of support (such as through the gatherings of the Hyogo Accord from 2005 to 2015), remains to date not only with few to-scale financial investments, but also with gaps in predictive science.

For community preparedness in environmental hazard mitigation a question is whether the most at-risk members of a group have sufficient motivation, or can be adequately empowered, for protective actions in advance. This is because, where poverty is a daily pressure, people focus on the immediate needs they experience, rather than longer term speculative ones. Community engagement in effective early warning requires inclusion of community development and well-being aspirations at the preparedness stage (IFRC 2009). This brings motivation that stimulates knowledge and capacity to be able to evaluate risk, albeit aided by indicator verification and appropriate technology, such that early warning is also early action where necessary. Furthermore, it is important to consider that the simplified representation of community involvement in early warning and development through a well-being driver takes place within culturally and politically specific contexts that vary from place to place and over time.

3.2 Reactions to Risk

Reactions to environmental risks vary from situation to situation dependent on experiences of previous events, perceptions of risks, personality, levels of dependency on resources, communication, and interpretational issues. The way risks have been conceptualized is notoriously influential on the way interventions may develop. Mass fatality is never inevitable should there be the *capability* of being in the right place, at the right time with adequate forms of protection (Collins 2009a). In this respect capacity is an overriding theme that tends to determine ultimately much of the progress in disaster reduction (Wisner et al. 2004). For the case of migrants "displaced" groups have frequently been associated with resourcefulness (Haines 2010), a phenomenon that extends to not only regulating their own livelihoods but also contributing to the opportunities of host communities (Jacobsen

2002). Other related approaches also resonate with the risk and resilience innovations outlined in Sect. 2.4 of this chapter using the language of social protection emphasizing social relations, local to global governance, justice and claims-based entitlements (Devereux et al. 2011), Adaptive Social Protection (IDS 2012), or Local Adaptive Capacity Framework (Jones et al. 2010), though are not yet applied to displacement and migration.

Across the variation of perspectives, at the institutional level, further challenges are in the moral, political, and scientifically guided decisions in relation to the level of impact of an instability and certainty about its occurrence (Table 8.2). Terms of engagement in these reactions and policy rationales vary across different contexts and institutional links to the crisis areas. It is important to note that Table 8.2 only represents a basic conceptualization of the decision-making domain where there is trade-off between reactions to higher and lower impacts under varying conditions of certainty. The point here is to demonstrate a range of potential reaction positions alongside implied and contrasting policy rationales. Analysts will recognize examples of disasters and displacement events as variously fitting parts of this representation, albeit subjective. It is also acknowledged that instability involving migration may at different stages of the unfolding events be represented by varying parts of such a matrix.

Table 8.2 Characteristic migration-related reactions and policy rationales under varying conditions of certainty and instability

	Higher potential impact	Lower potential impact
Higher certainty	*Characteristic reaction:* Focus on short-term and more visible migration impacts with limited investment in underlying vulnerability and development issues.	*Characteristic reaction:* Moral and political awareness of migrant and other issues submerged. Low intervention investment accompanied by non-action for "acceptable" risks of "minority" concern.
	Characteristic policy rationale: Mitigate likely impacts on known vulnerable migrant groups. Resilience building, well-being, poverty reduction, and development progress rarely an additional action when feasible.	*Characteristic policy rationale:* Use evidence base to interpret instability threat, and negotiate longer term adaptations, well-being, conflict, or poverty reduction.
Lower certainty	*Characteristic reaction:* Lack of motivation and precautionary investment to crucial areas of migration risk. Misdirected reactions due to uncertainty.	*Characteristic reaction:* Slower onset and "lower impact" instability is considered status quo with acceptable levels of unknown migration risk. Tendency to complacency.
	Characteristic policy rationale: Develop knowledge and understanding of complex impact on and consequences of migration. Opportunity for diplomacy, rights, and moralistic persuasion but characterized by political opportunism.	*Characteristic policy rationale:* Concord with available knowledge and development action only broadly mindful of underlying condition of resilience and well-being, poverty, development, and conflict risk reduction. Migration only lightly considered.

Source: Author

While the science of climate-related environmental hazards and effects remains at times patchy, albeit clear change is occurring, levels of certainty accompany environmental influences on human vulnerability. It is important to consider here that displaced people who may be economically impoverished often gather, or are gathered, in areas of higher environmental risk (Black 1998; Collins 1998). The poorest are more exposed to environmental hazards, such that vulnerability and poverty reduction mitigates disaster impact (Bankhoff et al. 2004; Collins 2009a; Pelling 2003; Wisner et al. 2004). The relationship between poverty and environmental impacts led to increasing convergence of disaster risk reduction as emphasized by the Hyogo Accord of 2005, poverty reduction as emphasized by the Millennium Development Goals, climate change adaptation (Collins 2009a; Pelling 2010), and what has followed these. International organizations increasingly recognize the interconnections including the U.N., World Bank, and national level strategies. Addressing poverty, disaster risk reduction, sustainable development, and climate change adaptation together is largely the conflation of aspects of the same agenda of survival. These are rights-based issues first and foremost. The ability to view environmental sustainability and human migration in this way potentially revitalizes global reactions to issues of poverty and well-being regardless of the levels of certainty that may be possible.

3.3 Disaster Communication

Communication mediates reactions to environmental shocks. State-of-the-art communications are those that are able to preserve the integrity of the message being communicated. At the site of potential displacement, those engaged in assessment and research have learnt to apply participatory technologies to be able to gather and interpret information (IFRC 2006, 2007, 2010). A further advance is the techniques of crisis communication and the ability to transmit information in real time and accessible to multiple actors, including migrants and displaced people. Interpretations of hazards, risks, vulnerabilities, adaptation processes, and the capacity to deal with environmental shocks and associated population movements are subject to the chain of communication. Disaster communication is in part dependent on who informs and who is informed, and what techniques are available to transmit the information. This is particularly important to the media and the reactions they are capable of inducing (Vasterman et al. 2005), particularly in contexts of migration during instability through public concern about where and in what conditions migrants may end up. For example, this is demonstrated through the current broadcasts of the plight of migrants attempting to enter Europe by boat across the Mediterranean, for which media attention has stimulated combinations of civil societal based relief, government interventions, and claims of negligence.

Since the way information is portrayed and received determines the reaction that may result, risk perception studies play an important role in understanding how people interact with risks (Adams 1995; Slovic 2000) and the distancing of risk through modernization (after Beck 1992). The need to communicate trust that can be developed from the basic ideas about trust (for example as defined by Longstaff and Yang 2008; Stephenson 2005) is central to many of the issues being addressed here. An analysis of the issue of communicating beyond the realms of experiential learning and incumbent certainty, that includes the requirement of more of trust, is addressed in Collins (2015b).

Additionally, information communication technology (ICT) allows environmental shocks to be known about instantly throughout the world. It is also a tool for those surviving instability to trace each other across dispersal landscapes. Some recent environmental emergencies, such as New Orleans, Haiti, Pakistan, and Japan, have all witnessed the adoption of mobile communication technology as a reaction mechanism. It is used to make contact between migrant or displaced relatives and in some instances to warn of further risks to come. The mobile phone and social media have ushered in a communication environment in which both those moving away from the location of environmental shock and observers can interact with each other and with the media in a manner previously not possible. Some relief packages of NGOs include communications services for survivors and migrant or refugee groups. Communication capacity changes the dynamic in the dialogue among those experiencing, those observing, and those reporting more formally on instability. For longer term displacements and migrations, once the main media interest has subsided, communication technology remains a tool by which mobile groups network their reactions and decision making.

3.4 Humanitarian Standards in Disaster Response and Migration

In the analysis of operational mechanisms displaced people are taken more into the framework of relief than they are prevention strategies. To improve the relief industry, measures have been taken to produce guidelines and minimum standards (i.e., MSF 1997; Oxfam 1995; Sphere 2004). These include standards for people's immediate primary subsistence needs following destitution, including nutrition, water, sanitation, shelter, and fuel. The case of Haiti revealed that security from theft and abuse needs to be added to this list together with criteria such as wider health requirements and communications facilities. Reference to the rights of refugees is apparent in the Humanitarian Charter, reiterated through Sphere in terms of the right to *non-refoulement* (i.e., not to be sent back), and also resounds with the current renewed calls for social protection. For example, the MSF and Oxfam specifically orient to refugee situations in terms of basic rights. However, minimum standards often remain unattainable in many humanitarian crises involving mass displacements both of rapid and slow onset. This is because for more rapid-onset events displaced people may be out of reach of immediate relief or not represented sufficiently as high priority, and for slow-onset events considered inevitable and "acceptable" aspects of under-development. As such, emergency response aspects of DRR can neither be assumed as sufficiently applied.

Minimum standards as portrayed by the agencies can be less appropriate if not locally derived. Consider the contrasting needs for minimum standards for the governance of displaced people and migrants in Somalia and within the UK, though each raises questions of representation. A further problem is that when migrations are also associated with contexts of political violence rather than environmental shocks alone, there may be deliberate worsening of standards when migrants end up on the wrong side of a political divide. Minimum standards in these contexts become more fundamentally about protecting rights through humanitarian action (IASC 2002). Consistent with aspired to DRR principles and operational mechanisms local players can manage standards and delivery of humanitarian aid effectively where suitable emergency management governance would be in place. This also would help address the problems of dependency and declining international interest once supporting agencies go home (Crisp 2005), and in recognizing displaced people's own adaptation strategies. However, humanitarian response is not yet well integrated with prevention and DRR principles. A growing problem is also that pledges for long-term reconstruction assistance following some disaster events are not being honored.

4 Conclusions

This chapter has analyzed principles and operating mechanisms apparent from a DRR perspective in terms of their applicability to migration associated with instability. It is hoped that the chapter has orientated an understanding of adaptation through migration processes in relation to DRR and draws attention to conceptual and practical implications informed by direct experiences of slow- and rapid-onset instability. Principles presented here are relevant to meeting the aspirations of those affected by environmental shocks to build back better, achieve well-being, or simply adapt to a place to live whether through migrating or in situ adaptation. This is applicable for those that move, those in host areas, and those who stay behind. It requires a vision for future actions beyond merely reacting to threats of mass migration, to realizing risk reduction in migration through rights and sustainable development.

References

Adams J (1995) Risk. UCL Press, London

Alam E, Collins AE (2010) Cyclone disaster vulnerability and response experiences in coastal Bangladesh. Disasters 34(4):931–953

Alexander DE (1997) The study of natural disasters, 1977-97: some reflections on a changing field of knowledge. Disasters 21(4):284–305
Bankhoff G (2001) Rendering the world safe: 'vulnerability' as western discourse. Disasters 25 (10):19–35
Bankhoff G, Frerks G, Hilhorst D (eds) (2004) Mapping vulnerability: disasters, development and people. Earthscan, London
Beck U (1992) Risk Society; towards a new modernity. Sage, London
Black R (1998) Refugees, environment and development. Longman Development Studies, Essex
Black R, Adger WN, Arnell NW, Dercon S, Geddes A, Thomas D (2011a) The effect of environmental change on human migration. Global Environ Change 21:S3–S11
Black R, Bennett SRG, Thomas SM, Beddington JR (2011b) Climate change: migration as adaptation. Nature 478:447–449
Cernea MM (1997) The risks and reconstruction model for resettling displaced populations. World Dev 25(10):1569–1587
Christie F, Hanlon J (2001) Mozambique and the Great Flood, African Issues Series. James Currey, Oxford
Collins AE (1998) Environment, health and population displacement: development and change in Mozambique's diarrhoeal disease ecology, Making of Modern Africa Series. Ashgate, Aldershot
Collins AE (2009a) Disaster and development. Routledge Perspectives in Development Series. Routledge, London
Collins AE (2009b) The people centred approach to early warning systems and the 'Last Mile'. International Federation of the Red Cross and Red Crescent Societies (IFRC), World Disasters report. pp 39–68
Collins AE (2013) Editorial—linking disaster and development: further challenges and opportunities, special edition. Environ Hazards 12(1):1–4
Collins AE (2015b) Beyond experiential learning in disaster and development communication. In: Egner H, Schorch M, Voss M (eds) Learning and calamities: practices, interpretations, patterns. Routledge, London, pp 56–76
Collins AE, Artur L (2012) Strengthening local risk management and mainstreaming disaster risk reduction 2008–2011, project evaluation. United Nations Development Programme, UNDP-Mozambique and New York
Collins AE, Jones S, Manyena B, Jayawickrama J (eds), (2015a) Hazards, risks and disasters in society, hazards risks and disasters series. Elsevier, Oxford
Crisp J (2005) No solutions in sight: the problem of protracted refugee situations in Africa. In: Ohta I, Gebre D (eds) Displacement risks in Africa: refugees, resettlers and the host population. Kyoto University Press, Kyoto, pp 17–52
Department for International Development (DFID) (2005) Disaster risk reduction: a development concern. DFID, London
Department for International Development (DFID) (2006) Reducing the risk of disasters: helping to achieve sustainable poverty reduction in a vulnerable world: a DFID policy paper. DFID, London
Devereux S, Béné C, Chopra D, Koehler G, Roelen K, Sabates-Wheeler R, te Lintelo D (2011) Social protection for social justice. IDS Bull 42(6)
Edgeworth R (2010) Self care for health in rural Bangladesh. PhD thesis, University of Northumbria
Edgeworth R, Collins AE (2006) Self-care as a response to diarrhoea in rural bangladesh: empowered choice or enforced adoption? Soc Sci Med 63:2686–2697
Erwin-Atwood L, Major AM (1998) Exploring the "Cry Wolf" hypothesis. Int J Mass Emerg Disasters 16(3):279–302
Frerks GE, Kliest TJ, Kirkby SJ, Emmel ND, O'Keefe P, Convery I (1995) A disaster continuum? Disasters 19(4):362–366
Foresight: Migration and Global Environmental Change (2011) Final project report. Government Office for Science, London

Goma Epidemiology Group (1995) Public health impact of Rwandan refugee crisis: what happened in Goma, Zaire, in July, 1994? Lancet 345:339–344
Haines DW (2010) Safe haven: a history of refugees in America. Kumarian Press, Sterling, VA
Harremoës P, Gee D, MacGarwin M, Stirling A, Keys J, Wynne B, Guedes V (eds) (2002) The precautionary principle in the 20th century: late lessons from early warnings. Earthscan and European Environment Agency, London
IASC (2002) Growing the sheltering tree: protecting rights through humanitarian action. United Nations, Geneva, http://www.reliefweb.int/iasc/rghr.html
IDS (2012) Adaptive social protection. Institute of Development Studies. http://www.eldis.org/go/topics/resource-guides/climate-change/key-issues/adaptive-social-protection. Accessed online May 2012
IFRC (2006) What is VCA? An introduction to vulnerability and capacity assessment. International Federation of the Red Cross and Red Crescent Societies, Geneva, http://www.ifrc.org/Global/Publications/disasters/vca/whats-vca-en.pdf
IFRC (2007) How to do VCA? A practical step-by-stop guide for Red Cross and Red Crescent staff and volunteers. International Federation of the Red Cross and Red Crescent Societies, Geneva, http://www.ifrc.org/Global/Publications/disasters/vca/how-to-do-vca-en.pdf
IFRC (2009) Focus on Early Warning, International Federation of the Red Cross and Red Crescent Societies. World Disasters Report, Geneva
IFRC (2010) Empowering communities to prepare for cyclones. Report of the Bangladesh ICRC. http://www.ifrc.org/Docs/pubs/disasters/resources/reducing-risks/194300-Empowering-communities-to-prepare-for-cyclones.pdf. Accessed January 2011
Ingleby D (ed) (2005) Forced migration and mental health: rethinking the care of migrants and displaced persons. Springer, New York
International Organization of Migration (IOM) (2009) Disaster risk reduction and climate change adaptation in IOM's response to environmental migration, Geneva
International Organization of Migration (IOM) (2010) Disaster risk reduction, climate change adaptation and environmental migration: a policy perspective, Geneva
Jacobsen K (2002) Livelihoods in conflict: the pursuit of livelihoods by refugees and the impact on human security of host communities. Int Migr 40(5):95–123
Jones L, Ludi E, Levine S (2010) Towards a characterization of adaptive capacity: a framework for analyzing adaptive capacity at the local level. Overseas Development Institute Background Note, December 2010
Jones S, Aryal K, Collins AE (2013) Local-level governance of risk and resilience in Nepal. Disasters 37(3):442–467
Kelman I, Mercer J, Gaillard JC (2012) Indigenous knowledge and disaster risk reduction. Geography 97(1):1–21
Kilby P (1993) Emergency relief programmes for pastoral communities. Dev Pract 3(2):92–102
Krüger F, Bankoff G, Cannon T, Orlowski B, Schipper ELF (2015) Cultures and disasters: understanding cultural framings in disaster risk reduction. Routledge, London
Longstaff PH, Yang S (2008) Communication management and trust: their role in building resilience to "surprises" such as natural disasters, pandemic flu, and terrorism. Ecol Soc 13(1):3 (online). http://www.ecologyandsociety.org/vol13/iss1/art3/
Lupton D (1999) Risk. Routledge, London
Manyena B (2006) The concept of resilience revisited. Disasters 30(4):433–450
McLeman RA (2011) Settlement abandonment in the context of global environmental change. Global Environ Change 21:S108–S120
McLoughlin D (1985) A framework for integrated emergency management. Public Admin Rev 45 (Special Issue: Emergency Management: A Challenge for Public Administration):165–172
MSF (1997) Médicins Sans Frontières refugee health: an approach to emergency situations. Macmillan Education Ltd., London
NCPTSD (National Centre for PTSD) (2005) Effects of traumatic stress in a disaster situation: a national centre for PTSD fact sheet

NICE (National Institute for Clinical Excellence) (2005) Post traumatic stress disorder (PTSD): the management of PTSD in adults and children in primary and secondary care, clinical guideline, No. 26. National Institute for Clinical Excellence, UK
Okada N (2004) Urban diagnosis and integrated disaster risk management. J Nat Disaster Sci 26 (2):49–64
Oxfam (1995) The Oxfam handbook of development and relief. Oxfam, Oxford
Palen L, Hiltz SR, Liu SB (2007) Online forums supporting grassroots participation in emergency preparedness and response. Commun ACM 50(3):54–58
Paton D, Johnston D (eds) (2006) Disaster resilience: an integrated approach. Charles Thomas Publishers, USA
Pelling M (2003) The vulnerability of cities: natural disasters and social resilience. Earthscan, London
Pelling M (2010) Adaptation to climate change: from resilience to transformation. Routledge, London
Practical Action (2010) http://practicalaction.org/disaster-reduction/livelihood_options. Accessed January 2011
Price L (2006) Treating the clock and not the patient: ambulance response times and risk. Qual Saf Health Care 15:127–130
Richards P (1985) Indigenous agricultural revolution: ecology and food crops in West Africa. Hutchinson, London
Slovic P (2000) The perception of risk. Earthscan, London
Smith K, Petley DN (2009) Environmental hazards: assessing risk and reducing disaster, 5th edn. Routledge, London
Sphere Project (2004) Humanitarian charter and minimum standards in disaster response. Sphere Project, Geneva
Stephenson M (2005) Making humanitarian relief networks more effective: operational coordination, trust and sense making. Disasters 29(4):337–350
UNDP (2004) Reducing disaster risk: a challenge for development. Bureaux for Crisis Prevention and Recovery, Geneva
UNDP (2009) Overcoming barriers: human mobility and development. Human development report. UNDP, New York
UNISDR (2005) Hyogo framework for action 2005-2015: building the resilience of nations and communities. United Nations International Strategy for Disaster Reduction, Geneva
UNISDR (2015) Sendai framework for disaster risk reduction. United Nations International Strategy for Disaster Reduction, Geneva
Vasterman P, Joris Yzermans C, Dirkzwager AJE (2005) The role of the media and media hypes in the aftermath of disasters. Epidemiol Rev 27:107–114
Williams L, Collins AE, Bauaze A, Edgeworth R (2010) The role of risk perception in reducing cholera vulnerability. Risk Manag Int J 12:163–184
Wisner B, Blaikie P, Cannon T, Davis I (2004) At risk: natural hazards, people's vulnerability and disasters, 2nd edn. Routledge, London
Zaman MQ (1993) Rivers of life: living with floods in Bangladesh. Asian Surv 33(10):985–996

Chapter 9
Linking Migration, Mobility, and Development for Strengthening Adaptation to Climate and Disaster Risks: Reflections from Nepal

Bishnu Raj Upreti and Gitta Shrestha

1 Setting the Context

> It is important to emphasize the positive contribution young migrants make to societies of origin, transit, and destination—economically and by enriching the social and cultural fabric. Most work hard to earn a living and improve their circumstances. The remittances they send to support families in their home countries are a major contributor to economies worldwide.
>
> -Ban Ki-moon, United Nations Secretary-General (UN 2013a).

The above quotation by the Secretary General Ban Ki-moon shows the importance of the contribution by migrants to the country of origin as well as the destination. The agenda for greater utility of migration to reduce poverty, combat disasters, and foster sustainable development has been growing, in high-level dialogues, i.e., the United Nations sustainable development agenda (SDGs), recognized in Cancun Adaptation Framework which was adopted at the United Nations Conference on Climate Change (COP 16) in 2010 (ADB 2012) and the post-2015 Hyogo Framework for Action. Despite its multifaceted potential for development, it is widely believed that migration has been ignored thus far in the framework of development agendas at national level (Ghimire and Upreti 2012).

The relation between migration and climate change is a contentious issue. Although the links between climate disasters and the consequences in Nepal are widely evident (Sudmeier-Rieux et al. 2012), their relation to migration is yet to be established. There exists a knowledge gap about the inter-linkages between migration, climate change, disasters, and development, especially in Nepal, where migration is as old as its history. Incidences of glacier depletion, shrinking water

B.R. Upreti (✉) • G. Shrestha
Nepal Centre for Contemporary Research, Kathmandu, Nepal
e-mail: bupreti@nccr.wlink.com.np

© Springer International Publishing Switzerland 2017

K. Sudmeier-Rieux et al. (eds.), *Identifying Emerging Issues in Disaster Risk Reduction, Migration, Climate Change and Sustainable Development*,
DOI 10.1007/978-3-319-33880-4_9

balances, disruptive and concentrated precipitation, heavy floods, landslides, soil erosion, droughts, and Glacier Lake Outburst Floods (GLOFs) have repeatedly been reported in Nepal. Changes in the fragile ecosystems of mountainous areas are affecting people's livelihoods, which are highly dependent on forestry and agriculture yet also aggravated by a lack of adaptive capacities (Oxfam International 2008).

Migration has thus been highly advocated as a key tool for increasing resilience by building human capital and empowering members through income diversification strategies (Ghimire and Upreti 2012). Migration therefore increases the longer term resilience of individuals and communities alike to the threats of environmental change (Foresight 2011). However, as advocated in the literature, there is still a need to develop and enhance proper knowledge and adaptation strategies to address climate change and disasters (Banerjee et al. 2011).

In recent years, this debate of whether migration could be harnessed as an adaptation strategy for climate change consequences has become more pronounced in the context of mountainous countries, which are more susceptible to climate change, conflict, and human mobility (Deshingkar 2012). In developing countries like Nepal, the increasing role of remittances in the annual gross domestic product (GDP) has changed the perceptions and outlook on migration and migrant workers (NIDS and NCCR 2012). With more and younger migrants in movement for better opportunities, migrants are increasingly recognized as a powerful agent of positive change and development through engaging, enabling, and empowering them to fulfill their development potential (UN 2013b). As the inter-linkages between climate change, disasters, migration, and development gain more traction worldwide, policy makers in Nepal are interested in exploring the potential of migration in combating the adverse impacts of climate change and its greater contribution to sustainable development.

This chapter discusses a contemporary positive outlook of migration, examined from a mobility and development perspective and as a response to climate change adaptation and disaster risk reduction. At present, Nepal's overall development process is passing through broader and closely inter-linked contemporary change processes ranging from climate change, disaster risks, migration, demographic dynamics, political change, livelihood strategies, and economic opportunities. The cases and examples of Nepal presented in this chapter will highlight the urgency of working from the broader perspective to address development challenges and inter-linkages with migration as a positive contribution toward solution for climate change adaptation (CCA) and disaster risk reduction (DRR).

2 Methodology

This chapter is the synthesis of research work undertaken by the authors during several years under the research collaboration with Swiss universities (Upreti et al. 2012a, b). Furthermore, the chapter draws on a systematic review of secondary sources from both scientific and available grey literatures, including secondary sources of data on returnee labor migrants. Qualitative research methods such as individual interviews, content analysis (mainly newspaper articles), and participatory appraisal of the relationship between migration, mobility, and development have also complemented other sources.

3 Conceptualizing Migration, Mobility, and Development

Worldwide, international migration has become a common multidimensional phenomenon (Shishido 2009) including in Nepal (Ghimire and Upreti 2012). It has increased steadily over the years, becoming an established feature of the contemporary social and economic landscape for many youths (Shishido 2009). Migration in the past has often been seen either from a humanitarian perspective or unidirectionally. However, the recent outlook toward migration is more focused on its possible role in progress toward sustainable development and thus also as a vehicle for CCA and DRR. Migrants can provide financial as well as social remittances, including innovative ideas, practices, identities, and social capital.

Our conceptual framework is based on the "mobility lens," or new ways to analyze migration issues with a specific focus on the development agenda in Nepal (Hannam et al. 2006; Urry 2007). According to Massey et al. (1993) studies of migration theories follow social theories and social theories follow social facts and social facts are related to development. The impacts of globalization and information technologies have diminished distances and resulted in the making of goods, people, ideas, and services as mobile entities and brought in a new perspective for examining development. The concept of mobility also includes corporate movement, infrastructure enabling mobility, capitalist spatial restructuring, tourism, and travel. These all have direct and indirect relations with development, disaster, and climate change issues.

4 International Migration from Nepal: A Brief Scenario

Although internal migration has been a historic phenomenon in Nepal, international migration has become more prominent and developed as an indispensible strategy for securing livelihoods for Nepali citizens, especially after the introduction of a multiparty democracy in 1990. Following the enactment of the Labor Act of 1985, greater incentives from foreign migration were recognized for the first time and consequently foreign employment destinations were diversified to include South-east and Far-East Asia and later the Middle East (CBS 2006). The scale of labor out-migration, especially to overseas countries (other than India), has however been staggering and unprecedented, with the result that remittances have become a veritable mainstay of the Nepali economy since the late 1990s. International migration has further expanded during periods of armed conflict and the post-peace agreement (2006) amid prolonged political instability (Upreti 2009).

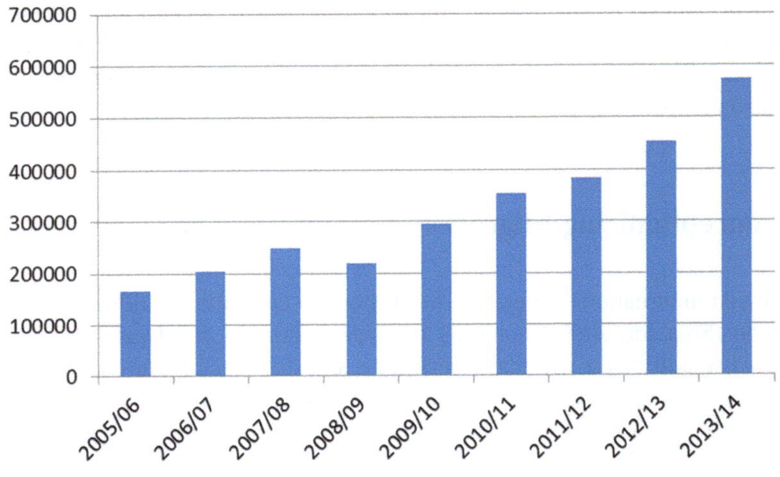

Fig. 9.1 Permits issued for overseas employment per fiscal year (except India). *Source*: DoF (2014)

According to National Population and Housing Census 2011 (CBS 2012), at least one member from each household is absent or is living out of country. The same census reports that the total number of absentee populations is 1,921,494 compared to 7,62,181 in 2001, and the highest proportion, i.e., 44.8 % of absent population, is youths aged 15–24. A large proportion of the young absentee population of Nepal is migrant workers (Sedhai 2012). It is estimated that on average, the number of individuals leaving from Nepal to foreign countries for employment accounts for 41,400 persons/month and 1,349 persons/day (Nepal News 2014). The fiscal year 2013–2014 has reported the highest issuance of permits (572,814) for overseas job placements, surpassing all past annual records (The Himalayan Times, 2014). A report by the Ministry of Labour and Employment 2014 shows that over the 6-year period from 2008/2009 to 2013/2014, a total of 2,226,152 labor permits were issued. According to the Department of Foreign Employment, between mid-April 2013 and mid-April 2014, 446,260 Nepalis went abroad for foreign jobs (DoFE 2014) (see Fig. 9.1).

In the past 6 years, labor migrants from Nepal have worked in as many as 131 countries. Migrants who sought jobs through recruitment agencies account for employment in a total of 31 countries between 1993 and 2000. Kuwait, Malaysia, Qatar, Saudi Arabia, and the UAE have been the most attractive destinations for Nepali migrants since 1993. From 1993/1994 to 2007/2008, these countries received nearly 97 % of all labor migrants from Nepal. Data shows that during the last 6-year duration until September 2014, Malaysia received 40.9 % of all male and female labor migrants, followed by Saudi Arabia (22.9 %), Qatar (20.3 %), UAE (11.2 %), and Kuwait (2.1 %). In addition to these top five destinations, Korea has been receiving more migrants since a ministerial agreement between Nepal and Korea in 2007. The top five destinations for female labor migrants through recruitment agencies are Malaysia (38 %), UAE (33.8 %), Qatar (7 %), Lebanon (6.4 %), and Cyprus (3.3 %) (DoFE 2014).

India accounts for 41 % of Nepali migrant workers (The World Bank 2009). Apart from the contradictory claims about the number and amount of remittances received from India (due to an open border policy between the two nations) India is the greatest work destination for Nepalis. According to the 2011 census, around 742,000 Nepalis were found to be working in India. Nepal receives an estimated amount of US$1,634 million in remittances per year, usually from its least wealthy migrants (NIDS and NCCR 2012). District-wise, Dhanusa, Mahottari, Jhapa, Morang, Nawalparasi, Saptari, Sunsari, Sarlahi, and Rupandehi were found as top ten districts with the highest number of migrants, many of which scored the lowest in terms of education on the UNDPs HDI.

Migrants are mostly male. According to the department of foreign employment (2013/2014), 95.1 % of labor permits were issued for men. Female migration is recent in Nepal, but it is on rise after a ban on women traveling to the Middle East was lifted in 2010. As of 2013/2014, 46,274 permits were issued to women going abroad to work. Recently, the government has also revised the minimum age for women to go for foreign employment. The limit was reduced from 30 to 25 years old. The percentage of female migrants has increased from 3.9 % in 2008/2009 to 5.6 % in 2013/2014 (Sedhai 2014). A majority of migrants who migrate to Gulf countries and who go through Department of Foreign Employment are from middle and lower economic classes.

5 Remittances

Nepal is one of the most heavily remittance-based economies in the world, mainly from labor migration (World Bank 2011). A recent study by the World Bank shows that Nepal ranked third in the world in terms of the contribution of remittances to the GDP (Global Times 2014). In 2012/2013, remittances aggregated US$4.93 billion, a rise by 11 % compared to US$4.41 billion in 2011/2012 (SAARC 2014). Likewise, the remittances to GDP ratio increased from 10.7 % in 2000/2001 to 13.8 % in 2006/2007 and further to 25.5 % in 2012/2013 and a 29 % share during the first 8 months of 2013/2014. With such an increasing trend in foreign employment migration, the annual GDP growth has increased from 3.4 % in 2011 to 4.9 % in 2012 (World Bank 2014). In the fiscal year of 2012/2013, Nepal's GDP was recorded at 28.8 % (World Bank 2013) accounting for US$4.44 billion (through the tenth month of Nepali fiscal year 2013/2014) (Global Times 2014).

About half of the remittances come from Gulf countries. According to the Nepal Living Standard Survey (NLSS) 2010/2011, the share of remittances received from abroad is the highest from Qatar (21.3 %), Malaysia (19.2 %), Saudi Arabia (14.9 %), India (13.4 %), the UK (2.2 %), and other countries (29 %) (Table 9.1). The World Bank (2011) reports that a Nepali household with a migrant worker in India receives an average of NPRs 62,000 (roughly US$620 at the time of printing) in remittances annually. According to NLSS 2010/2011, around 55.8 % of the households receive remittances, which is almost double that of 1995/1996 NLSS

(23 %). The average per capita remittance received was NPRs 9,245 in 2010/2011 (Kharel 2011). With export declining over the years, money sent home by migrant workers has become the main source of foreign exchange for Nepal (Table 9.1).

Around 79 % of remittances are spent on daily consumption, 7 % to repay loans, 4.5 % on household property, 3.5 % on education, 2.4 % on capital formation, 0.5 % in new business development, and only 0.6 % for savings (CBS 2012) (see Table 9.2).

According to a survey from the Nepal Rastra Bank in 2008, remittances in 2005/2006 were used largely to buy land and a house (49 of the respondents), followed by repayment of the debt (25 %), savings in a bank (11), education, health and other (9 %), social work (3 %), and other investments (2 %) (see Table 9.3).

Because of high consumption and the consequent low gross domestic savings (6.7 % of GDP in 2010/2011), the gross domestic savings-investment (gross capital formation, GCF) gap is negative and high (−23.6 % of GDP in 2010/2011).

Table 9.1 Share of remittances from major migrant-receiving countries

Country	Percent (%)
Qatar	21.3
Malaysia	19.2
Saudi Arabia	14.9
India	13.4
UK	2.2
Other countries	29

NLSS (2010/2011)

Table 9.2 Use of remittances according to CBS 2012

Use of remittances	Percentage (%)
Daily consumption	79
Repay loans	7
Household property	4.5
Education	3.5
Capital formation	2.4
New business development	0.5
Savings	0.6

Source: CBS (2012)

Table 9.3 Use of remittances according to Nepal Rastra Bank 2008

Use of remittances	Percentage (%)
Buy land and a house	49
Repayment of debt	25
Savings in a bank	11
Education, health, and other	9
Social work	3
Other investment	2

Source: Nepal Rastra Bank (2008)

However, remittance inflow, even as it fuels consumption, has been so large that the gross national savings (GNS) have exceeded GCF in all but 2 years since 2000/2001 (Kharel 2011).

6 Migration and Development

With increasing international migration, the volume of remittances has exceeded the volume of development aid in many developing countries like Nepal. Lucas (2005), for instance, argues that remittances are more reliable capital flows than official ODA and FDI. The positive effects of remittances are observed on food security, access to basic needs in times of livelihood shocks such as drought, finance, the acquisition of human, social, physical, and natural capital, and increased demand and growth in local production. Research indicates that families with labor migrants have better adaptation abilities during livelihood crises than those that do not.

Remittance has positively contributed to poverty reduction in the last 15 years in Nepal. It has been estimated that the headcount poverty ratio has decreased from 42 % (1995/96) to 31 % (2003/4) to 25 % (2010/11) and the poverty ratio has declined by 2 percent points per year (CBS 2012). Lokshin et al. (2007) show that international migration has contributed about 14.5 % in poverty reduction.

Poverty reduction in the place of remittances origin is reported by Upreti et al. (2012a, b). Sunam (2014) also shows the changed economic status in migrant households. According to UNDP's "Human Development Report 2010," remittances have been one of the factors behind Nepal's remarkable success in human development in the last 40 years. Indeed, Nepal has emerged as one of the world's most dynamic countries in the Human Development Index (HDI) since 1970, coming third among 135 countries studied. The same has been reported by BTI 2014 Nepal Country Report, which informs a rise in Nepal's HDI from 0.449 to 0.458 between 2009 and 2011. The report also outlines improved development indicators: literacy, school enrolment, and average life expectancy rates and child and maternal mortality rates. Additionally, Aubriot (2010) reports the positive impact of international migration on local life in the Gulmi district in terms of changes in social organization, family, and land tenure. She found that migration has strongly supported marginalized families, considering the fact that migrating has generally yielded higher incomes than staying in one place. Migrants have started financing their children's education from cities including girls' education, metal roofs, concrete houses, telephones, televisions, and furniture.

Katseli et al. (2006, cited in Barnett and Webber 2010) demonstrate that migration also boosts incentives to pursue education as this is a determinant of how mobile one may be. Migration thus indirectly increases the educational attainment of migrant populations. This has indeed been the case for Nepal. Similar to the other studies indicated above (such as Aubriot 2010; Sunam 2014), Gautam (2005) finds that migration has brought improvements in the literacy and educational status of many villagers. According to his study, 5 % of migrants improved themselves in terms of literacy while living and working in India and other countries.

Improvements have also been noticed in the Gender Inequality Index (GII). BTI 2014 Nepal Country Report shows improved female-male literacy ratio—67.2 % (male) to 70.2 % (female) at primary school level, 26.9–26.6 % in lower secondary, and 15.7–14.2 % in secondary school. Similarly, the NLSS 2010–2011 indicates that labor force participation is similar for men and women, around 80 % for both. It has helped women to escape gender-based violence (GFMD 2010).

Similarly, despite the increased feminization of migration and the associated negative impact on women, the literature also argues that migration has contributed to the empowerment of women, particularly when dealing with public and social issues (Aubriot 2010; ICIMOD 2014; Sunam 2014). For example, in Jumla, the Shree Kanjiroba Mahila Bachat Cooperative run by women is financed by remittance deposits. It has provided loans to over 400 women at 14 % interest and paid interest of 8 % on deposits, which is higher than many other banks. They also provide training in entrepreneurship and skill development in areas such as chip-making, stitching, and vegetable farming. The cooperative did not require collateral, as their bad debt was low because they had a strong emphasis on loan recovery. They used strategies like social boycotting and the seizure of debtors' houses to its credit. Because of this, it has also been used as a safe channel to finance migration (Sherpa 2010).

Interestingly, migration has also made a broader developmental social impact in unequal rural society. For instance study of Sunam (2014) shows that Dalits (known as "low caste") have started purchasing land in the village from higher caste/higher income groups, who move to urban areas or the Terai areas. A similar study by Gautam (2005) found that cash from remittances has helped lower caste people buy small pieces of fertile grass land, meaning that higher caste people have started treating them with more respect.

Furthermore, international migration has also contributed to an increase in non-agricultural incomes (through increases in remittance-induced demand) and agricultural wages (through a tightening of the labor market). It has also changed the social relationships between tenants and landlords. Tenants with earnings from migration are now able to make stronger claims when seeking legal remedies in cases of exploitation by landlords.

Despite the many positive consequences of migration, there are also certain drawbacks. A major impact of migration in Nepal has been in the agricultural sector. It has caused shortages of labor in villages with migrant households because more of the remittance savings are channeled to urban areas for better infrastructure, health, and education. Gautam (2005) argues that in the long run, this causes a shortage in the necessary manpower for developmental activities of the nation. He also highlights the negative impact on migration in the decreased agricultural productivity because of reduced manure. Due to a lack of manpower, livestock has been reduced per household and therefore has a negative impact on pastoral activities. Similar findings have been found by Schwilch et al. (Chap. 11), who highlight the decrease in manure due to less livestock which is causing a negative impact on land management.

Apart from agriculture, migration has a deeper social impact on family members that are left behind. As indicated above, since a major proportion of the migrants are moving to urban centers, there is no one to look after the old family members. Migration has also been reported to cause family breakdown. Wives' elopement and polygamy are some of the negative impacts of changing relationship in Nepal society due to migration (Gautam 2005). In addition to an increasing trend of divorce, less attention is paid to the elderly, scarcity of agricultural labor and feminization of agriculture, psychological problems, trafficking of young females, and an increasing trend of HIV/AIDS, STDs, heart attack, and accidental injuries and death (Gautam 2005; Rijal 2013).

The below section attempts to link this evidence in relation to CCA and DRR strategies.

7 Migration as Potential Adaptive Strategy Against Climate Change Impacts and Disasters

Despite contention around the issue of climate change-driven migration, changing variations in climate and its impact on population and resources have been well documented in Nepal. Land degradation for instance, due to the abandonment of land, reduction in soil fertility, and changing rainfall variation, has been a factor forcing mountain populations to seasonally migrate to many parts of India and elsewhere (Massey et al. 2007; Schwilch et al. Chap. 11). Mobility, as is found in other parts of the world (Tacoli 2009; Warner and Afifi 2014), has been a major adaptive mechanism adopted by migrants in many rural hills of Nepal to cope with food insecurity. It has also helped migrant households to add to economic and social resources which in turn help their resilience against future risks and environmental disasters. Resilience to risks and disasters including climate change thus requires increased adaptive capacity, which depends on a wide range of factors, including but not limited to financial resources, governance, information, social resources, infrastructure, and technology. If on the one hand it is believed that improvement in these factors can reduce the need for people to migrate, on the other hand, migration has also been argued as a way to improve these factors (ADB 2012).

Despite the fact that the adaptive capacity of migrating communities has increased due to a decrease in resource demand from sending regions (ADB 2012), newer understandings about migration *consider* remittances as not only a major contributor to "economic development" but also as a major contributor to household risk reduction and decreased vulnerability at the household level. It has been found that remittance-receiving households are less disrupted by environmental disasters. For instance, a study by Banerjee et al. (2011) found remittances to be a strategy toward disaster preparedness and a safety net in the aftermath of disasters. They found that financial remittances have immediate and direct impacts on household responses to water hazards in the Hindu Kush region by facilitating irrigation facilities in drought-affected households, buying boats, and improving or strengthening housing quality in households affected by floods.

Social remittances from migration such as skills, networks, knowledge, and information enhance the prospects of local businesses and simultaneously provide a safety net in times of disaster (Tacoli 2009). In the long run, it has been argued that migration can be instrumental in fostering resilience at the household level, as it allows for income diversification and better access to information and social networks (ADB 2012). It can increase the acquisition of new technologies, and migrants and their families are often early adopters of information and communication technology. Use of technology at the local level, such as mobile phones to obtain information on available resources and new adaptation technologies, has also been identified as sound adaptive practices. Expanded local networks have also been found to mobilize technical and financial resources for adaptation (ICIMOD 2014). This has been confirmed by a recent study by Jaquet et al. (2015) which shows that migration has several positive consequences including the strengthening of coping capacities, household income, access to education and employment, expansion of opportunities for marginalized low-caste groups in Nepal, and the empowerment of women. Banerjee et al. (2011) show that social remittances such as certain skills such as carpentry, masonry, and tailoring have all led to the income and sectoral diversification of livelihoods in migrant households in the Hindu Kush Himalayan region. Such diversification ultimately contributes to the enhancement of a household's adaptive capacity.

As stated above, the greatest visible contribution from social remittances has been the development of entrepreneurial endeavor among migrants, which make them a potentially important resource that can enhance the capacities of hosting communities to adapt to risks including climate change. Scholars (i.e., Black 1994) argue that migrants often bring resources with them, including skills and labor, and use these to build their livelihoods. Nepali media report several such cases where a migrant, upon his return, engaged in entrepreneurship such as commercial agriculture and shops (grocery, tailor, tea, etc.). A local newspaper reported that Suresh Chaudhary, a returnee from Malaysia, was able to set up a profitable vegetable farming business.[1] He was able to earn more than double of what he would earn in the Gulf. Similarly, Srikishna Adhikari, a 30-year-old returnee from Israel, was able to utilize his knowledge about deep irrigation systems that he learned while working in Israel. He has set up a commercial vegetable farming venture in Hetauda while also counseling and supporting 23 other vegetable farm projects (Ghimire 2014). Holmelin (2014) reports that young returnee migrants from Dolakha are moving back to the countryside to cultivate new crops like cardamom, tea, garlic, cauliflower, potatoes, peas, tomatoes, chilli peppers, and ginger, as well as

[1] Republica National Daily, April 25, 2014.

promoting traditional forms of food production. Some are focusing on medicinal plants, such as the cancer medicine Himalayan Yew (Lothsalla) or honey production, while others are experimenting with kiwifruit, pears, and oranges. This has been seen as an example of fostering local development. Local innovation and experiments to cultivate new crops and improve farming methods in turn make villagers better equipped to face a future with increased climate uncertainty.

Similar reports have been published on women returnee migrants. For example, it has been widely reported by the national newspapers that foreign employment has helped women turn into small-scale entrepreneurs. As Gautam (2014) reports, Manju Shrestha is one such woman, who has been able to establish herself as a successful entrepreneur after coming back from Kuwait. Similarly, Pratima Ghimire, a 26-year-old returnee from Oman, has established a tailoring business. Likewise, Muna Gautam, after working in a couple of Gulf states, returned to run a travel agency jointly with her husband (Gautam 2014). To add more, Khatiwada (2012) highlights how remittances from women migrants have helped to start new businesses such as auto rickshaw rentals in Nepal, which have in turn become positive role models for others.

Remittances from migration have become a key contributor for development, economy, livelihoods, and service sector growths. Remittances contribute to higher local wages and a higher demand for locally produced goods and services (CBS 2006). Migrants, especially returnees, have positively contributed to the building of assets such as financial capital (savings) and human capital (training and work experience), as well as securing livelihoods and enhancing adaptive capacities. Based on the experiences of the past two and half decades of high migration rates, migrants have added value to development by bringing financial resources and technical skills like vegetable or animal farming and assisted the government to opt for agricultural reforms. They have also helped to develop drinking water and access road/trails, and contributed to generate local employment for unemployed youths.

Such gains from migration such as women's empowerment, Dalit empowerment, changed social relations, agricultural and land reforms, improvement in education, and awareness could, we argue, indirectly contribute to the improvement of the adaptive capacities of society's most marginalized in facing the adversities of life, including during natural hazard events. Migration has not only proved as a strategy to diversify incomes, but it also helps to transport skills, knowledge, awareness, and the formation of agencies. This could lead to the establishment of participatory governance (which as mentioned before has already become more inclusive due to exposure to knowledge and awareness), a cohesive and just society (due to increase in networks), and free flows of information (due to access to modern technologies), all which greatly aid in addressing unanticipated climate risks and disasters.

Nevertheless, as seen from the case of Nepal, the benefits from international migration can only be maximized and sustained when the migrants themselves originate from favorable environments. It has been argued that in the absence of favorable support, many migrants are caught within a cycle of re-migration. The lack of favorable support such as counseling centers on entrepreneurship and proper market value for goods produced in the country of origin are major hindrances to starting an enterprise in Nepal. Returnee migrants face obstacles such as a lack of counseling centers, training and difficulties to lease land, and a lack of collateral and start-up money while trying to start a business. Due to the unfavorable business environment of Nepal, it is difficult to implement the vision they bring from overseas. Moreover, a recent study by the World Bank (2013) argues that entrepreneurship for migrants is more of an economic necessity than a desire to use the new skills they learned overseas. The study shows that the returnee did not find the skills learned overseas relevant while starting business in Nepal. Similar to other studies, it also finds that proper support such as business management advice and investment capital to returnee migrants is needed in order to realize entrepreneurship as a critical component of poverty reduction and development strategies.

Maharjan et al. (2013) show how migration provides opportunities to make significant contributions to the agricultural sector, and how farming households unwilling to invest in subsistence farming can be motivated to shift to commercial farming if a suitable environment existed. Similarly, Barnett and Webber (2010) argue that migration could contribute toward the development of both receiving and sending countries. Receiving countries create wealth through received labor in the context of labor shortages. The migrants working in the host countries contribute positively to growth and the provision of public goods by paying taxes. The ability and willingness of local and national governments to assist migrants are therefore critical to maximizing the benefits they can bring back to the origin. It is thus important to pave the way for safer and more cost-effective migration, by assisting migrants in the migration process from their villages to their destinations and back again to their origin through creating and providing proper opportunities that adapt well to local cultural and natural settings.

Conceptually, migration and development are closely interrelated (Ghimire and Upreti 2012). Hence, migrants in Nepal are increasingly seen as agents of development, because their exposure and earning make them a medium for the exchange of information and resources between places of residence and origin communities, even though they are not primarily responsible for the development. As described above, migrants can therefore accelerate development through remittance expenditure and investment in development, climate change, and DRR activities. However, their individual interest in rural development through collective investments in services and infrastructure is not high. The expenditure and investment choices made by individual migrants primarily reflect their legitimate individual interests in improving the livelihoods of their own families and, to a certain extent, their localized communities (Sunam 2014). The contribution to development of migrants who are still away is still notable as they send money to their family members to invest in their basic infrastructures. In this context, returning is not necessarily a condition for development; however, a conductive policy framework is essential to the promotion of development through migrant remittances.

8 Conclusions

As argued elsewhere in the chapter, migration is a proven development strategy to maximize the needs of local communities (De Haan 1999) as well as adaptation strategies to climate risks and disasters (Agrawal 2008). Adger et al. (2007) highlight that the capacity of a system to respond to climate change to moderate or avoid its negative consequences is a function of a number of properties, including financial resources (to pay for adaptation); governance (how well society can steer the adaptation process and how legitimate that process is); information (to anticipate climate risks, devise appropriate adaptations, and learn from their implementation); social resources (networking and bonding among people and groups so that social responses to climate change are cohesive, equitable, and robust); infrastructure; and technology (tools and crafts that help adapt). Migration can therefore make significant positive contributions to many of these determinants of adaptive capacity. In Nepal, remittances have increased financial resources; migrants have increased communities' access to information, expanded social networks, and contributed to climate-sensitive local infrastructure development like rainwater harvesting, solar panel, plastic ponds, drinking water taps, resting place, and feeder roads.

This chapter concludes that migration is multidimensional and a proper understanding of the relationship between migration, mobility, development, and risks emerging from disaster or climate change requires a comprehensive analytical framework, in-depth analysis of the phenomenon, and broader understanding of the emerging issues and challenges. This exploratory study on the interrelationship between migration, mobility, development, climate, and disaster risks in Nepal is in the rudimentary stages and needs further strengthening to be properly understood. However, the preliminary research works from different researchers and organizations are indicating important positive and negative links between the different dimensions.

It is important to link the emerging research to policy and develop a convergence of strategies and programs that work toward addressing challenges brought by the global change process such as disasters, climate variability, migration, development, and demographic changes. It is important to understand and reconcile local trends vis-a-vis global demand and pressure before the development of any adaptive measures. Remittances from migration of unskilled and semi-skilled Nepali people are far more than the contribution of the skilled people. However, the Government of Nepal needs to work more effectively to develop adaptive capacities of individual migrants, local communities, and governments through a more effective use of remittances.

References

ADB (Asian Development Bank) (2012) Addressing climate change and migration in Asia and the Pacific. Asian Development Bank, Mandaluyong City, Philippines

Adger WN, Agrawala S, Mirza M, Conde C, O'Brien K, Pulhin J, Pulwarty R, Smit B, Takahashi K (2007) Assessment of adaptation practices, options, constraints and capacity. In: Parry M, Canziani O, Palutikof J, Van der Linden P, Hanson C (eds) Climate Change: impacts, adaptation and vulnerability, contribution of Working Group II to the fourth assessment report of the inter-governmental panel on climate change. Cambridge University Press, Cambridge and New York, pp 717–743

Agrawal A (2008) The role of local institutions in adaptation to climate change. Paper prepared for the social dimensions of climate change. Social Development Department, The World Bank, Washington, DC

Aubriot O (2010) International and national migrations from a village in Western Nepal: changes and impact on local life and impact on local life. Eur Bull Himalayan Res 35–36:43–61

Banerjee S, Gerlitz JY, Hoermann B (2011) Labour migration as a response strategy to water hazards in the Hindu Kush-Himalayas. International Centre for Integrated Mountain Development, Kathmandu

Barnett J, Webber M (2010) Accommodating migration to promote adaptation to climate change. Policy research working paper 5270. The World Bank

Black R (1994) Refugee migration and local economic development in Easter Zambia. Tijdschr Econ Soc Geogr 85(3):249–262

CBS (Central Bureau of Statistics) (2006) Statistical Pocket Book of Nepal, Kathmandu

CBS (Central Bureau of Statistics) (2012) National population and housing census 2011 (national report), vol 1. CBS, Kathmandu

DoFE (2014), Labour Migration for Employment A Status Report for Nepal: 2013/2014. Kathmandu: Department of Foreign Employment.

De Haan A (1999) Livelihoods and poverty: the role of migration—a critical review of the migration literature. J Dev Stud 36(2):1–31

Deshingkar P (2012) Environmental risk, resilience and migration: implications for natural resource management and agriculture. Environ Res Lett 7(1):015603. doi:10.1088/1748-9326/7/1/015603

Foresight (2011) Migration and global environmental change: final project report. The Government Office for Science, London

Gautam T (2005) Causes and impact of migration: a sociological study of emigration from Kandebash, Baglung, Nepal. Dhaulagiri 1:146–163

Gautam L (2014) Relations cost high when remittance steals away wives. www.thehimalayantimes.com. Accessed 27 Aug 2014

GFMD (Global Forum for Migration and Development) 2010 background paper, Roundtable 2: human mobility and human development. RT session 2.2: migration, gender and family, Mexico

Ghimire S (2014) Changes on the farm. My Republica. April 24, 2014

Ghimire A, Upreti BR (2012) Migration, mobility and development: enhancing our understanding in Nepal, SAs RCO discussion paper 6. South Asia Regional Co-ordination Office of National Centre of Competence in Research North-South and Nepal Centre for Contemporary Research, Kathmandu

Global Times (2014) Contribution of remittance to GDP reaches 25 pct in Nepal. http://www.globaltimes.cn/content/866600.shtml. Accessed 25 Dec 2014

Hannam K, Sheller M, Urry J (2006) Mobilities, immobilities, and moorings. Mobilities 1(1):1–22

Holmelin N (2014). Dolakha farmers learnt to adapt Nepali Times. http://nepalitimes.com/article/nation/dolakha-farmers-learn-to-adapt,1260

ICIMOD (International Centre for Integrated Mountain Development) (2014) Women's empowerment at the frontline of adaptation: Emerging issues, adaptive practices and priorities in Nepal. Working paper 2014/3. Kathmandu, Nepal

Jaquet S, Schwilch G, Hartung-Hofmann F, Adhikari A, Sudmeier-Rieux K, Shrestha G, Liniger HP, Kohler T (2015) Does outmigration lead to land degradation? Labour shortage and land management in a western Nepal watershed. Appl Geogr 62:157–170. doi:10.1016/j.apgeog.2015.04.013

Kharel P (2011) Remittances as development resource: the experience of Nepal. South Asia watch on trade. Economics and Environment (SAWTEE), Kathmandu, Nepal

Khatiwada PP (2012) Migration in Middle East: a gender lens. In: NIDS, NCCR (eds) Nepal migration year book. NIDS and NCCR, Kathmandu, Nepal

Lokshin, M, Bontch-Osmolovski M, Glinskaya E (2007) Work-related migration and poverty reduction in Nepal. World Bank policy research working paper 4231, May 2007. The World Bank, Washington, DC

Lucas R (2005) International migration and economic development: lessons from low-income countries. Expert group on development issues. Ministry for Foreign Affairs Sweden, Stockholm

Maharjan A, Bauer S, Knerr B (2013) Migration for labour and its impact on farm production in Nepal. CESLAM, Kathmandu

Massey DS, Arango J, Hugo G, Kouaoci A, Pellegrino A, Taylor JE (1993) Theories of international migration: a review and appraisal. Popul Dev Rev 19(3):431–466

Massey D, Axinn W, Ghimire D (2007) Environmental change and out-migration: evidence from Nepal. Population studies centre report, 07-615. Population Studies Centre, University of Michigan, Ann Arbor, MI

Nepal News (2014) Over 492,000 persons left Nepal for foreign jobs last year. My Republica. Published on 2014-05-11 20:19:02. See more at: http://www.myrepublica.com/portal/index.php?action=news_details&news_id=74537#sthash.QuBA7dEu.dpuf

Nepal Living Standard Survey NLSS (2010/2011) Statistical report volume 2. Central Bureau of Statistics, Kathmandu

Nepal Rastra Bank (2008) Household budget survey Nepal (Mid November 2005-Mid November 2006)

NIDS (Nepal Institute for Development Studies) and NCCR (Nepal Centre for Contemporary Research) (2012) Migration yearbook 2011. NIDS and NCCR, Kathmandu, Nepal

Oxfam International (2008). Climate change adaptation in practice community-based adaptation and advocacy in coastal Pakistan [Case study on a project of Oxfam GB]. Retrieved from CAKE

Rijal Y (2013) Socio-economic and Health Impact of Labour Migration in Nepal. ISA Symposium for Society, Germany

SAARC (2014) Role of remittances in economic and financial development (SAARC/IGEG.07/27). SAARC Secretariat, Kathmandu

Sedhai R (2012) Youth's majority in absentee population alarming. The Kathmandu Post 11 Dec 2012

Sedhai R (2014) Government to lower migrant women age bar to 25. The Kathmandu Post 19 Dec 2014

Sherpa D (2010) Labour migration and remittances in Nepal. ICIMOD, Kathmandu, Nepal

Shishido H (2009) Large-scale migration and remittances: benefits and costs. World Bank, Washington, DC

Sudmeier-Rieux K, Jean-Christophe G, Sundar S, Dubois J, Jaboyedoff M (2012) Floods, landslides and adapting to climate change in Nepal: what role for climate change models? In: Kelman I, Lamadrid A (eds) In climate change modelling for local adaptation in the Hindu Kush-Himalayan region community, vol 11. Environment and disaster risk management. Emerald Group Publishing Limited, UK, pp 119–140

Sunam R (2014) Marginalized Dalits in international labour migration: reconfiguring economic and social relations in Nepal. J Ethnic Migr Stud. doi:10.1080/1369183X.2014.948393

Tacoli C (2009) Crisis or adaptation? Migration and climate change in a context of high mobility. In: Guzmán JM, Martine G, McGranahan G, Schensul D, Tacoli C (eds) Population dynamics and climate change. UNFPA/IIED, New York/London, pp 104–118

UN (United Nations) (2013a) International youth day: focus on rights and potential of young migrants. http://www.un.org/en/development/desa/news/population/marking-youth-day.html. Accessed 12 Aug 2013

UN (United Nations) (2013b) Youth migration and development. The Himalayan Times, 11 Aug 2013

Upreti BR, Zimmermann AB, Debele B, Cissé G, Contributions from the NCCR North-South Regional Coordinators (2012a) Partnerships in development-oriented research: lessons learnt and challenges ahead. NCCR North-South, South Asia Coordination Office, Kathmandu, Nepal

Upreti BR, Sony KC, Mallett R, Babajanian B, Pyakuryal K, Ghimire S, Ghimire A, Sharma SR (2012b) Livelihoods, basic services and social protection in Nepal. Working paper 7. ODI, London

Upreti BR (2009), Nepal from War to Peace: Legacies of the past and hopes for future. New Delhi: Adroit Publishers

Urry J (2007) Mobilities. Polity Press, Cambridge

Warner K, Afifi T (2014) Where the rain falls: evidence from 8 countries on how vulnerable households use migration to manage the risk of rainfall variability and food insecurity. Clim Dev 6(1):1–17

World Bank (2009), Nepal Migration Survey. Kathmandu: The World Bank.

World Bank (2014) Nepal: country snapshot. http://wwwwds.worldbank.org/external/default/WDSContentServer/WDSP/IB/2014/04/25/000442464_20140425132556/Rendered/PDF/868110REVISED00Spring0Meetings02014.pdf

World Bank (2011) Large-scale migration and remittance in Nepal: issues, challenges and opportunities. World Bank, Washington, DC

World Bank Group (2013) Migration and entrepreneurship in Nepal with a focus on youth: an initial analysis. Kathmandu, Nepal

Chapter 10
Overcoming Land Tenure Barriers in Shelter and Other Reconstruction Activities in Post-disaster Settings

Ina Rehema Jahn, Lorenzo Guadagno, Ethel Gandia, and Valentin Bonnefoy

1 Introduction

Disaster-induced movements take place along a continuum of more or less forced and more or less voluntary decisions (Hugo 2008). However, the loss of personal safety, assets, housing, infrastructures, and services that often result from the impact of a hazard on a vulnerable setting can greatly erode the options available in situ to affected communities, and therefore often results in displacement out of affected areas.

Displacement, in turn, in particular if inadequately managed, can deplete people's assets and reduce their access to capital, including local support and assistance networks, hence strongly constraining the well-being of those on the move. The inflow of displaced populations can also put pressure on host communities' access to essential assets, services, and opportunities, and ultimately have negative spillover effects on their wellbeing (Zetter and Deikun 2010). The latter is often the case for those displaced in urban areas, who necessarily rely on and interact with local service providers, markets, and institutions for their survival (Weiss Fagen 2014). As a result, displacement can consolidate and compound losses induced by natural hazards upon affected communities, including by eroding resilience in the long term. In fact, the longer the displacement period, the more likely the production of these negative direct and indirect, short and long-term impacts (Gütermann and Schneider 2011).

I.R. Jahn (✉) • L. Guadagno • E. Gandia • V. Bonnefoy
Department of Operations and Emergencies (DoE), International Organization for Migration (IOM), Geneva, Switzerland
e-mail: ijahn@iom.int

© Springer International Publishing Switzerland 2017

K. Sudmeier-Rieux et al. (eds.), *Identifying Emerging Issues in Disaster Risk Reduction, Migration, Climate Change and Sustainable Development*,
DOI 10.1007/978-3-319-33880-4_10

In the aftermath of events that disrupt the physical and built environment, addressing displacement as part of recovery efforts necessarily involves (re-)constructing and rehabilitating housing and revitalizing affected settlements. When shelter and other reconstruction actors carry out their relief and recovery efforts, they are likely to be confronted with a broad range of issues related to land rights and access to land. Consideration of these issues is critical for actions aimed to support temporary resettlement, return, and relocation of displaced persons, in order to allow for "do no harm"[1] principle-compliant reconstruction operations. Insufficient consideration of these issues can result in an increasing number of land conflicts, hampering the recovery process (Mitchell 2011; Sherwood et al. 2014). Land disputes can also lead to displaced persons being evicted from informally occupied areas (Haysom 2013; Jacobsen and Howe 2008), and ultimately in situations of protracted or repeated displacement that result in increased vulnerability (Esnard and Sapat 2014).

In addition, addressing land and property issues can help tackle a key factor of insecurity and disaster risk: land tenure regimes contribute to determining people's settlement and livelihood options, their (and the authorities') ability and willingness to invest to increase safety and security, and the kind of support and assistance they receive in the aftermath of disasters (Mitchell 2011). Access to secure land tenure is therefore a key element of risk reduction policies, but also of development efforts at large.

The main purpose of this chapter is to describe three technical tools that proved particularly effective in addressing land-related barriers for reconstruction and recovery and achieving increased overall tenure security in the aftermath of the 2010 Port-Au-Prince earthquake: (a) data-gathering and mapping of areas affected by the earthquake and needing reconstruction efforts; (b) creation of the Shelter Program Legal Team; and (c) adherence to the Due Diligence Guidelines from the Global Shelter Cluster. The use of these tools will be highlighted through an analysis of the International Organization for Migration's (IOM) "Overcoming Land Tenure Barriers" (OLTB) project in Port-Au-Prince (2010–2012).

These tools aim to operationalize the "Due Diligence" principle, support shelter and shelter-relevant actors in paying the needed attention to potential land issues emerging from their interventions. These issues, as will be outlined in this chapter, are of particular importance for the effectiveness and sustainability of post-disaster recovery interventions.

[1] The "do no harm" principle refers to the fact that all actors must strive to minimize the negative consequences they may be producing on direct beneficiaries and other stakeholders through the mere fact of being present and providing assistance. Humanitarian actors need to be aware of this and take steps to minimize the potential sources of harm (UNICEF's Humanitarian Principles 2003).

2 The Overcoming Land Tenure Barriers Project: OLTB Project

The devastating Mw 7.0 earthquake that struck Haiti on 12 January 2010 resulted in an estimated 200,000–250,000 deaths, US$8 billion in damages (Calais et al. 2010) and the destruction of more than 300,000 buildings. In the aftermath of the earthquake, 1.5 million people were displaced into camps, while an estimated 630,000 had to find shelter outside camps (including host communities in the city and other locations within and across national borders)—equivalent to about 20 % of the population of the whole country (IOM Haiti undated). The need for a massive reconstruction effort to house the homeless appeared evident from the immediate aftermath of the earthquake. Widespread disruption compounded pre-earthquake land-related grievances, including prevalent land tenure insecurity, weak land administration and management institutions, unequal distribution of land and property, lack of enforcement of land and property laws, as well as lack of urban planning (ECHO 2010; Sherwood et al. 2014).

It is important to highlight that, even before the 2010 earthquake, the large urban centers in Haiti (Port-au-Prince and Cap Haitian) were suffering from a chronic and serious housing shortage primarily caused by decades of unmanaged population movements from impoverished rural areas to peri-urban and urban centres (Levine et al. 2012). In particular since the 1970s, economic decisions favoring the development of low-wage productions for exporting cheap products to North American markets, and a widespread urban bias in investments related to basic services and infrastructure, resulted in massive inbound population flows, as people sought to access opportunities not available in their communities of origin. As a consequence, Port-au-Prince's population grew from 1.2 million in 1982 to about 3.2 million in 2003. An average of 75,000 migrants moved to the metropolitan region every year, despite increasing scarcity of employment opportunities. In a context characterized by a complete lack of land-use planning, this resulted in explosive, unregulated urban growth, with people building makeshift dwellings on available empty spaces. This, in turn, translated into widespread informality of land and property relations, despite the presence of a plethora of legal rules supposedly dedicated to govern land tenure in Haiti (Tobin 2013).

These conditions considerably complicated effective shelter responses, and overall reconstruction and recovery, in the aftermath of the disaster (CBS 2010; Kushner 2015). In an effort to effectively address the land-related challenges arising out of this context, IOM together with the United States Agency for International Development (USAID) developed the Overcoming Land Tenure Barriers (OLTB) project in the immediate aftermath of the earthquake.

The initiative was implemented in the neighborhoods of *Delmas 32* and *Carrefour Feuilles* of Port-au-Prince. Both areas have seen rapid expansion and development over the last decades and today host a young yet diverse population. The built environment, while largely developed informally, includes only a limited number of substandard dwellings. However, structural and social vulnerability is

Table 10.1 Overview of building status, type, and use in the project area

	Carrefour		Delmas		Total Project area	
	No.	%	No.	%	No.	%
Total plots surveyed	4,550		6,145		10,695	
Total households	5,887		8,238		14,125	
Total individuals	21,798		31,801		53,599	
Building status						
No damage	935	21 %	2,107	34 %	3,042	28 %
Rebuilt/repaired	1073	24 %	1,707	28 %	2,780	26 %
Damaged but in use	964	21 %	1,148	19 %	2,112	20 %
Destroyed or unrepairable	684	15 %	1,138	19 %	1,822	17 %
Repair/construction in progress	103	2 %	40	1 %	143	1 %
Not applicable	791	17 %	5	0 %	796	7 %
Type of building						
Low house	2181	48 %	3,070	50 %	5,251	49 %
Multistory and apartment buildings	691	15 %	1,876	31 %	2,567	24 %
Temporary shelter	820	18 %	466	8 %	1,286	12 %
Substandard	327	7 %	54	1 %	381	4 %
Other	531	12 %	679	11 %	1,210	11 %
Use of building						
Residence only	3304	73 %	4,246	69 %	7,550	71 %
Residence and other	180	4 %	417	7 %	597	6 %
Business	63	1 %	125	2 %	188	2 %
Empty house	399	9 %	743	12 %	1,142	11 %
No data	604	13 %	614	10 %	1,218	11 %

Source: IOM OLTB Survey, 2012

still prevalent, and the 2010 earthquake damaged two thirds of all local buildings in these neighborhoods (Table 10.1).

Project partners included the *Bureau de Monétisation des Programmes d'Aide au Développement* (BMPAD), the *Comité Interministériel d'Aménagement du Territoire* (CIAT), the *Direction Générale des Impôts* (DGI), the Mayors of both neighborhoods, as well as local communities. The latter were involved through the *Conseil du Projet de Développement Participatif* (COPRODEP) in *Delmas 32* and through a professional, nonprofit international solidarity and cooperation association called *GRET* in *Carrefour Feuilles.*

This initiative aimed to address some of the challenges faced by post-earthquake housing, infrastructure and other reconstruction actors as a result of the lack of clarity of the land tenure system in Haiti, with a specific focus on the tenure situation in areas considered for shelter projects. To reach this objective, the OLTB project included three main components:

1. Data gathering and validation
2. Establishment of a land tenure database and data sharing mechanisms
3. Land dispute mediation.

2.1 *Data Gathering, Validation, and Sharing*

2.1.1 Data Collection and GIS/Mapping Exercise

Due to the widespread lack of formal registration of land titles and the loss of much of the existing land documentation in the earthquake, shelter and reconstruction actors were presented with the risk that securing land for shelter and supporting the sustainable return of the displaced would be complicated by innumerable land disputes.

Hence, the first step of the project was to gather data on the local land and property situation in the neighborhoods of *Delmas 32* and *Carrefour Feuilles* in Port-au-Prince. This allowed assessing the earthquake's impacts on buildings and infrastructures and the availability of locations for the temporary resettlement of the displaced, and provided all stakeholders with a better understanding of the land tenure system at the neighborhood level. The results of the exercise are highlighted in Table 10.1. This exercise cleary showed existing high levels of informality—largely dependent on decades of informal land occupation processes as a result of urban migration, and only made worse by the disruption brought about by the earthquake, including through the loss of land registries—and the potential for future land disputes.

The information was collected and consolidated using a GIS platform. All buildings, empty spaces, streets, infrastructures, and other physical characteristics of the two target neighborhoods were identified and plotted. Following the consolidation of GIS data, teams of community representatives and IOM staff were deployed directly in the neighborhoods to gather household-level information. Data on status of the buildings, household demographics and land tenure were collected and geographically referenced. In order to obtain accurate and first-hand data on land and property tenure, OLTB experts adapted their methodology to the Haitian context, so the collected data was based on the knowledge of community members, and not necessarily on official records.

Comprehensive maps were then produced for each of the two neighbourhoods, which allowed to qualify individual plots as being not damaged, rebuilt, under reparation, damaged or destroyed, and to identify the degree of land conflict potential for each of them (Fig. 10.1). These details enabled shelter actors to obtain a clearer idea of priorities and options for addressing displacement in the area, in compliance with the "do no harm" principle.

2.1.2 Establishing the Tenure Situation

In order to obtain an accurate picture of the land tenure situation of a planned shelter construction site, it is necessary to fully clarify the current occupancy status of the land, who holds what type of rights over the land and the (potential) competing claims over the land.

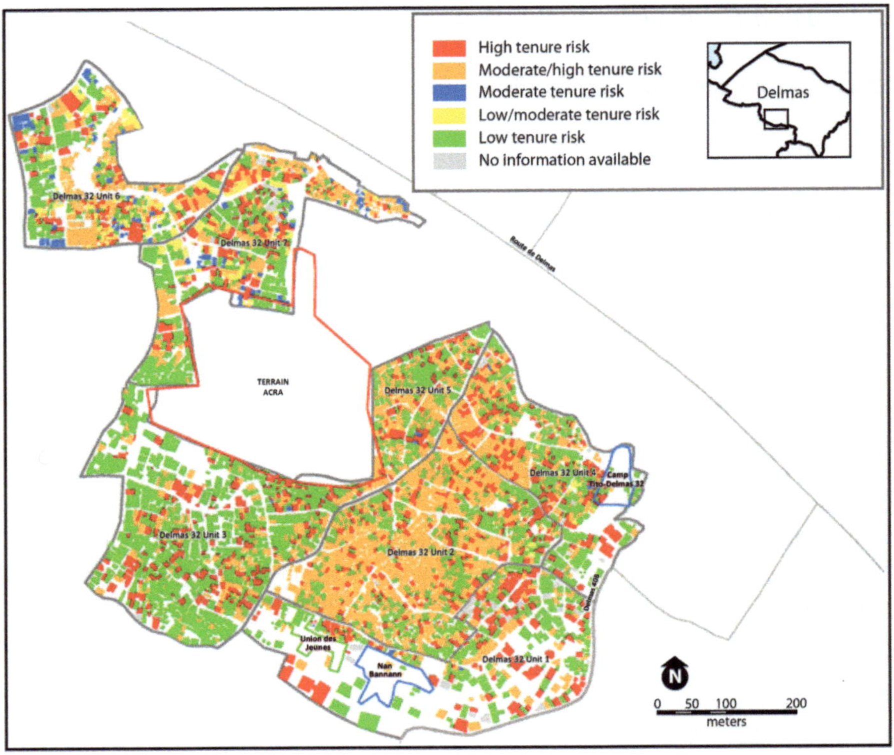

Fig. 10.1 Map of land tenure situation in Delmas 32. *Source*: IOM OLTB Survey, 2012

In the cases where land targeted for shelter (re-)construction was public, the relevant local authorities were approached to provide supporting land tenure documents prior to construction. If authorities were unable to provide the required documentation, the legal team verified the status of the land further with the Haitian General Tax Office. In cases where the land targeted for shelter (re-)construction was privately owned, the legal team still coordinated and obtained approval or endorsement from the local government prior to construction. In such cases, approval was also sought from the presumed (self-declared) landowners. Once endorsement from the local government and landowners was secured, a thorough land tenure investigation was launched.

In addition to research regarding the formal evidence of the property rights, the legal team carried out an investigation in the neighborhood of the construction site in order to assess the possible existence of other owners or potential claimants. The legal team also verified with the General Tax Office that the land had not been the subject of multiple sales; that it was not the object of a land conflict; and that it was not under sequestration (IOM 2012).

2.1.3 Community Validation

As a next step, a community validation exercise was conducted to verify findings from the tenure survey. In contexts such as Port-au-Prince, where formal land frameworks coexist with informal systems, the perception of the local population about land ownership as well as the grassroots knowledge of who owns what plot of land in a given neighborhood may be more accurate than more formal sources. Decisions by reconstruction actors involving land tenure aspects in these areas therefore require close consultation with community representatives. The participatory process itself has to be agreed upon by the local population, in order to ensure a better level of implementation due to an increased feeling of ownership and acceptance of the initiative by the population (ICG 2010).

OLTB staff therefore undertook community validation of all the data previously gathered on the two sites. A meeting was organized to present the household-level information to community representatives, who were asked to confirm reported position and occupancy status of identified buildings and to validate existing claims. Ultimately, the process of data collection and validation helped determine the relationship between residents and occupied land and property tenure for about 10,700 plots and over 14,000 households. In addition to serving as a mean to adapt the evidentiary standards to the Haitian context, the community validation process also discouraged people from providing false information.

The process enabled reconstruction actors to obtain a precise understanding of who owned what parcel of land and the conditions and legal framework applicable in the different cases. Figure 10.2 highlights how land and property conditions varied widely within the target areas. In any given neighborhood, it was common to find several kinds of ownership or tenure situations, together with different levels of damage and extremely diverse family and individual backgrounds. As a consequence, it was necessary to adopt a parcel and household-specific approach to design and implementation of reconstruction and recovery programs in order to facilitate the sustainable return of displaced persons and the reconstruction of their neighborhoods, as will be highlighted in the following sections.

This component also provided a comprehensive catalogue of the related informal, community-based agreements that govern land use and building occupancy, advancing national institutions and international actors' understanding of land tenure dynamics in Haiti before the earthquake. For example, one of the main findings in *Delmas 32* was that the vast majority of residents had lived in the neighborhood since 1986. This illustrated that the targeted neighborhoods benefited from a high level of *de facto* land tenure security. Land and property transactions were occurring, seemingly without great difficulties, often based on non-authentic documentation perceived as proof of property right. However, many of the *Delmas 32* inhabitants appeared to unknowingly own their plots of land through a concept of "*possession utile*" (adverse possession) under Haitian law (as a consequence of 20 years of peaceful and uninterrupted occupation) and could hence oppose that right to anyone wanting them to leave their homes (IOM 2012).

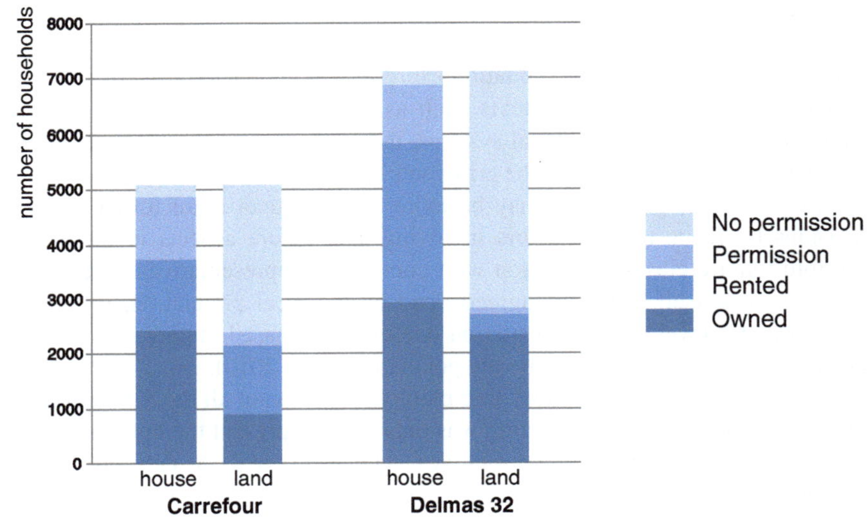

Fig. 10.2 House and land tenure situation in the project area. *Source*: IOM OLTB Survey, 2012

2.1.4 Public Information and Sharing

Throughout the whole OLTB project and during each of the phases presented above, public information and awareness raising initiatives about the project were held with the affected communities and wider public. Clarifying what land policies and legal framework were applicable to the various land issues and what the objectives of the project looked like resulted in the target areas being largely spared from land grabbing attempts, speculative informal land sales and other types of conflicts over land, which became increasingly common in the aftermath of the earthquake (Levine et al. 2012; Protection Cluster 2010).

Land issues can also be exacerbated by the fragmentation of responsibilities among different actors including different Ministries, local governments and authorities, international organizations, civil society actors, community-based organizations, international financial institutions and donors (Mitchell 2011). Transparent coordination between all reconstruction and humanitarian actors is therefore absolutely essential to avoid the duplication of efforts. In order to ensure coordination, the data collected and validated under the OLTB project was shared with all relevant national and international actors. A Web-based and publicly accessible platform was designed to host the land tenure database in order to discourage possible malpractices in the management of collected data. Ownership of the data was then transferred to the Haitian Government, which was trained on how to properly use the database along with other relevant actors.

2.1.5 Preventing and Mediating Land Conflicts Throughout the Reconstruction Phase

As part of its recovery effort after the 2010 earthquake, IOM developed the "Shelter program," which, as of March 2012, had successfully constructed 11,248 housing units across Port-au-Prince (IOM 2012). The program was designed to support durable solutions for the displaced through activities promoting return, relocation, and local integration. Shelter efforts are still ongoing more than six years after the disaster: as of 2014, an estimated 65,000 individuals still remained displaced, and even those who had formally left displacement sites still encountered significant obstacles in securing their housing, land and property rights and accessing essential services (Sherwood et al 2014).

Land tenure disputes always depend on the context, and tailoring reconstruction interventions to local specificities is critical not only to recovery but also to tackling the root causes of disaster vulnerability. In a context as complex as post-earthquake Haiti, adequate consideration of land issues has proved essential to the effectiveness and sustainability of post-disaster shelter and housing solutions. Data gathered throughout the project showed how the land tenure situation differed widely from neighborhood to neighborhood, and even from household to household. Responses therefore needed to be adapted to the specificities of the micro-level context.

This required particular flexibility from the legal perspective. Understanding the land tenure situation specific to target neighborhoods and devising adequately informed reconstruction strategies was arguably fundamental to successfully preventing or resolving land disputes and allowing for a swift reconstruction process.

To this aim, the IOM Shelter Program established a Shelter Legal Team, which focused on investigating the land tenure situations in construction areas and provided technical and legal advice on how to efficiently handle land issues that were likely to hinder shelter and housing (re-)construction.

2.1.6 Prevention of Land Conflicts

The Legal Team's activities were based on the identification of potential land conflicts that may arise before, during and after the construction of emergency and/or more transitional shelters, based on the results of the household-level survey (Fig. 10.3). In the project areas' dense, prevalently informal urban context, the team resorted to a number of measures to try to prevent displacement, return and resettlement of displaced persons from resulting into widespread conflict over land, which could have delayed or undermined the shelter reconstruction program.

The most common land-related barriers to the (re-)construction of shelter included: (1) the lack of information on who had what rights to what land and/or property; (2) the (perceived) prevalence of conflict and competing claims over land and property; (3) the refusal by landowners to allow that land be used for the

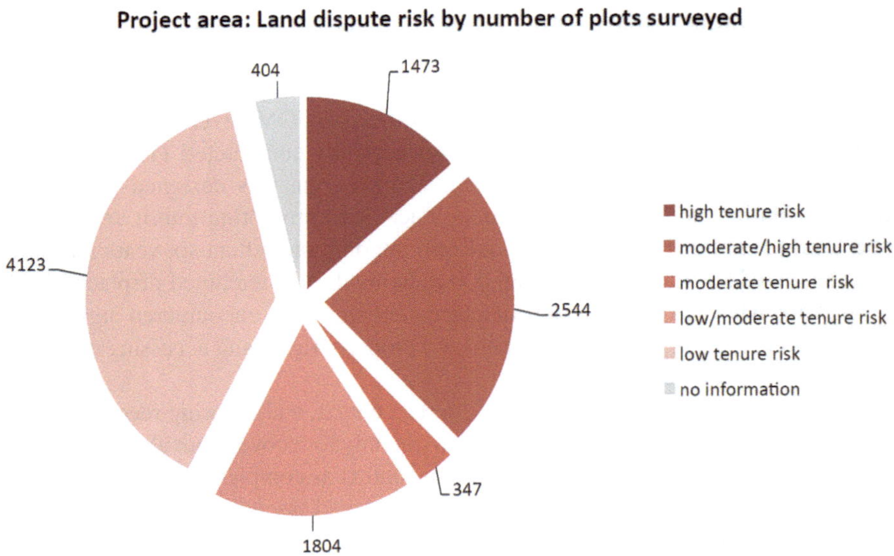

Fig. 10.3 Land dispute risk in the project area, by number of plots surveyed. *Source*: IOM OLTB Survey 2012

construction of shelter; (4) the lack of available public land that could be allocated for the purpose of shelter construction; and (5) the lack of legal security.

In addition to the data collection and validation activities described above, these problems were addressed by employing the following principles:

- Safeguarding land and property registers (whenever they existed and to the extent possible)
- Recording property claims of, and providing legal assistance and documentation to, IDPs and returnees
- Identifying situations where property dispute resolution mechanisms were needed
- Keeping records of shelter construction milestones.

In cases in which the land tenure situation was unclear and in order to prevent potential disputes and conflicts over land, the Legal Team designed a number of standard forms to organize and keep records of, among others, requests for shelter construction, authorization to begin construction and agreement letters between land owners and shelter beneficiaries.

Clarifying the land and property context was instrumental to identifying land free of claims or disputes that could be used to host displaced populations without fear of future evictions—a situation which, elsewhere in Haiti, forced population already displaced by the earthquake to move in marginal areas prone to landslides and flooding (Sherwood et al 2014). Identifying sites close to people's land and livelihoods, safe from hazards and well-served by essential services was also key to prevent forced movements from further eroding the affected people's resilience and wellbeing.

2.1.7 Mediating Land Disputes

As mechanisms put in place to prevent land conflict cannot entirely eliminate the risk of disputes, the Shelter Legal Team also developed a proactive methodology to resolve conflicts arising as a result of shelter construction.

The most frequent causes of conflict handled by the team were:

1. End of farm lease: Tenants had legitimately built houses on formally leased land, and had then failed paying the landowners for legal occupation of the property. Once their houses were destroyed by the earthquake, land owners prohibited (re-)construction until past debts were paid.
2. Land boundary problems: Neighboring landowners initially agreed to the construction work. However, during construction, and in particular in denser urban areas, disputes about the land boundaries arose in due course, delaying or preventing the construction process.
3. Claims to right of ownership: The person who approved shelter construction was not the legal owner and the real owner came to claim his right once the shelter intervention had already been agreed on. This created particularly serious issues in cases where construction had started based on the available documentation and community-based information.

A mechanism to refer land dispute cases to the Haitian Judicial authorities was established but never activated, as no such dispute arose during the implementation of the project. The approach of the legal team was to encourage as many amicable settlements as possible and, to the extent possible, to avoid court cases. Amicable settlements were identified as the best way to resolve disputes, as they enabled the parties to maintain an open dialogue and avoid paperwork, evidentiary requirements, expenses, time consumption and difficulties of implementation of the sentences related to land disputes before the Court.

For example, in an “End of Farm Lease” conflict case, the IOM legal team had to deal with violations of pre-earthquake agreements between the land owner and the tenants. The land owner refused to allow shelter construction for the reason that in the neighborhood many tenants’ lease contracts had been expired since over 10 years. In response, the legal team updated the terms of payment between tenants and land owners in order to respect both parties’ rights. This method allowed solving about 50 similar disputes.

Legal negotiation and mediation proved crucial in preventing an escalation of micro-level conflicts. The IOM legal team applied flexible evidentiary standards, providing an alternative to the strict evidentiary requirements of legal land claims. Shelter and other reconstruction actors could not avoid accepting some degree of “legal uncertainty” in the post-disaster context of Haiti, and such a procedure provided a way to effectively use a sufficient amount of information, including that collected and validated through community-based efforts. In total, the legal team addressed 138 conflicts, some of which involved more than 300 shelter units at the same time. In *Carrefour-Feuilles 59*, the disputes were solved through mediations (in 40 cases), negotiation (2) and agreement (17). In two cases, the conflict could not be solved and the construction process was interrupted (IOM 2012).

2.1.8 Lessons Learned

The work under the OLTB project showed that investing resources in preventive measures, working closely with local and national stakeholders, the community and with all actual and potential right holders, and establishing effective dispute resolution systems proved effective in avoiding obstacles to reconstruction and recovery related to land tenure. These measures might also reduce future conflicts, and help pursue solutions to displacement that are durable and that comply with the "do no harm" principle, guiding humanitarian work.

Shelter and other reconstruction actors are usually the ones in charge of addressing land issues in situations of post-disaster population displacement. Nonetheless, in most cases, shelter construction projects do not include a strong land and property component, and often lack staff with the needed expertise. By adopting a set of systemic responses such as outlined above, shelter and other reconstruction actors can prevent, mitigate, or resolve all kinds of land issues that can emerge before, during or after a shelter project and are likely to obstruct reconstruction efforts. These efforts should be complemented with training of all project staff and partners in the Land and Shelter Due Diligence Standard further outlined below.

3 The Due Diligence Standard

Land represents a crosscutting issue relevant to many areas, including livelihoods, shelter, and agriculture, and should therefore be a key concern in disaster recovery efforts. Shelter and housing actors, in particular, should take an active interest in the rights existing on the land they intend to build on, in order to ensure:

- Protection of beneficiaries in the medium and long term
- Reduction of the risk of tensions and conflicts around land
- Prevention of tensions or land conflicts in the communities where shelters are constructed
- Respect of the "do no harm" principle
- Accountability vis-à-vis the donor community
- Higher level of legal certainty
- Ability to implement shelter and housing projects timely and efficiently.

To this end, IOM together with the Global Shelter Cluster developed the Land and Shelter Due Diligence Standard (Shelter Cluster 2013), a document which aims to operationalize the Due Diligence approach to land in shelter interventions and which is now part of the Global Shelter Cluster tools.

The Due Diligence principle refers to the application of the "do no harm" approach, and hence provides guidance on how to prevent causing or increasing tensions and conflicts around land by ensuring the greatest possible level of legal certainty about land rights of beneficiaries throughout shelter and reconstruction programs. The "do no harm" approach further requires shelter actors to consider the impact of their proposed activities on any given community—sometimes inaction might be the best option available if all other alternatives lead to increased overall vulnerability.

In general, taking land into consideration at every stage of the response helps to avoid future evictions of the beneficiaries, provide safe shelters in secure settlements and contribute to durable recovery. The Guidelines queried shelter actors with the necessary guidance to establish a systematic process that anticipates, mitigates and responds to land rights issues (potentially) emerging before, during or after the construction of shelter or housing. For example, such a process requires the close involvement of government entities, community leaders, civil society organizations, and humanitarian and recovery actors based on the specific land governance system(s) prevalent in a given country or region (both statutory and customary).

Throughout the reconstruction process, it is crucial that shelter actors: (1) understand the context of how land and property relations are organized; (2) ensure sufficient certainty to the extent possible (assess the legal certainty of rights to a given land plot); (3) ensure that shelter actors are prepared to react to land issues at all stages of the shelter construction project (how to react and who to refer to); and (4) reduce the risk for beneficiaries to suffer from land rights-related issues (what can reasonably be undertaken to address land issues).

Adaptation to the local context is of key importance. Local resources, institutions, expertise, and knowledge exists, even in severely damaged human settlements, and must be taken into account to properly understand disaster impacts, existing needs and opportunities, and to guide response and recovery activities. The households and community validation processes illustrated above provide a good example of the operationalization of this principle. Participatory approaches should be adopted throughout the subsequent site selection and shelter reconstruction activities. By increasing local ownership over the whole process, shelter actors are likely to ensure a better acceptance of their activities and minimize risks of conflicts arising during the construction or after shelter actors have left.

Another concern to be taken into consideration in a shelter construction context is the hosting phenomenon, understood as the shelter provided (most of the time informally) to disaster survivors by family and friends. Frequently, this hosting will occur before international actors arrive, and it tends to continue until after the latter have left. The medium/long term nature of this phenomenon may bring a high conflict-potential with it, that is, between displaced, host communities and the Government, and even within a same family if appropriate mitigation and/or resolution measures are not developed and implemented immediately (Brookings 2013).

In addition, specific attention should be given to the situation of those who were renting before the disaster (both formally and informally). Concentrating exclusively on the situation of homeowners during reconstruction efforts might lead to overlooking tenants whose conditions, often already more precarious in the first place, can be worsened in the inflated housing market resulting from scarcity of shelter options in the post-disaster phase (Hooks and Miller 2006; Masozera et al. 2006).

It is of crucial importance that these principles are taken into account by all relevant actors, and not only those dealing with shelter and housing reconstruction, through shared approaches and coordination of efforts. During the reconstruction phase, working with the existing land legal framework is identified by the UN as one of the major land administration requirements that need to be complied with. That is the main objective of the Land and Shelter Due Diligence Standard.

Shelter and other reconstruction actors should contribute to restoring capacity of the land administration system, in particular when it is inefficient or suffers from lack of capacity in the aftermath of a disaster. This will allow for an equitable and legal process to be followed in acquiring land for the construction of public buildings and infrastructure, and help ensure that a land plot is indeed cleared for construction purposes. Being attentive to land issues in shelter activities is hence indispensable for a sustainable settlement solution and resumption of livelihood activities. Support given to the local land administration system is also crucial in order to provide secure land tenure and coordinate the registration of all land-related activities in the long term. This is especially important for addressing the situation of people displaced for long periods of time and for those returning to their land wanting to build permanent housing.

4 Conclusions

Land disputes do exist almost everywhere in the world, in particular in contexts characterized by weak legal systems and high levels of informality. Land issues are likely to be compounded in disasters, and in particular as a consequence of displacement situations. Preventing and addressing conflicts - whether preexisting to or resulting from a given disaster - is often key to reducing long-lasting impacts of hazards on displaced persons, host communities and other affected populations through effective and sustainable reconstruction and recovery. Indeed, land rights and land use planning help manage and address population displacement and contributes to increased resilience to future disasters (USAID 2013).

In the post-earthquake Haitian context, most disputes were due to: lack of information on who has what right to what land, property, and housing; competing claims over land and property; the refusal by land owners to allow land use for construction of housing and shelter; the lack of available public land to allocate for construction of housing and shelter; and the lack of legal security for reconstruction actors. The OLTB project showed how paying due attention to land in shelter interventions was an essential step to adress and most often overcome these disputes and enable recovery.

The main recommendations for addressing disaster-induced displacement, in particular if characterized by high prevalence of land disputes, are as follows:

1. The creation of a legal team, largely comprising local staff cognizant of the specific context, proved extremely useful in elucidating the actual tenure situation on the ground before proceeding with land allocation for shelter construction.
2. The community validation mechanisms employed to directly consult those affected by land access and ownership disputes in verifying official tenure records supplied by local authorities proved highly effective in preventing land grievances.
3. Support to the (re-)establishment of land rights for all members of the affected population, including those who never held any formal land title, is crucial for inclusive recovery and for building back better.
4. In addition to preempting potential conflicts in the immediate aftermath of disaster, the benefits of investing in extensive data gathering and validation exercises often go beyond the immediate shelter reconstruction, as they can be strong evidence in supporting land tenure system reform or land laws enforcement, thereby tackling one of the main obstacles to addressing displacement situations and one of the main underlying causes of vulnerability to disasters.
5. Coordination and information-sharing with communities, local organizations and authorities, and the whole range of humanitarian, reconstruction and recovery actors is essential to avoid duplication of efforts and ensure adequate follow-up to one-off projects activities.

Paying due attention to land and property aspects when addressing disasters and displacement and supporting the recovery of affected populations is essential to support communities to better anticipate, withstand, and cope with future hazards. The entire set of activities undertaken by shelter and reconstruction actors in a post-disaster context, therefore, cannot be divorced from their larger political context and must go beyond the immediate time frame of emergency disaster response. Failing to do so can result in protracted situations of displacement, which in turn can compound disaster impacts and result in long-lasting vulnerabilities.

References

Brookings (2013) Under the radar: internally displaced persons in non-camp settings. Brookings-LES Project on INTERNAL DISPLACEMENT, Washington

Calais E, Freed A, Mattioli G, Amelung F, Jonsson S, Jansma P, Hong SH, Dixon T, Prepetit C, Momplaisir R (2010) Transpressional rupture of an unmapped fault during the 2010 Haiti earthquake. Nat Geosci 3(11):794–799

CBS (2010) Fights over Land Stall Haiti quake recovery. CBSNEWS. http://www.cbsnews.com/news/fights-over-land-stall-haiti-quake-recovery/. Accessed 11 July 2010

ECHO (2010) Haiti before the earthquake. http://ec.europa.eu/echo/files/aid/countries/Haiti_paper_01102010.pdf

Esnard AM, Sapat A (2014) Displaced by disaster: recovery and resilience in a globalizing world. Routledge, New York

Gütermann N, Schneider E (2011) The earthquake in Haiti. In: Gemenne F, Brucker P, Glasser J (eds) The state of environmental migration 2010. Institut du développement durable et des relations internationale and International Organization for Migration, Paris and Geneva

Haysom S (2013) Sanctuary in the city? Urban displacement and vulnerability—final report. Overseas Development Institute, London

Hooks JP, Miller TB (2006) The continuing storm: how disaster recovery excludes those most in need. Cal W L Rev 43(1):21–54

Hugo G (2008) Migration, development and environment. International organization for migration research series. IOM, Geneva

ICG (2010) Haïti: stabilisation et reconstruction après le tremblement de terre. International Crisis Group, Rapport Amérique/Caraïbes n.32. www.crisisgroup.org/fr/regions/amerique-latine-caraibe/haiti/032%20Haiti%20Stabilisation%20and%20Reconstruction%20after%20the%20Quake.aspx

IOM (2012) Dealing with land barriers to shelter construction in Haiti: the experience of the IOM Haiti legal team. International Organization for Migration, Geneva

IOM Haiti. Displacement Tracking Matrix (DTM) portal. http://iomhaitidataportal.info/dtm/

Jacobsen K, Howe K (2008) Internal displacement to urban areas: the tufts-IDMC profiling study; case 3: Santa Marta, Colombia

Kushner J (2015) Who owns what in Haiti?. Pulitzer Centre for Crisis Reporting. http://pulitzercenter.org/reporting/who-owns-what-haiti-land-rights-tourism-mining-ile-a-vache-carnival

Levine S, Bailey S, Boyer B, Mehu C (2012) Land, institutions and humanitarian action in post-earthquake Haiti. Humanitarian Policy Group Working Paper

Masozera M, Bailey M, Kerchner C (2006) Distribution of impacts of natural disasters across income groups: a case study of New Orleans. Ecol Econ 63:299–306

Mitchell D (2011) Assessing and responding to land tenure issues in disaster risk management: training manual. Food and Agriculture Organization, Rome

Protection Cluster (2010) Policy options and recommended approach to address forced evictions in post-earthquake Haiti. Protection Cluster Housing, Land and Property Working Group

Shelter Cluster (2013) Land rights and shelter: the due diligence standard. https://www.sheltercluster.org/References/Pages/Housing-Land-and-Property.aspx

Sherwood A, Bradley M, Rossi L, Gitau R, Mellicker B (2014) Supporting durable solutions to urban, post-disaster displacement: challenges and opportunities in Haiti. Brookings Institution and International Organization for Migration, Washington and Geneva

Tobin KA (2013) Population density and housing in Port-au-Prince: historical construction of vulnerability. J Urban Hist 39:1045

UNICEF (2003) UNICEF humanitarian principles. UNICEF, New York

USAID (2013) Land tenure and property rights framework. USAID, Washington, DC

Weiss Fagen P (2014) Flight to the cities: urban options and adaptations. In: Martin SF, Weerasinghe S, Taylor A (eds) Humanitarian crises and migration: causes consequences and responses. Routledge, New York

Zetter R, Deikun G (2010) Meeting humanitarian challenges in urban areas. Forced Migr Rev 34:5

Chapter 11
Impacts of Outmigration on Land Management in a Nepali Mountain Area

Gudrun Schwilch, Anu Adhikari, Michel Jaboyedoff, Stéphanie Jaquet, Raoul Kaenzig, Hanspeter Liniger, Ivanna M. Penna, Karen Sudmeier-Rieux, and Bishnu Raj Upreti

1 Introduction

1.1 Study Objectives

Worldwide land users face the challenge of producing from land while minimizing natural resource degradation (Foley et al. 2011). Land management in mountains poses particular problems, constrained by steep slopes, inaccessibility, remoteness, and often harsh climatic and weather conditions. Working on sloping land hampers mechanization and therefore requires more manual labor; it also requires special care to manage run-off caused by heavy rains, and related soil erosion and mass movements. Mountain areas also depend on roads to access markets and centers, which are difficult to construct. There are, however, advantages, such as access to sufficient water, higher (although sometimes excessive) rainfall than in lowlands, and often very adapted, traditional, and rich sociocultural systems that have evolved over thousands of years. These systems are neither fixed nor isolated from

G. Schwilch (✉) • S. Jaquet • H. Liniger
Centre for Development and Environment CDE, University of Bern, Bern, Switzerland
e-mail: gudrun.schwilch@cde.unibe.ch

A. Adhikari
International Union for Conservation of Nature IUCN, Kathmandu, Nepal

M. Jaboyedoff • I.M. Penna • K. Sudmeier-Rieux
Group of Risk Analysis – Institute of Earth Sciences (ISTE), University of Lausanne, Lausanne, Switzerland

R. Kaenzig
Institute of Geography, University of Neuchâtel, Neuchâtel, Switzerland

B.R. Upreti
Nepal Center for Contemporary Research NCCR, Kathmandu, Nepal

© Springer International Publishing Switzerland 2017

K. Sudmeier-Rieux et al. (eds.), *Identifying Emerging Issues in Disaster Risk Reduction, Migration, Climate Change and Sustainable Development*,
DOI 10.1007/978-3-319-33880-4_11

globalization, and outmigration has always been an adaptation strategy of mountain societies.

The mountain ecosystem can be particularly sensitive to the dynamics of human activities, but also to changing climatic conditions. Although climate change and environmental degradation as drivers of migration have been widely studied (Piguet 2013; Warner et al. 2010), there is a gap in the literature on how migration impacts land management, and how climate change and increasing disaster occurrence affect sustainable land management practices. While migration research is often focused on those who migrate or where they migrate to (Black et al. 2011; Findlay 2011; Piguet 2013), land management studies, on the other hand, tends to consider land user families as a complete and stable unit (de Graaff et al. 2008; Gisladottir and Stocking 2005; Hurni 2000). Additionally, earlier land management studies focused more on whether population growth increased or decreased land degradation (Carswell et al. 1998; Thapa and Weber 1995; Tiffen et al. 1994; Warren 2002), but not on the reverse phenomenon—population decline—which has been observed more recently. It is therefore crucial to examine in detail how land management is sustained under changing population and climatic conditions, as failing to care for land could promote natural hazards such as floods or landslides. Nonetheless, outmigration may also have positive impacts on the land. Remittances, for example, can improve living conditions or help finance new land management measures. This chapter focuses on the impacts—both positive and potentially negative—of outmigration on land management.

The aim of the study was to identify how land management in mountains is affected by migration. The main study site was in Nepal, in the Panchase area of the Western Development Region; a sister study site was also selected in Bolivia, in the Quillacollo District of Cochabamba. Through the assessment of land use, land degradation, sustainable land management practices, migration patterns, use of remittances, and concerns of the population left behind, the study contributes to an improved understanding of the population and land dynamics in those areas. This chapter presents preliminary findings from the fieldwork, with a main focus on results from the Nepali case study, as not all field assessments were also conducted at the Bolivian site. Nevertheless, the results from Bolivia are briefly presented in a box, as they still give important insights into differences and similarities across the two sites on two continents.

1.2 Conceptual Framework

The study is based on two complementary conceptual frameworks: the sustainable livelihoods framework and the sustainable land management (SLM) framework. The sustainable livelihoods framework (DFID 1999; Scoones 1999) defines livelihoods as access to five basic capitals or resources (social, human, economic, natural, and physical) and is considered as the standard approach to understanding livelihoods and vulnerability (Upreti and Muller-Boker 2010). The SLM framework (Schwilch et al. 2011; UNU-INWEH 2011) is a holistic and integrated concept which provides an overview of the cause–effect interactions of land degradation and SLM on environment and human well-being. SLM is considered the "response" to the "drivers," "pressures," and "states" of degradation, and it enhances the provision of ecosystem services and thus improves human well-being and reduces poverty. The World Overview of Conservation Approaches and Technologies (WOCAT) network and program defines SLM "as the use of land and water resources, including soils, water, animals and plants, for the production of goods to meet changing human needs, while simultaneously ensuring the long-term productive potential of these resources and the maintenance of their environmental functions" (WOCAT 2014). SLM has the capacity to directly address all three UN environmental conventions. It helps to prevent desertification (UNCCD), increase biodiversity (UNCBD), and make people less vulnerable to the effects of climate change (UNFCCC). Although many of these positive effects of SLM are known, there is insufficient evaluation of how SLM is impacted by migration in a context of disasters and climate change.

The two frameworks are therefore combined to explore the impact of migration on land management and livelihoods. Access to resources, environmental degradation, mountain hazards/climate change, and changing demographics are identified drivers of change. They can be in a causal relationship and create—alone or combined—more vulnerability and strains on livelihood strategies. These strategies then lead to a change in land management or increased outmigration. Adapted livelihood and land management strategies can in turn have an impact on resource access or environmental degradation and hazard events, especially in fragile environments such as mountains.

1.3 Natural Hazards and Climate Change Impacts

Natural hazards and climate change are adding a new dimension to land management in mountainous areas. Climate change models are uncertain at best, but data on monsoon rains point towards more intense rainfall patterns, longer drought periods, and new higher ranges for crops (Sudmeier-Rieux et al. 2012). Data analyzed by Petley (2010) show the South Asian Seasonal Monsoon Index (SASMI) correlated with a higher incidence of landslides since 1995. Drought spells are now more common and crops can grow at increasingly higher elevations. According to inhabitants from different regions in Bolivia, the incidence of hail storms and drought has increased (OXFAM 2009). Such natural hazards, also including wind storms, frost, landslides, and floods, can further aggravate the vulnerability of mountain communities and mountain farming (Kollmair and Banerjee 2011). The uncertain climate conditions, noticeable over the past 10 years, have created more uncertainty for managing land, especially as mountain farming communities are now heavily impacted by labor loss due to migration.

1.4 Migration Impacts

In Nepal, migration is a common strategy to improve livelihoods, especially among mountain people. Migrants are mostly men and they migrate to other Asian countries, most commonly India, followed by the Gulf countries (NIDS and NCCR 2011). In recent years the number of migrants has increased, for reasons related to rural poverty, environmental degradation, enhanced communication means, higher education, changing values, and rising economies attracting thousands of migrant workers, such as Qatar and other gulf countries in the case of Nepal (Andersen 2002; Seddon et al. 2002).

Migration-induced demographic changes and economic factors such as remittances influence the natural resource management of those left behind (Gartaula et al. 2012; Maharjan et al. 2012). Land abandonment, for example, is a growing consequence of outmigration (Khanal and Watanabe 2006; Paudel et al. 2012), but it is not yet clear what other impacts that migration may have on land management. It is this aspect upon which we aim to shed light. Migration leaves behind fragmented families (often a wife and children) and the elderly to deal with managing the land in addition to daily life challenges (Black et al. 2011; Grau and Aide 2007). Having less labor may be accompanied by a reduction in land management expertise on cropland, grazing, forest, and mixed land (e.g., maintenance of terraces and irrigation canals, or stabilization of slopes), and is further challenged by changing climate conditions and growing environmental threats (Gray 2009; Valdivia et al. 2010). On the other hand, remittances might help households invest in more sustainable agricultural practices in addition to supporting livelihoods through the purchase of food, goods, and better schooling (Davis and Lopez-Carr 2010; Nijenhuis 2010).

2 Methods

This study integrates biophysical and socioeconomic data through a case study and a mapping approach. From the social sciences, we drew upon well-tested qualitative methodologies typically used in vulnerability and capacity analyses, such as semi-structured interviews with key informants, focus group discussions, and transect walks. The biophysical analysis of the current environmental conditions determining hazards and structural vulnerability was obtained from remote sensing analysis, fieldwork studies, and GIS analysis. Some of the methods are outlined in more detail below.

In Nepal, the assessment of the consequences of migration on the area of origin was linked with a mapping and appraisal of land management practices (Schwilch et al. 2011). The WOCAT mapping tool makes it possible to capture major land degradation as well as land management practices and technologies, their spread, effectiveness, and impact within a selected area (Liniger et al. 2008). Data drawn from a variety of sources is compiled and harmonized by a team of experts comprising land degradation and conservation specialists working in consultation with land users from various backgrounds. Knowledge provided by specialists and land users is combined with existing datasets and documents (maps, GIS layers, high-resolution satellite images) and jointly verified in workshops. The workshops are designed to build consensus among the participants regarding the assessment of land degradation and SLM. This process is also referred to as participatory expert assessment or consensus mapping. In the Panchase area in Western Nepal, the WOCAT mapping tool was applied in the Harpan river subwatershed in the Kaski district, an area of 36 km^2. A land use map of the Phewa catchment area, of which the Harpan watershed is part, served as a base map. It was produced by the local NGO (Machhapuchhre Development Organization) using *rapid eye* satellite images of 2012. With the help of local experts and field visits, each land use type of 48 administrative units (12 wards with further subdivisions along landscape features) was assessed individually regarding land use trend, land degradation, and land management, by filling in the attribute tables of the WOCAT mapping tool.

Finally, the mapping was used together with other socioeconomic data to combine information about migration types and land management issues. To obtain these data, a household survey was conducted in six communities within the watershed. Swiss and Nepali researchers jointly designed and tested the questionnaire based on previous knowledge and research conducted in Nepal (Sudmeier-Rieux 2011). The questions were centered on household demographics and socioeconomic status, housing and infrastructure, land use and land management, damage from disasters, the migration situation and remittances, and other institutional and social issues. Between 10 and 30 % of the households were thus surveyed using a systematic sampling method. The resulting data were complemented through interviews, focus group discussions, and expert consultation. GIS and descriptive statistical analysis, as well as a content analysis of the qualitative data, was used to understand, quantify, and visualize those links. As research and analysis is still underway, not all data collected was used in this chapter.

3 Study Area

The study area was selected considering the following criteria—an area with a high level of outmigration, proximity to an urban center, harsh environmental conditions, and land management issues. We obtained information about outmigration from national statistics. To identify harsh environmental conditions, we analyzed relief and geomorphology, referring to processes shaping the landscape, causing intense erosion or occurs at a rate that challenges the capacity of people to deal with their dynamics. Information about land management included available knowledge about SLM practices from within the local partners and previous projects (e.g., NEPCAT,[1] EbA[2] projects from IUCN in Nepal).

The study area is located in the Panchase area, or more specifically, the Harpan river subwatershed in the Kaski District, Western Nepal, near the city of Pokhara. With a precipitation of 4,000 mm per year and altitudes ranging from 800 m to 2,517 m.a.s.l., the area is considered the water tower of the Phewa watershed and lake (GoN 2013). The climate is subtropical and soils are usually thick. Based on data from the Central Bureau of Statistics (GoN 2012), the estimated number of households in the subwatershed is 894, with 3,330 persons in total and 93 persons/km^2. Major nonclimate pressures include overexploitation or unsustainable use of resources, overgrazing, and pest infestation, while climate-induced pressures include habitat destruction, biodiversity loss, an increase in invasive alien plant species, and a degradation of ecosystem functioning (EbA 2014). The Panchase region is one of the most heavily affected by landslides in Nepal and has high outmigration, creating a severe and underreported impact on food security and sustainable local development (Sudmeier-Rieux 2011).

4 Results

4.1 Migration: People's Concerns and Use of Remittances

Of the 58 households surveyed, 81 % have at least one member, mostly men and young people, living somewhere else in Nepal or abroad. However, due to cultural and legal barriers, only few women migrate abroad, and those who can afford it prefer to move to main Nepali cities (Ghimire and Upreti 2012). People interviewed in the different areas prioritized their main concerns according to a list of proposed choices. The top priority was unemployment, which appears logical given the significant outmigration rates in those villages (see Fig. 11.1). Health and sanitation, road access, and education also featured prominently. Landslides and flooding are only deemed minor problems, mainly mentioned by families most affected by these in Harpan and Ghatichhina. Of lower concern were agricultural and farming issues as well as access to land and water.

This may not be surprising considering that agriculture has become a smaller part of household income: 78 % of respondents said off-farm income comprised more than half their earnings. Many households depend mainly on remittances, whether in the form of financial or material support. Remittances are used primarily for food (63 % of responses) and to an extent on education (9 % of responses). The

[1] Nepal Conservation Approaches and Technologies, http://www.icimod.org/nepcat

[2] Ecosystem Based Adaptation Programme, Nepal, http://ebaflagship.org/ecosystems/mountains/nepal

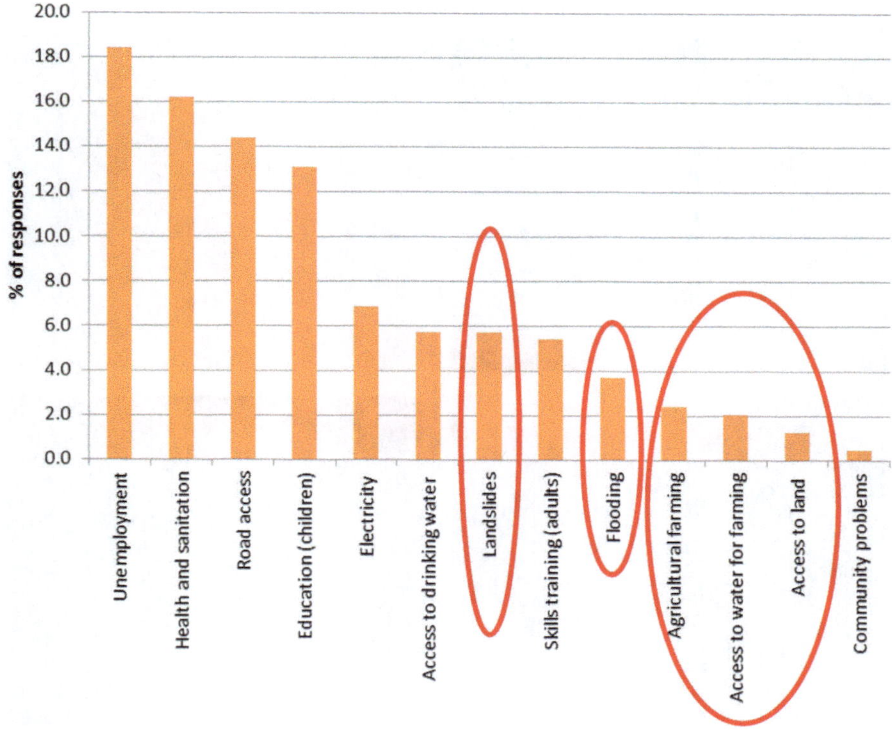

Fig. 11.1 Main concerns mentioned by 58 household respondents in Panchase, Nepal (the five main concerns were weighted by their rank). *Red circles* highlight environmental concerns

analysis revealed that remittances are hardly used for agricultural purposes (2 % of responses).

4.2 Land Degradation and Land Abandonment

Forest covers more than 70 % of the watershed, mainly in its upper western and southern reaches. Since 1996 forest cover has shown a net increase of over 12 %, due to the abandonment of cropland (Jaquet et al. 2015). The irrigated (*Khet*) and non-irrigated (*Bari*) agricultural fields make up 26 % of the watershed and are located in the central and lower eastern parts. According to Jaquet et al. (2015), almost 22 % of the cropland area is currently abandoned.

The WOCAT mapping revealed that only 1.3 % of the area and 2.8 % of the cropland is affected by soil erosion (loss of topsoil and gullying), with the trend decreasing slightly on average. Favorable ecosystem recovery conditions mean that abandoned terraces are rapidly covered by grasses, bushes, and trees. A woman from Upper Makawanpur says: "*Abandoned land is stronger now through trees, not*

Fig. 11.2 Terraces are usually well maintained. Damages on terraces, whether abandoned or not, are rare. (**a**) animal trampling on abandoned terraces, (**b**) mice holes in terrace riser, (**c**) application of manure, (**d**) terrace riser reinforced with stones (Photos: G. Schwilch)

fragile anymore." Even when grazed, there were very few signs of damage on the terrace risers such as those shown in Fig. 11.2. The damages seen in Fig. 11.2a, b represent rare exceptions and most interview partners confirmed this impression from the field survey. With fewer animals in the villages, there was also less deterioration from grazing on abandoned terraces. The terraces still in use for agriculture are usually well maintained, i.e., manure is applied (Fig. 11.2c) and, where necessary, terrace risers are enforced by stone walls (Fig. 11.2d).

It is often assumed that less land management and abandoning terraces would lead to more soil erosion (see, e.g., Gerrard and Gardner 2002; Harden 1996; Khanal and Watanabe 2006). However, the results of this study show the contrary or at least a more differentiated picture. The field survey and mapping, interviews, and focus group discussions revealed no increased soil erosion problems from agricultural or forest land.

Road construction, however, did cause increased erosion and shallow landslides. Within the Harpan subwatershed, damage from road construction (e.g., soil deposits on the land just below the new road) was visible in a few cases only, but a detailed assessment was not done. However, related studies (Raya and Sharma 2008; UNDP 2012) in the surrounding areas refer to huge problems with badly planned and negligently implemented road construction. A majority of rural earthen roads are funded partly by local government authorities (Village Development Committees) and partly by communities themselves. They are usually constructed using a local bulldozer contractor with no technical or geological expertise. Such roads are commonly wiped out during heavy monsoon rains, requiring costly clearing with heavy equipment and increasing landslide risk and impacts to settlements, forests, water sources, agriculture lands, and infrastructure (Oven et al. 2008).

The results from the WOCAT mapping show that the main type of degradation in the area is biological. More specifically, this has been caused by the increased spread of invasive alien plant species, which have affected 39.2 % of the studied area. The two main species are Nilo Gandhe (*Ageratum Houstonianum*) and Banmara (*Ageratina adenophora*), plants which appeared only 5 years and 10 years ago respectively. Nilo Gandhe is toxic for animals and both are difficult to remove. People participating in the mapping clearly linked this degradation type to outmigration and land abandonment, as illustrated by these two statements: "*If there were more people in the village, we would have less invasive species, as these people would use the land*" (man in Kuiredanda village) and "*The more Banmara is cleared, the more it grows; we lack the labor to clear it better*" (woman in Upper Makawanpur). The problem is that Banmara can spread vegetatively and could worsen if care is not taken when weeding or plant pieces are dropped randomly. Thus, with land abandonment and less labor available for weeding, both species spread increasingly (see also Jaquet et al. 2015).

Fertility decline and reduced organic matter content is a problem on 8.4 % of the area, and in particular, within 32.1 % of the cropland. Those participating in the mapping, as well as the interview partners, attributed this to outmigration, as this has reduced the number of animals and thus the availability of manure, along with the manpower to distribute the manure. On the other hand, it was also said that "*fertility has increased because land is used less*" (focus group participant in Upper Sidane), referring to the labor shortage-induced reduction from two crops a year to one. Leaving out the second crop, e.g., maize in case of irrigated rice-based cropping fields (*Khet*) or millet in the rainfed maize-based system (*Bari*), increases fertility as fewer nutrients are extracted from the soil. However, this might not be significant and would need to be further investigated. Around Ghatichhina, the village to which some upstream households migrate, cropland suffers from declining fertility as the growing population increases production from two to three crops per year. There is no land degradation on just over half (50.6 %) of the total area. In the forest, this value is as high as 61.9 %, whereas only 23.7 % of cropland is unaffected by degradation.

If we consider the total extent of degradation per map unit (see Fig. 11.3a), pressure on natural resources is higher in the lower, more densely populated sectors of the watershed in the west as well as in most cropland units. Upstream areas and the eastern part of the catchment are much less affected. As mentioned above, it is mainly the forest units that show very little degradation, with less than 10 % affected. A similar picture emerges when showing the extent of SLM measures

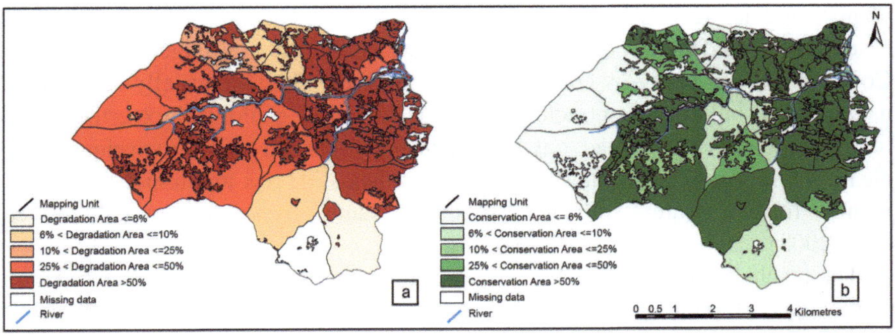

Fig. 11.3 Comparison of the total extent of degradation (**a**) and of SLM measures (**b**) per map unit, showing that areas with a high extent of degradation often also have a high extent of SLM measures

applied (Fig. 11.3b). The map units with a high extent of degradation often also have a high extent of SLM measures. Based on the interviews, this can be explained by the higher population density in downstream areas, reinforced by outmigration from upstream or remote locations to these lower sectors of the watershed. This supports the hypothesis that abandoned land resulting from outmigration does not increase land degradation problems (see also discussion and conclusions).

4.3 The Social Impacts of Outmigration and Land Abandonment

More than half (54.3 %) of the households interviewed have abandoned land, with the main reason mentioned for this as a lack of labor availability (Jaquet et al. 2015). Lack of labor means that some of the cropland can no longer be cultivated, but gender issues also play a role. Asked about problems when women manage the land, a representative from Lower Makawanpur responds: "*Yes, it is a big problem for the women to plough. Some of the abandoned land is from these women. It is abandoned because they cannot cultivate it.*" The women left behind are overburdened with additional male work, such as plowing or roofing, which they are traditionally not allowed to do. Although women face problems cultivating land because one or more of their male family members have migrated, women are also empowered in terms of overall land management, as they are forced to make decisions and deal with different actors such as the Village Development Committee (VDC) and government officials. Many of the inhabitants see no positive side to land abandonment. They regret that the land is no longer cultivated and complain that "*wild animals hide there.*" Social life in the villages also suffers, with several villagers expressed really feeling like left-behinds. As a woman from Lower Sidane put it: "*If they have money, they migrate—we don't have money, so we are here.*"

4.4 Sustainable Land Management (SLM)

Forest management (afforestation, forest protection) is applied to 34 % of the area (50.7 % of the forest), and terraces on 14 % of the area (52.4 % of cropland). Community forest management has a long tradition in Nepal (Adhikari et al. 2004). Families have access to these community forests, which are steered by community forest user groups with specific rules about use and management. Additionally, there is the Panchase Protected Forest, which has been a protected area since 2011 (Panchase Protected Forest Program, Ministry of Forest and Soil Conservation). According to several interview partners, forest protection is effective in reducing landslide problems in the forest, and thus also reducing damage to the adjacent cropland. Overall, forest cover has increased by 12 % between 1996 and 2012 (Jaquet et al. 2015), during which time pressure on forest resources has decreased. According to a woman from Upper Sidane, this illustrates the link with outmigration and land abandonment: "*With abandoned land, we have access to fodder and don't need to go to the forest. Also with better stoves we use less firewood from the forest.*" However, conserving the forest resource appears to have some disadvantages for the remaining population and their cropland, with several people reporting increased problems with monkeys, rats, and other wild animals affecting the crops.

Almost all crops are grown on terraces. When asked about SLM practices applied in the area, people often omit to mention terraces, just because they are so normal in Nepal. However, they are a highly effective measure against land degradation, especially soil erosion. Besides maintaining traditional terracing, only few new SLM practices, such as vegetable farming, mulching, and agroforestry, are emerging. Overall, 53 % of the land in the Harpan subwatershed has some kind of SLM practice (49.3 % of forest, 59.7 % of cropland). In addition to forest management and terracing, these include nutrient management, protection against natural hazards (such as gabion walls with vegetative measures, 1 %), and grazing land management (0.6 %) (Jaquet et al. 2015).

Box 1. Summary of results from the sister study conducted in Bolivia

The second study site for part of the same research is the Jatún Mayu-Pankuruma watershed from Sipe Sipe municipality in the Quillacollo district (Cochabamba Department), central Bolivia. This is one of the poorest areas in the Cochabamba Department, with high outmigration rates and harsh environmental conditions. The area suffers from severe land degradation, mainly due to landslides, fluvial erosion, wind erosion, and gully formation, which leads to a loss of land and productivity. Furthermore, it is becoming increasingly difficult to predict extreme climatic events such as occurrences of frost and hail. In Bolivia, mountain communities are highly mobile—a traditional

(continued)

and well-established way in Andean livelihoods to reduce vulnerability to both environmental and nonenvironmental risks (Kaenzig and Piguet 2014). Consequently, most households are involved in some type of migration; either long term (abroad), temporary (to another department), commuting and multiresidency (to neighboring cities), or return migration. Despite this high degree of mobility, migrants do not become disconnected with their community of origin. Instead, they usually stay in close touch with their mountain villages and return on a regular basis, either to maintain their plots of land or to fulfill their share of the responsibilities and roles that are divided among the community members. However, migration is causing schools to close or become centralized, traditional and rural knowledge is lost, and conflicts over land tenure emerge. In turn, these impacts themselves become drivers and push factors of migration.

Figure 11.4 shows that environmental hazards in Bolivia are high on people's list of major concerns, the most important being landslides and weather-related farming difficulties (e.g., crops threatened by hail and frost). Health and sanitation, unemployment, electricity, and education are also important concerns in this area. Of the five communities in the watershed, only one primary school is still operational, and in order to access secondary school or university, students have to move downstream to Sipe Sipe or Cochabamba. The lack of basic education is increasingly putting pressure on people, pushing them to move and leave the mountains.

SLM measures practiced in the area are terraces, gabion walls, and grass strips. The impacts of migration on land management practices are not yet

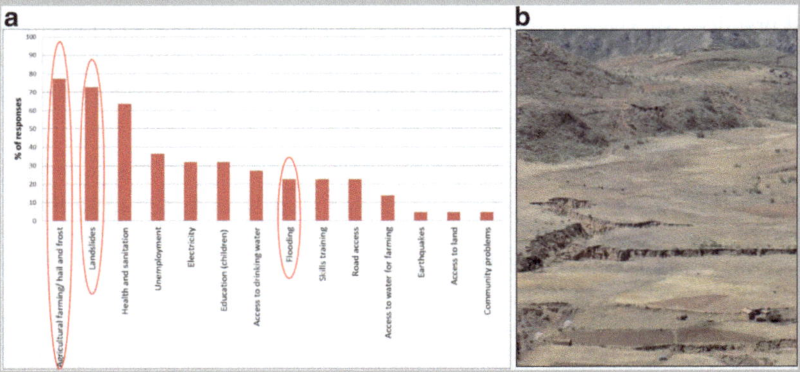

Fig. 11.4 (**a**) Main concerns mentioned by 22 household respondents in three communities from the Jatún Mayu-Pankuruma watershed, Quillacollo district, Bolivia. *Red circles* highlight environmental concerns. (**b**) Agricultural land with gully erosion (Photo: I. Penna)

(continued)

fully clear, but some important aspects are emerging from the information obtained during the interviews. Double-residence migrants still use their land, coming back on some days to work on it. However, multiresidency leads to a tendency to reduce land management work or to change the way land is used. Less able to exploit land due to labor force shortages, people farm mainly on the most accessible and/or productive parcels. Whether land is being completely abandoned is unclear, as people are reluctant to speak about land tenure with outsiders like researchers. Inhabitants have only obtained the right to own their land and crops since an agrarian reform in 1952 and land property has remained a very sensitive issue amid people's fears of losing their rights again.

Contrary to our initial hypothesis, we observed that there was no substantial process of feminization in the study area in Bolivia. Instead, entire families are migrating. Most families migrate to the valley of Cochabamba or within Bolivia, and the men (sometimes with their wives) return to the mountain to maintain fields or participate in community meetings. Remittances are not common and most of the time family members return with food or consumer items. The main phenomenon observed in Bolivia relates to an aging population. Migration leads to a loss of traditional farming knowledge, but it also leads to changes in the traditional plantation calendar, which is itself threatened by increasingly uncertain climate predictions. Our analysis shows that younger farmers have less knowledge about natural weather indicators and corresponding agricultural measures, leading to a loss in agricultural production. Additionally, the decrease in labor force is leading to a lack of maintenance of agricultural lands and increased problems with land management, possibly leading to degradation. However, some important questions remain: what causes such intense gullying? Is grazing a major cause of degradation? Do people abandon their land at all and if so, what impact does this have on the land? In Bolivia, it is difficult to distinguish fallow land from abandoned land, as land abandonment is a sensitive issue.

5 Discussion

This study has assessed land use, land degradation, SLM practices, migration patterns, use of remittances, and concerns of the people left behind, to achieve a better understanding of how migration affects land management in mountainous areas. By bringing together these two issues—land management and migration—it goes further than previous studies, which have focused either on migrants' livelihoods or evaluated land management without taking into account migration-induced changes in family patterns.

Our study sheds light on the strong link between migration and land management in the Panchase area of Nepal. Even within a rather small watershed like the

Harpan river subwatershed, there are areas of outmigration as well as in-migration. While they are linked and interdependent through the people migrating from one place to the other, the consequences on the land are very different.

We have identified five causal consequences of **outmigration**, which are all in a cause–impact relationship:

1. ***Less human population*** results in less livestock in villages. This means that less manure is produced to fertilize the soils.
2. The population decrease-induced ***labor shortage*** has several impacts. First, reducing production to only one crop per year leads to reduced income (as more food has to be bought). Second, cropland terraces are no longer cultivated and thus abandoned. Third, invasive alien plant species are increasing on abandoned terraces.
3. ***Land abandonment*** provides better access to more fodder, initially from better access to grazing land and forage grass, later from fully grown woodland. Fodder thus no longer needs to be sought from the forest.
4. ***Less land degradation*** due to less intensive use of resources. An important exception here is that there is a massive increase in invasive alien plant species. Soil erosion thus generally decreases, while vegetation cover increases.
5. ***Expansion of forest***, with a likely increase in biodiversity and wildlife (again, partly threatening cropland).

We found hardly any literature differing from or confirming these causal consequences, which we believe verifies our statement that the impacts of outmigration on land management in the area of origin are not well researched. Our results do not confirm those of earlier studies in mountainous areas that reveal increased landsliding, soil erosion, and a higher susceptibility to slope failures on less-managed cultivation terraces and on grassland (abandoned terraces) (Gerrard and Gardner 2002; Harden 1996), although long-term monitoring might change the picture (Khanal and Watanabe 2006).

However, our study did confirm the finding of Paudel et al. (2012) that the labor shortage resulting from outmigration is the major constraint to agricultural activities, leading to changes in the cropping patterns. Labor shortages are causing farmers either to shift towards a single crop per year or even to abandon their land in favor of wage labor, providing them with less-risky and more immediate benefits. The observation of Davis and Lopez-Carr (2014) that remittances are either invested to expand agricultural land, intensify its management (e.g., through chemical fertilizer and pesticides), or shift to more cattle-based systems, however, was not confirmed in our study, where remittances are hardly invested in agriculture at all.

As there are also areas of **in-migration** in the Harpan subwatershed, especially the downstream village of Ghatichhina and surroundings, we were also able to identify and consider these impacts:

1. ***More population*** leads to more pressure on land resources in in-migration villages.

2. More people on ***less land*** requires the cultivation of an additional crop, i.e., an increase from two to three crops per year.
3. ***Increasing the intensity of land use*** leads to decreasing soil fertility and to potentially less careful land management, such as cultivating crops that require less postharvest processing (e.g., potatoes). Potatoes potentially increase soil erosion due to the heavy soil disturbance during harvest.

These impacts of in-migration on the lowland area confirm the findings of Gautam et al. (2003), where the authors reported an increased fragmentation of lowland agricultural areas due to urbanization and increased crop diversification in the remaining lowlands.

Comparing these results from Nepal with the sister study done in Bolivia (see Box 1) reveals two very different contexts, although in both cases migration is spurred by unemployment and lack of educational opportunities. In Nepal, migrants (mainly male) leave behind fragmented families and the elderly to manage the land, and those left behind rarely use remittances to invest in agriculture. In Bolivia, people move dynamically between urban areas and their communities of origin in the mountains, returning home at intervals to work on their lands. In a major difference between the two contexts, environmental threats are of greatest concern in Bolivia, while in Nepal these figured only in seventh place and below. In both locations, climate change is increasing the frequency and magnitude of extreme weather events, shifting crops (and invasive species) upslope and rendering less certain the traditional planting season due to more erratic temperatures and rainfall. In light of these trends, the challenge for the populations studied will be how to manage land in the face of changing demographics, uncertain climatic conditions, less labor available, and lower food production in mountain areas. There is an obvious shift from local food production to a greater reliance on remittance income and imported foods.

Lessons drawn from this research are that multiple pathways of outmigration, land abandonment, ecological impacts, and corresponding livelihood opportunities exist and thus need to be carefully investigated for each situation (Munroe et al. 2013). Evaluating these pathways can provide evidence to develop adapted land management and livelihood strategies.

6 Conclusions

The findings of this study suggest that a more differentiated, context-specific view is required when looking at the impact of migration on land management. There was no evidence to prove the often assumed negative impact of outmigration on land degradation. On the contrary, some types of degradation such as soil erosion or landslides were even reduced, and overall vegetation and forest cover had increased. But other types of land degradation have emerged, such as the increase in invasive alien plant species coverage, making farming more difficult for those left behind. The main negative effect is sociocultural, as people's livelihoods are not only affected by the environmental state of the land they live on, but much more by the absence of their family members and neighbors. This threatens the social fabric and leads to a loss of traditional knowledge and cultural customs. A feminization of agriculture has been observed in the Nepali case study, while in the Bolivian case study, whole families generally migrate together and it is mostly the men who temporarily return to manage the land.

Even within small rural areas, topographic location and specific migration patterns determine how land is used and managed, and the consequences on land and livelihoods thus vary considerably from place to place. It is this interaction of livelihoods and land, which characterizes the high dynamism of these mountain areas and, when evaluated carefully and within its migration context, offers opportunities for sustainable development of both the environment and people's well-being.

Acknowledgements This 2-year study (2012–2014) was funded by the Swiss Network for International Studies, based in Geneva. We would like to send sincere thanks to the communities and people who helped us during the fieldwork. We are grateful to our partners in Nepal, G. Shrestha, S. Devkota, G. Subedi, and P. Baral; and in Boliva, E. Cuba and A. Machaca. We also thank Tina Hirschbuehl for language editing this chapter.

References

Adhikari B, Di Falco S, Lovett JC (2004) Household characteristics and forest dependency: evidence from common property forest management in Nepal. Ecol Econ 48(2):245–257. doi:10.1016/j.ecolecon.2003.08.008

Andersen LE (2002) Rural-Urban migration in Bolivia: advantages and disadvantages. Documento de Trabajo, 5

Black R, Bennett SRG, Thomas SM, Beddington JR (2011) Climate change: migration as adaptation. Nature 478(7370):447–449

Carswell GM, Lindblade KA, Tumuhairwe JK (1998) Mitigating the relationship between population growth and land degradation: land-use change and farm management in Southwestern Uganda. AMBIO 27(7):565–571

Davis J, Lopez-Carr D (2010) The effects of migrant remittances on population–environment dynamics in migrant origin areas: international migration, fertility, and consumption in highland Guatemala. Popul Environ 32(2-3):216–237. doi:10.1007/s11111-010-0128-7

Davis J, Lopez-Carr D (2014) Migration, remittances and smallholder decision-making: implications for land use and livelihood change in Central America. Land Use Policy 36:319–329. doi:10.1016/j.landusepol.2013.09.001

de Graaff J, Amsalu A, Bodnár F, Kessler A, Posthumus H, Tenge A (2008) Factors influencing adoption and continued use of long-term soil and water conservation measures in five developing countries. Appl Geogr 28(4):271–280. doi:10.1016/j.apgeog.2008.05.001

DFID (1999) Sustainable livelihoods guidance sheet. UK Department for International Development, London

EbA (2014) Ecosystem and ecosystem services of panchase: an overview. EbA, Kathmandu

Findlay AM (2011) Migrant destinations in an era of environmental change. Global Environ Change 21(Suppl 1):S50–S58. doi:10.1016/j.gloenvcha.2011.09.004

Foley JA, Ramankutty N, Brauman KA, Cassidy ES, Gerber JS, Johnston M, Mueller ND, O'Connell C, Ray DK, West PC, Balzer C, Bennett EM, Carpenter SR, Hill J, Monfreda C, Polasky S, Rockstrom J, Sheehan J, Siebert S, Tilman D, Zaks DPM (2011) Solutions for a cultivated planet. Nature 478(7369):337–342, http://www.nature.com/nature/journal/v478/n7369/abs/nature10452.html#supplementary-information

Gartaula HN, Visser L, Niehof A (2012) Shifting perceptions of food security and land in the context of labour out-migration in rural Nepal. Food Secur 4(2):181–194. doi:10.1007/s12571-012-0190-3

Gautam AP, Webb EL, Shivakoti GP, Zoebisch MA (2003) Land use dynamics and landscape change pattern in a mountain watershed in Nepal. Agr Ecosyst Environ 99(1–3):83–96. doi:10.1016/S0167-8809(03)00148-8

Gerrard J, Gardner R (2002) Relationships between landsliding and land use in the Likhu Khola Drainage Basin, Middle Hills, Nepal. Mt Res Dev 22(1):48–55. doi:10.1659/0276-4741(2002)022[0048:rblalu]2.0.co;2

Ghimire A, Upreti BR (2012) Migration, mobility and development: enhancing our understanding in Nepal, SAs RCO Discussion Paper 6. South Asia Regional Coordination Office of National Centre of Competence in Research North-South and Nepal Centre for Contemporary Research, Kathmandu

Gisladottir G, Stocking M (2005) Land degradation control and its global environmental benefits. Land Degrad Dev 16(2):99–112. doi:10.1002/ldr.687

GoN (2012) National population and housing census 2011. GoN, Kathmandu

GoN (2013) District and VDC profile of Nepal. GoN, Kathmandu

Grau HR, Aide TM (2007) Are Rural–Urban migration and sustainable development compatible in mountain systems? Mt Res Dev 27(2):119–123. doi:10.1659/mrd.0906

Gray CL (2009) Environment, land, and rural out-migration in the Southern Ecuadorian Andes. World Dev 37(2):457–468

Harden CP (1996) Interrelationships between land abandonment and land degradation: a case from the Ecuadorian Andes. Mt Res Dev 16(3):274–280. doi:10.2307/3673950

Hurni H (2000) Assessing sustainable land management (SLM). Agr Ecosyst Environ 81 (2):83–92. doi:10.1016/S0167-8809(00)00182-1

Jaquet S, Schwilch G, Hartung-Hofmann F, Adhikari A, Sudmeier-Rieux K, Shrestha G, Liniger HP, Kohler T (2015) Does outmigration lead to land degradation? Labour shortage and land management in a western Nepal watershed. Appl Geogr 62:157–170. doi:10.1016/j.apgeog.2015.04.013

Kaenzig R, Piguet E (2014) Migration and climate change in Latin America and the Caribbean. In: Piguet E, Laczko F (eds) People on the move in a changing climate, vol 2, Global migration issues. Springer, Dordrecht, pp 155–176. doi:10.1007/978-94-007-6985-4_7

Khanal R, Watanabe T (2006) Abandonment of agricultural land and its consequences. Mt Res Dev 26(1):32–40

Kollmair M, Banerjee S (2011) Drivers of migration in mountainous regions of the developing world: a review. In: Project F (ed) Migration and global environmental change. UK Government, London

Liniger HP, Van Lynden G, Nachtergaele F, Schwilch G (2008) Questionnaire for mapping land degradation and sustainable land management (QM), WOCAT, LADA, DESIRE. CDE Bern, FAO Rome, ISRIC Wageningen

Maharjan A, Bauer S, Knerr B (2012) International migration. Evidence from Nepal. International Migration, Remittances and Subsistence Farming, Kathmandu

Munroe DK, van Berkel DB, Verburg PH, Olson JL (2013) Alternative trajectories of land abandonment: causes, consequences and research challenges. Curr Opin Environ Sustain 5 (5):471–476. doi:10.1016/j.cosust.2013.06.010

NIDS, NCCR (2011) Nepal Migration Year Book 2010. Nepal Institute of Development Studies (NIDS), National Centre of Competence in Research (NCCR) North-South South Asia Coordination Office, Kathmandu

Nijenhuis G (2010) Embedding international migration: the response of Bolivian local governments and NGOs to international migration. Environ Urban 22(1):67–79. doi:10.1177/0956247809356182

Oven KJ, Petley DN, Rigg JD, Dunn CE, Rosser NJ (2008) Landslides, livelihoods and risk: vulnerability and decision-making in Central Nepal. Paper presented at the First World Landslide Forum, Tokyo, Japan

OXFAM (2009) Bolivia: climate change, poverty and adaptation. OXFAM International, Bolivia

Paudel KP, Dahal D, Shah R (2012) Study report on abandoned agriculture land in mid-hills of Nepal, Status, causes and consequences. Australian Centre for International Agriculture Research, World Agroforestry Centre, Forest Action, IUCN, Nepal

Petley DN (2010) On the impact of climate change and population growth on the occurrence of fatal landslides in South, East and SE Asia. Q J Eng Geol Hydrogeol 43(4):487–496. doi:10.1144/1470-9236/09-001

Piguet E (2013) From "Primitive Migration" to "Climate Refugees": the curious fate of the natural environment in migration studies. Ann Assoc Am Geogr 103(1):148–162. doi:10.1080/00045608.2012.696233

Raya RB, Sharma S (2008) Impact of land use in water quality of Phewa Lake, Pokhara, Nepal. Paper presented at the World Lake Conference

Schwilch G, Bestelmeyer B, Bunning S, Critchley W, Herrick J, Kellner K, Liniger HP, Nachtergaele F, Ritsema CJ, Schuster B, Tabo R, van Lynden G, Winslow M (2011) Experiences in monitoring and assessment of sustainable land management. Land Degrad Dev 22 (2):214–225

Scoones I (1999) New ecology and the social sciences: what prospects for a fruitful engagement? Annu Rev Anthropol 28:479–507

Seddon D, Adhikari J, Gurung G (2002) Foreign labor migration and the remittance economy of Nepal. Critical Asian Stud 34(1):19–40. doi:10.1080/146727102760166581

Sudmeier-Rieux K (2011) On landslide risk, resilience and vulnerability of mountain communities in Central-Eastern Nepal. UNIL, Lausanne

Sudmeier-Rieux K, Gaillard J-C, Sharma S, Dubois J, Jaboyedoff M (2012) Chapter 7 Floods, landslides, and adapting to climate change in Nepal: what role for climate change models? In: Climate change modeling for local adaptation in the Hindu Kush-Himalayan region. Emerald Group Publishing, Bingley, pp 119–140. doi:10.1108/S2040-7262(2012)0000011013

Thapa G, Weber K (1995) Status and management of watersheds in the Upper Pokhara Valley, Nepal. Environ Manag 19(4):497–513. doi:10.1007/bf02471963

Tiffen M, Mortimore M, Gichuki F (1994) More people less erosion: environmental recovery in Kenya. John Wiley, London

UNDP (2012) Multi Hazard Risk Assessment Study and Development of Strategic Implementation Plan Phewa Watershed Area, Kaski

UNU-INWEH (2011) Guidelines for the Preparation and Reporting on Globally-relevant SLM Impact Indicators for Project-level Monitoring. United Nations University, Hamilton, ON

Upreti BN, Muller-Boker U (2010) Livelihood Insecurity and Social Conflict in Nepal. South Asia Regional Coordination Office, Swiss National Centre of Competence in Research (NCCR) North-South., Kathmandu

Valdivia C, Seth A, Gilles JL, Garcia M, Jimenez E, Cusicanqui J, Navia F, Yucra E (2010) Adapting to climate change in Andean ecosystems: Landscapes, capitals, and perceptions shaping rural livelihood strategies and linking knowledge systems. Ann Assoc Am Geogr 100:818–834

Warner K, Hamza M, Oliver-Smith A, Renaud F, Julca A (2010) Climate change, environmental degradation and migration. Nat Hazards 55(3):689–715. doi:10.1007/s11069-009-9419-7

Warren A (2002) Land degradation is contextual. Land Degrad Dev 13(6):449–459. doi:10.1002/ldr.532

WOCAT (2014) Sustainable land management. https://www.wocat.net/en/about-wocat/vision-mission/sustainable-land-management.html. Accessed September 20, 2014

Part III
Climate Change Adaptation

Chapter 12
Reflections on Disaster Diplomacy for Climate Change and Migration

Ilan Kelman

1 Introduction

Migration has always been part of human existence, supporting and inhibiting sustainable development while being a cause and consequence of disaster vulnerability, as amply demonstrated by other chapters in this volume. Migration as a cause and consequence of disaster vulnerability has led to both cooperation and conflict, most notably between (1) migrants and those remaining in the home community and (2) migrants and hosts. Some aspects of migration-related conflict have become populist in recent years because the speculations help to serve certain political agendas such as supporting militarisation and being anti-immigration.

Contemporary climate change is caused in part by anthropogenic influences, namely emissions of greenhouse gases and land use changes such as deforestation (IPCC 2013–2014). One possible consequence from the responses to climate change, or lack thereof, is migration. Migration could occur at any point along the forced-voluntary continuum, although reasons linked to climate change must be viewed in the context of multiple other reasons which people use, and have always used, for migrating and for not migrating (Foresight 2011; Hugo 1996; Petersen 1958).

Nonetheless, cases occur where the residents would rather remain in their homes, but have decided to move involuntarily, solely due to climate change. Residents of the Carteret Islands of Papua New Guinea have already left their homes and moved elsewhere, seemingly due to sea-level rise (Connell 1997). Communities in coastal Alaska are planning to move inland, as they face increasingly rapid coastal erosion, partly occurring because less ice on the ocean permits

I. Kelman (✉)
University College London, London, UK

University of Agder, Kristiansand, Norway
e-mail: ilan_kelman@hotmail.com

© Springer International Publishing Switzerland 2017

K. Sudmeier-Rieux et al. (eds.), *Identifying Emerging Issues in Disaster Risk Reduction, Migration, Climate Change and Sustainable Development*,
DOI 10.1007/978-3-319-33880-4_12

more severe storms to occur (Bronen and Chapin III 2013; GAO 2003). In each case, the migrants have experienced both conflict and cooperation with the various parties involved, with some recognising the challenges faced and so being willing to help, while others decline to help or oppose the migration.

Many frameworks and models exist for analyzing how different elements within sustainable development, including dealing with climate change, lead to and support cooperation or conflict. One framework which has not been explored extensively for climate change and migration is "disaster diplomacy". This chapter is a conceptual exploration of a disaster diplomacy perspective for climate change and migration and thus provides reflections rather than results, further helping to indicate the importance of diplomacy for those involved in disaster-related activities.

In a strict and fundamental definition, "diplomacy"—building on Bull (1977) while acknowledging that this citation comes from the English school of international relations, and that concepts apply beyond it—is generally considered to involve sovereign governments and their entities conducting peaceful and official business with each other. This view has widened considerably to apply to many more aspects of international and non-international affairs and relations alongside peace and cooperation activities. These activities can be undertaken by any form of entity including individuals, non-sovereign jurisdictions, the private sector, and all forms of non-governmental organizations including universities, foundations, international organisations, cultural groups, schools, and charities (which are not mutually exclusive categories; for instance, media could be stand-alone for-profit, stand-alone not-for-profit, or embedded within any other entity).

Davidson and Montville (1981) formalized the differences in two complementary diplomacy tracks, with "Track-One Diplomacy" referring to official and formal governmental interactions and "Track-Two Diplomacy" referring to unofficial and unstructured interactions involving any entities. Diamond and McDonald (1993) expanded those into "Multi-Track Diplomacy" comprising nine interacting tracks: (1) government; (2) professional conflict resolution; (3) business; (4) private citizens; (5) research, training and education; (6) activism; (7) religious; (8) funding; and (9) public opinion/communication. Other entities such as international organizations and the media are implicitly included in these categories.

In these tracks and activities, many linked words and phrases emerge to describe diplomacy-related activities, including "peace", "cooperation", "collaboration", "conflict reduction", and "conflict resolution" amongst many more. These terms are not synonymous and can diverge in theory, policy, and practice. Yet with such vast literature and detailed discussions, overlaps and fuzziness will inevitably occur in trying to understand and describe "diplomacy". Disaster diplomacy work (e.g. Kelman 2012) has lacked rigor in using these terms, explicitly accepting loose definitions and not adopting any specific school of thought. Consequently, sometimes the terms are used interchangeably depending on the demands of each case study and the literature on it. This chapter does similarly, accepting multi-track diplomacy and using "diplomacy" loosely to apply to interactions related to conflict, cooperation, and other related terms.

With this understanding of diplomacy, the next sub-section in this chapter describes the disaster diplomacy framework to move into the context of climate change diplomacy. It provides background for why disaster-related activities, including those involving climate change, rarely bring people together over the long-term, further explaining the importance of diplomacy for disaster professionals due to the number of assumptions about disasters that are shown to be challengeable. The third part of this chapter reflects on that knowledge and background to make climate change-related migration a disaster diplomacy case study. The final part suggests an agenda for moving forward through providing thoughts on the questions on which this book focuses.

2 Disaster Diplomacy

2.1 Background to Disaster Diplomacy

Disaster diplomacy (http://www.disasterdiplomacy.org) investigates how and why disaster-related activities do and do not influence conflict and cooperation (Kelman 2012). "Disaster-related activities" cover (1) pre-disaster efforts including prevention, preparedness, planning, and mitigation and (2) post-disaster actions including emergency response, reconstruction, and recovery. The disaster diplomacy literature (e.g. Kelman 2012) defines the process to relate to disaster-related activities causing wider diplomacy-related activities; that is, establishing a unidirectional causal link between disaster-related activities and further diplomacy-related activities. Disaster diplomacy does not compile all instances where entities, including but not limited to sovereign states, work together for disaster-related activities, nor does disaster diplomacy as currently defined explore diplomatic processes leading to disaster-related activities. An example of this is the 2015 World Conference on Disaster Risk Reduction producing a voluntary framework for action (UNISDR 2015).

With respect to climate change, climate change diplomacy may be relevant for all climate change-related activities, including mitigation and adaptation, through all mechanisms, including science and policy negotiations. For science, the main international mechanism is the Intergovernmental Panel on Climate Change (IPCC 2013–2014) while for international policy, the main international mechanism is the United Nations Framework Convention on Climate Change (UNFCCC). In these fora, member states as well as other entities are involved in different ways, but no diplomatic processes or outcomes have yet been identified which emerged beyond or exclusively from these fora. In contrast, some authors identify degrees of conflict fomented by them, even though that conflict does not preclude all forms of cooperation (Beck 2012; Halvorssen 2007–2008). The processes, outcomes, and interactions are mixed and complicated, without indication or suggestion that they can or would lead to wider diplomacy beyond climate change. An example which has been investigated is renewable energy systems being set up across international borders of countries which are not on good terms (e.g. Mostafaeipour and Mostafaeipour 2009), but that activity alone will not bring these countries closer together. More work is needed to identify and analyze other examples.

This conclusion is the same for other disaster diplomacy cases. All evidence so far suggests that while disaster-related activities do not create fresh diplomatic opportunities, they sometimes catalyse action. Such catalysis occurs only in the short-term, not in the long-term. In the short-term—on the order of weeks and months—disaster-related activities can, but do not always, impact diplomacy. They have the potential for influencing it, for spurring it on, and for affecting it, as long as a pre-existing basis existed for that influence. This pre-existing basis might be culture or trade links or secret negotiations.

One example occurred on 26 December 2004 (Beardsley and McQuinn 2009; Gaillard et al. 2008; Le Billon and Waizenegger 2007) when a powerful earthquake off the west coast of Indonesia generated tsunamis around the Indian Ocean. Sri Lanka and the Indonesian province of Aceh each witnessed tens of thousands of fatalities against the backdrop of each having a long-running internal conflict which had become particularly violent over the previous 30 years. In Aceh, a peace deal was reached months after the tsunami which has so far held. In Sri Lanka, humanitarian emergency and international aid were used to exacerbate the conflict. Within a few years, Sri Lanka's military had defeated the fighters.

Yet the tsunami did not create the peace deal in Aceh. Secret negotiations had started just 2 days before the earthquake and tsunami disaster (Gaillard et al. 2008) and those negotiations formed the basis for the peace deal (see also Le Billon and Waizenegger 2007). Nevertheless, the earthquake and tsunami disaster had a significant influence on the peace deal. The devastation provided a space in which peace negotiations could be successful, if the parties involved wanted it, but other factors had to exist for desiring the peace prior to the tsunami. The parties all had non-tsunami reasons for achieving peace, so they were able to use the tsunami as one of several excuses to make peace work. Thus, the tsunami provided the opportunity to reach peace, but was not the cause of it (see also Enia 2008; Klimesova 2011).

This conclusion is corroborated by the Sri Lanka example (Wickremesinghe 2006). One rebel commander was facing corruption charges, so had an incentive to keep the fighting going. Many Sri Lankans opposed involving the main fighting groups in humanitarian aid (Enia 2008; Kelman 2012; Klimesova 2011; Le Billon and Waizenegger 2007). On 17 November 2005, less than a year after the tsunami, Sri Lanka elected a pro-war anti-negotiation President, Mahinda Rajapaksa, whose party also won major victories in the April 2006 local elections, endorsing Rajapaksa's approach to the conflict. The main parties involved had non-tsunami reasons for keeping the violent conflict going, mainly deeply engrained political antipathy against the other side, so they were able to use the tsunami as one of several excuses to make peace fail. The tsunami provided an opportunity to overcome the root causes of the conflict, but that opportunity was deliberately not grasped so that the conflict would continue.

The catalysis effect of disaster-related activities, as seen in Aceh, tends to emerge in the short-term only, on time scales of weeks and months (Kelman 2012). Over the long-term—such as years—non-disaster factors tend to take over diplomacy. Examples for perpetuating conflict are a leadership change, distrust, belief that an historical conflict or grievance should take precedence over present-day humanitarian needs, or priorities for action other than conflict resolution and diplomatic dividends (see also Pelling and Dill 2010). The latter two were seen in post-tsunami Sri Lanka. Examples of longer-term influences for overcoming a conflict are a leadership change, conflict weariness, a feeling of military disadvantage, or prioritising peace as a principle. All were seen in post-tsunami Aceh to varying degrees.

These same conclusions are visible for climate-related case studies. In the context of droughts, storms, and climate, Glantz (2000) details Cuba–USA relations when Fidel Castro led Cuba. Glantz (2000) describes how several opportunities arose for climate-related and weather-related disaster diplomacy between the two countries, but when deals seemed feasible, one side or the other inevitably came up with an excuse to scuttle the emergent diplomacy. Kelman (2012) extends that analysis to events after 2000 including the 11 September 2001 terrorist attacks in the USA, Hurricane Michelle in 2001, and the 2005 hurricane season including Hurricanes Dennis, Katrina, and Wilma. Glantz (2000) and Kelman (2012) agree that the reason that disaster diplomacy did not work for the Cuba–USA case study under Fidel Castro is that both sides benefitted from having the other as an "enemy". That animosity was played up, irrespective of any ongoing cooperation. Castro needed a monolithic power that claimed to be against the Cuban people in order to prop up his regime. American politicians in Washington, D.C. could not engage with Cuba for fear of upsetting the vocal Cuban-American lobby groups who had influence in the key electoral swing state of Florida. With Raúl Castro now leading Cuba, economic advantages are seen in engaging more with the USA and the American President Barack Obama has responded carefully to the overtures. Cuba–USA diplomacy is thawing, but not because of disaster-related activities.

The overall conclusion from disaster diplomacy research so far across multiple case studies including climate change (Kelman 2012) is that disaster diplomacy sometimes, but not always, appears in the short-term, if the parties involved have non-disaster reasons for enacting diplomacy. In the long-term, non-disaster factors supersede the influence of disaster-related activities. This result holds beyond international politics—that is, beyond diplomacy at the bilateral or multilateral level (e.g. see Gaillard et al. 2009 and Klimesova 2011 for intra-state disaster diplomacy in the Philippines)—and also beyond modern instances (e.g. Segalla 2012 describes a 1959 disaster diplomacy incident between Morocco and the USA over an oil spill).

Since this chapter focuses on the global challenge of climate change and migration across international borders, the focus of the disaster diplomacy framing remains at the international level for contemporary times. Consequently, reflections upon sub-national and historical disaster diplomacy are not detailed, although they are mentioned when relevant.

2.2 Explaining Disaster Diplomacy's Failure

Why does disaster diplomacy usually not work? The principal reason is that people make active choices regarding the politics that shape the active choices against diplomacy or against enacting disaster-related activities, either before or after a disaster. Reconciliation is not necessarily an important objective, despite the potential for joint life-saving actions, as seen for Cuba and the USA (Glantz 2000). As Glantz (2000) notes, plenty of scientific and technical cooperation occurs without the politicians knowing about it, such as researching and monitoring hurricanes, but this cooperation most likely occurs because the governments do not know about it—meaning that they do not use the research. Active political choices were made to avoid engaging with each other, because the enmity suited political goals.

Similarly, inertial prejudice, misgivings, and mistrust can hinder disaster diplomacy efforts. On 26 December 2003, an earthquake shattered the World Heritage city of Bam, Iran, killing 25,000 people (for background to this case study, see Kelman 2012; Warnaar 2005, 2013). Iran stated that the country would accept aid from any country apart from Israel. The disaster could not overcome Iran's bias. US aid, however, was accepted and led to suggestions that Iran-US disaster diplomacy might result. Media hype and lack of political forethought derailed possible good intentions. The US State Department's position vis-à-vis Iran had not actually changed, even though the press presented it as being an opening to *rapprochement*. The US government then tried to send a high-profile emissary with aid supplies to Iran, but it appears as if the US government did not fully clear that desire with the Iranian government beforehand. Iran declined, squashing any hope for disaster diplomacy. Simultaneously, the leaders in each country had domestic political reasons, related to the 2004 elections in each country, to avoid reconciliation, indicating a possible worry that disaster diplomacy could lead to too much diplomacy.

The fear of disaster diplomacy can not only harm diplomacy, but can also harm the process of dealing with disaster. As Kelman (2012) describes, with lingering memories of the failed 2003 earthquake diplomacy, Iran declined an American offer of aid following the February 2005 earthquake which killed hundreds. Iran stated that the country could handle the disaster domestically, even though aid was accepted from several other countries and several international organisations. Disaster diplomacy became a spectre to avoid, even at the cost of humanitarian aid.

The fundamental conclusion from disaster diplomacy is that saving lives is not necessarily important for political decision-making. That is not an innovative or surprising result. Instead, that is a fundamental basis for politics which is further seen in climate change diplomacy in the UNFCCC with the absence of an international, legally binding, enforced treaty that makes substantive contributions towards climate change mitigation and adaptation (see also Helm 2012).

2.3 *Is There Hope for Disaster Diplomacy?*

Given this evidence and analysis, is there any hope for disaster diplomacy, especially for climate change? Since few disaster diplomacy successes are seen (Kelman 2012), a question emerges about whether or not disaster-related activities (including those related to climate change) should be deliberately used to induce or force cooperation amongst adversarial states? Both "yes" and "no" are possible answers (e.g. Anderson 1999; Ferris 2011; Fox 2001; Hannigan 2012). Some who claim "no" believe that extensive effort occurs to divorce disasters from politics, such as through neutrality and impartiality. New mechanisms for relating disasters and politics are not needed. Instead, encouraging further separation is preferable. Some who claim "yes" explain how disasters are inherently political, so it is naïve to think otherwise. According to this argument, the more positive outcomes which could be fostered through disaster-related activities, the better. One corollary is that such outcomes should be actively pursued. Nevertheless, being aware of this viewpoint does not necessarily lead to disaster diplomacy success.

An example of a climate-related situation, although not necessarily climate change, is when media and humanitarian organisations tried to push drought diplomacy during the Eritrea-Ethiopia war from 2000 to 2002 (Kelman 2012). The suggestion was that Ethiopia needed food aid due to a drought and that aid could be offloaded using Eritrean ports and then sent to Ethiopia overland through Eritrea. Eritrea agreed, but Ethiopia declined, extolling numerous excuses for why aid should not go through Eritrea. Two examples are Ethiopia's claim that Eritrea would siphon off aid before it reached Ethiopia and that it was more efficient to use non-Eritrean ports, with Ethiopia suggesting that Eritrea wanted the food aid and the business for the ports rather than to help Ethiopia. The charges are not necessarily unfounded since Eritrea struggles with corruption and does need income (Desta 2006)—but that does not necessarily mean that selfishness was the only driver of Eritrea's offer. Fundamentally, despite active aims to effect drought diplomacy, both countries felt that they could win the war and so preferred to avoid drought diplomacy—or any form of diplomacy.

These case studies, and others (Kelman 2012), yield bleak prospects for hope regarding climate change diplomacy. In many instances, there is active effort to support disaster diplomacy, as with Ethiopia–Eritrea and Cuba–USA. In other instances, such as the 2004 tsunami, disaster diplomacy had strong prospects, but was not a deciding factor. Climate change being simply one hazard or hazard driver amongst many which affects locations facing conflict may not yield anything substantively different from other disaster diplomacy cases.

Nonetheless, a fundamental tenet in research is that "absence of evidence is not evidence of absence". Disaster diplomacy case studies so far yield an absence of evidence for disaster-related activities unilaterally and unidirectionally causing new diplomatic endeavours. That does not rule out future studies identifying a successful example of new diplomacy based on only disaster-related activities, either from history or in the future. Many possible examples have not yet been thoroughly investigated, so they are undecided regarding disaster diplomacy. As an example with implications for climate change, Dinar et al. (2010) found that variability in precipitation and runoff did display a tendency to increase cooperation surrounding international waterways in some instances, although their analysis did not drill down significantly into prior conditions or root causes. Nonetheless, it provides some hope that climate change affecting weather and weather variability has the potential for supporting cooperation and reducing conflict.

That is particularly the case since different forms of conflict exist (e.g. Binns 1977) which may lead to differing prospects for disaster diplomacy, although this realm has not yet been studied thoroughly. Even within violent conflict, differences occur amongst physical violence, psychological violence, structural violence, and various combinations. Conflicts which have been running for decades or centuries, such as in pre-tsunami Aceh and Sri Lanka, might have different responses to disaster-related activities than conflicts which exist mainly due to a disaster, such as if hosts are hostile to forced migrants. Since neither hostility nor dislike inevitably leads to any form of conflict, interventions for conflict prevention can make a difference, yielding another area which has not been explored extensively in disaster diplomacy literature.

In fact, no matter how unsuccessful disaster diplomacy seems to be based on current publications, those analyses have gaps and superficialities. One possibility to explore further is considering climate change migration as a disaster diplomacy case study.

3 Climate Change-Related Migration as a Disaster Diplomacy Case Study

Much of the rhetoric on climate change-related migration constructs it as being threatening or as a significant driver for major conflicts. A film by Michael Nash, *Climate Refugees* (2010), uses these points as central themes. The film's main downfall is aiming to frighten viewers about migrants. Dramatic red arrows leap from poorer countries to converge on richer locales. Accompanying commentary decries that "our" children might die fighting to protect "us" from "them". With limited evidence and critique, the film's start and finish frame climate change-related migration as being about "national security", presumably to appeal to American patriotism.

Bettini (2013), Hartmann (2010), and Nicholson (2014) deconstruct and critique this discourse. They argue that it plays into the hands of those who wish to securitise and militarise migration, seeking an excuse to create confrontation and fear for supporting melodramatic, militaristic measures that will apparently stem the mythical flows of dangerous people. These papers point to little empirical evidence that masses of people will suddenly migrate due to climate change, suggesting instead that the ideas of "climate refugees" and "climate change refugees" are politically constructed. Meanwhile, authors such as McNamara and Gibson (2009) garner viewpoints from the countries of potential migrants, focusing on Pacific islands, demonstrating how many accept the likelihood—perhaps inevitability—of moving due to climate change, but they do not wish to be labelled as "refugees" and would instead prefer to have international support to control the movement on their own terms in their own way.

None of this discussion precludes conflict emerging from migration including climate change linked migration. Instead, it notes that no reason exists to assume the inevitability of conflict and it queries why any migration associated with climate change is presented as being conflictual—especially when conflict prevention interventions might succeed. In fact, the history of migration, including due to natural hazards and hazard drivers such as the changing climate, involves different degrees of conflict and cooperation.

Numerous countries, such as Norway, welcome large numbers of refugees and asylum seekers, providing many resources for assessing their cases and, for the successful ones, supporting their settlement and adjustment (Government of Norway 2009). As a political conflict example, when the volcano on the tiny island of Montserrat (a U.K. overseas territory) in the Caribbean first erupted in 1995—an eruption still ongoing today—more than 2/3 of the population left, with political conflict resulting. Montserratians complained that the U.K. government was not doing enough to assist them (Pattullo 2000). Regarding cultural conflict, some Montserratians settling in London were upset over what their children were taught at school, which differed from the morals of their island life (Shotte 2006). For violent conflict, post-apartheid South Africa is an example of locals becoming increasingly abusive towards migrants from outside the country (Crush 2001). In most migration instances, different hosts can display diverse responses, along the entire spectrum from loathing to open arms.

Since different forms of conflict and cooperation are evident for migration cases throughout history, the expectation is that it could be similar for climate change. Could measures be implemented in advance to try to ensure that cooperation is a more likely outcome than conflict; that is, could climate change diplomacy be actively enacted? Issues which need to be resolved for migrants—and which can lead to different forms of conflict and cooperation—include specific locations for resettlement; who pays for moving and resettling; and the laws and norms which would govern the resettled people, for instance territorial and jurisdictional issues (Gerrard and Wannier 2013; Yamamoto and Esteban 2014). The possibilities range from complete assimilation, whereby the migrants adopt the laws and customs of the hosts, to complete sovereignty, in which the migrants run their own independent state on the hosts' land or on reclaimed land.

With this range of options being tabled, the disaster diplomacy question is whether or not all the migration-related decisions would or could proceed without problem. If migration occurs and is attributed principally to climate change, would or could some form of climate change diplomacy manifest so that people can resettle without problem, because compromises will be reached and the migrants will be assisted? Disaster diplomacy work so far demonstrates that the answer is not likely to be positive.

Without extensive evidence for successful disaster diplomacy so far, little reason exists to expect that climate change diplomacy for migration will succeed. Entirely precluding the possibility would be erroneous, for there is a chance that such migration would yield the first successful disaster diplomacy case study so far, especially given that this form of case study has not yet been well-analyzed. Nevertheless, with climate change diplomacy so far not being successful (the IPCC and UNFCCC are examples), relying on its success for migration might be overly optimistic. In fact, the empirical case studies so far, such as the Carteret Islands and Alaska mentioned earlier, demonstrate the high level of contentiousness which these cases evoke, although violent conflict has not yet emerged.

Any thoughts of seeking cooperative agreements regarding migration in advance of the people moving also seem to be misplaced. The reality appears to be similar to the situation experienced in other forced migration case studies involving environmental influences (e.g. Forced Migration Review 2008): significant action is most likely to be taken only once people are forced to migrate. Rather than planning ahead, it appears that anyone forced to migrate due to climate change will suffer the same squalor which so many others have been forced to suffer. That includes temporary shelters and temporary social accommodations, which are usually wholly inadequate, because those with the power and resources to prepare in advance actively choose not to do so.

Again, this situation matches the other disaster diplomacy case studies which have been examined. Few plans are made ahead of time, meaning that cooperation in advance is limited. Instead, ad hoc measures are taken only when forced by circumstances. There seems to be little impetus to learn from the past in order to try to prepare for this migration beforehand, while balancing that approach with the need to avoid expending too many resources in case the migration does not manifest. Consequently, the people affected will suffer in the short-term—and then in the long-term as they aim to rebuild their own lives without adequate external support.

This bleak view is based on past disaster diplomacy analyses with climate change diplomacy thus far demonstrating significant similarities. Optimism should nonetheless not be eschewed, because further work might reveal substantive differences in the migration and potential conflict forms emerging from climate change. These investigations require further research, especially to better integrate conflict and cooperation theories with multi-track diplomacy and disaster diplomacy analyses. The hope always exists that in any particular case, whether general such as the climate-migration nexus or specific such as a named community migrating to a named location due to a specific climate change impact, the situation will be different. That difference could be that genuine climate change diplomacy emerges, which provides a clear-cut case of disaster diplomacy.

4 Implications for Sustainable Development

This chapter has reviewed the disaster diplomacy case study of potential migration linked to climate change. Climate change is not in itself a new influencer on migration choices, although the human input to contemporary climate change's causes is new whilst the combination of rapidity and planetary-scale of the changes might be unique in human experience. Nevertheless, contemporary climate change in itself does not engender entirely new, unprecedented migration situations, and in fact, has the possibility for encouraging some populations not to migrate while supporting others in migrating. Meanwhile, many responses to climate change might deliberately or inadvertently encourage migration or non-migration, leading again to the question of causality: were migration linked to climate change, does it lead to conflict or cooperation; is the conflict or cooperation caused by climate change or by the human decisions related to responding or not responding to climate change? Disaster diplomacy analyses suggest that it is the latter because choices exist in the implementation of responses to hazards or hazard drivers such as climate change, while simultaneously with choices on pursuing diplomacy or otherwise.

The reflections tend to move towards the conclusion that this case study will not produce much difference from other disaster diplomacy case studies, irrespective of any differences in conflict types and migration types. These differences should never be ignored, yet a balance is needed because they should not be overplayed. Migration, in different combinations of forced and voluntary, is not and never has been confined to contemporary climate change, despite much populism over climate change causing migration. Even climate change-related migration has precedents.

Nunn et al. (2007) summarize potential migration from Pacific island communities due to climate-related sea-level fluctuations in the fourteenth century; however, Nunn et al.'s (2007) analysis shows a clear correlation but not necessarily indisputable causation. Precedents that are not climate-related include volcanoes. Lewis (1979) describes the 1946 eruption of and subsequent evacuation from Niua Fo'ou, Tonga. Tobin and Whiteford (2006) detail evacuations around the eruption of Tungurahua, Ecuador. Additionally, a long history exists of communities migrating or being moved, followed by them rebuilding or integrating, for so many different reasons. Residents of St. Kilda in the U.K. were forced to abandon their community and settle on the mainland, with claims that it was important for economic viability (Steel 2011). Along the Canada-U.S. border, the creation of the St. Lawrence Seaway forced several villages to move in 1958 so that the land could be flooded (Cox et al. 1999). These experiences from history can be used to plan for potential future migration in advance of climate change-related decisions.

Plenty is known about the climate change and migration nexus, across many disciplines such as law, political science, geography, migration studies, and development studies. Dozens of books, journal papers, and special issues have been published in the last several years, notwithstanding decades of prior work going to back to El-Hinnawi (1985) and earlier. No claim is made that knowledge usually leads to action or that policy and practice are necessarily based on science or on science alone, although there are ways in which the links could be improved (Jasanoff 2004). Instead, it is yet another statement in a long litany of publications on this topic to use science as one input into thinking ahead of crisis, which would support sustainable development interests better than waiting for a crisis to manifest.

Of particular importance is that climate change discussions do not always fully highlight the truism that migration has always been, and continues to be, a life strategy for humanity, supporting rather than inhibiting sustainable development. Migration leads to the rich exchange, cooperation, and advancement for humanity. Simultaneously, history and the world today are rife with examples of people being forced to move against their will, which does not support sustainable development.

Overall, people move or stay, and have always moved or stayed, for numerous environmental and social reasons, including responding to short-term and long-term environmental changes, responding to short-term and long-term social changes, education, joining family, adventure, and seeking different livelihoods (Foresight 2011; Hugo 1996; Petersen 1958). Such migration and non-migration is sometimes entirely voluntary, is sometimes entirely forced, and is most frequently somewhere along the voluntary-forced continuum, with conflict, cooperation, and neutrality being present to different degrees.

This reality does not justify the forcing of people to move against their will due to reasons such as climate change. The key is having the choice to migrate or not and having the resources to successfully carry out that choice so that sustainable development is pursued, which includes dealing with any present conflicts and preventing others into the future. In a sense, rather than climate change diplomacy, the suggestion here is for wider contexts to seek "migration diplomacy" so that neither the migrants, the hosts, nor those staying behind experience conflict due to the migration, but instead support the decisions which people make and gain positive outcomes from it.

This hope sounds almost utopic, but having the ability and resources to select one's location and livelihood, without causing problems for oneself or others, is part of the wider definitions of "sustainable development" and "diplomacy" processes. That does not obviate practical reasons for limiting or controlling migration, such as citizenship requirements and a location's resource limitations. Those with authority, power, and resources have the ability to analyze the situation and make choices for sustainable development which avoid conflict and support cooperation for climate change-related migration; that is, the active implementation of disaster diplomacy before a crisis manifests, with methods described in Kelman (2012). Such methods, however, so far have rarely been enacted, indicating that advance planning is unlikely to be witnessed because disaster diplomacy is not especially desired by those with decision making power. Disaster diplomacy therefore has the possibility for emerging as climate change diplomacy or migration diplomacy, with the two topics interacting to support sustainable development, but this appears unlikely.

References

Anderson MB (1999) Do no harm: how aid can support peace—or war. Lynne Rienner Publishers, London

Beardsley K, McQuinn B (2009) Rebel groups as predatory organizations: the political effects of the 2004 Tsunami in Indonesia and Sri Lanka. J Conflict Res 53(4):624–645

Beck S (2012) Between tribalism and trust: the IPCC under the "Public Microscope". Nat Cult 7 (2):151–173

Bettini G (2013) Climate Barbarians at the gate? A critique of apocalyptic narratives on 'climate refugees'. Geoforum 45:63–72

Binns D (1977) Beyond the sociology of conflict. St. Martin's, New York, NY

Bronen R, Chapin III FS (2013) Adaptive governance and institutional strategies for climate-induced community relocations in Alaska. Proc Natl Acad Sci U S A 110(23):9320–9325

Bull H (1977) The anarchical society: a study of order in world politics. Columbia University Press, New York, NY

Connell J (1997) Papua New Guinea: the struggle for development. Routledge, London

Cox HM, DeMelle BG, Harris GR, Lee CP, Montondo LK (1999) Drowning voices and drowning shoreline: a riverside view of the social and ecological impacts of the St. Lawrence Seaway and Power Project. Rural Hist 10:235–257

Crush J (2001) The dark side of democracy: migration, xenophobia and human rights in South Africa. Int Migr 38(6):103–133

Davidson WD, Montville JV (1981) Foreign policy according to Freud. Foreign Policy 1981–82:145–157

Desta Y (2006) Designing anti-corruption strategies for developing countries: a country study of Eritrea. J Dev Soc 22(4):421–449

Diamond L, McDonald J (1993) Multi-track diplomacy: a systems approach to peace. Institute for Multi-Track Diplomacy, Washington, DC

Dinar A, Blankespoor B, Dinar S, Kurukulasuriya P (2010) Does precipitation and runoff variability affect treaty cooperation between states sharing international bilateral rivers? Ecol Econ 69(12):2568–2581

El-Hinnawi E (1985) Environmental refugees. United Nations Environment Programme, Nairobi

Enia J (2008) Peace in its wake? The 2004 Tsunami and internal conflict in Indonesia and Sri Lanka. J Public Int Affairs 19:7–27

Ferris EG (2011) The politics of protection: the limits of humanitarian action. Brookings Institute, Washington, DC

Refugee Studies Centre (2008) Climate change and displacement. Force Migr Rev 31 (October):1–80

Foresight (2011) Migration and global environmental change. Report for the Government Office for Science, London

Fox F (2001) New humanitarianism: does it provide a moral banner for the 21st century? Disasters 25(4):275–289

Gaillard JC, Clavé E, Kelman I (2008) Wave of peace? Tsunami disaster diplomacy in Aceh, Indonesia. Geoforum 39(1):511–526

Gaillard JC, Kelman I, Orillos MF (2009) US-Philippines military relations after the Mt Pinatubo Eruption in 1991: a disaster diplomacy perspective. Eur J East Asian Stud 8(2):301–330

GAO (2003) Alaska native villages: most are affected by flooding and erosion but few qualify for federal assistance. GAO (United States General Accounting Office), Washington, DC

Gerrard MB, Wannier GE (eds) (2013) Threatened island nations: legal implications of rising seas and a changing climate. Cambridge University Press, Cambridge

Glantz MH (2000) Climate-related disaster diplomacy: a US–Cuban case study. Cam Rev Int Affairs XIV(1):233–253

Government of Norway (2009) St.meld. nr. 25: Lokal vekstkraft og framtidstru. Om distrikts- og regionalpolitikken. Government of Norway, Oslo

Halvorssen AM (2007–2008) UNFCCC, the Kyoto protocol, and the WTO - brewing conflicts or are they mutually supportive. Den J Int Law Policy 36:369–379
Hannigan J (2012) Disasters without borders. Polity Press, Cambridge
Hartmann B (2010) Rethinking climate refugees and climate conflict: rhetoric, reality and the politics of policy discourse. J Int Dev 22(2):233–246
Helm D (2012) Climate policy: the Kyoto approach has failed. Nature 491:663–665
Hugo G (1996) Environmental concerns and international migration. Int Mig Rev 30(1):105–131
IPCC (2013–2014) IPCC fifth assessment report. Geneva: IPCC (Intergovernmental Panel on Climate Change)
Jasanoff S (ed) (2004) States of knowledge: the co-production of science and the social order. Routledge, London
Kelman I (2012) Disaster diplomacy: how disasters affect peace and conflict. Routledge, Abingdon
Klimesova M (2011) Using carrots to bring peace? Negotiation and third party involvement. PhD Dissertation, Charles University in Prague, Prague, Czech Republic
Le Billon P, Waizenegger A (2007) Peace in the wake of disaster? Secessionist conflicts and the 2004 Indian Ocean Tsunami. Trans Inst Br Geog 32(3):411–427
Lewis J (1979) Volcano in Tonga. J Admin Overseas XVIII(2):116–121
McNamara KE, Gibson C (2009) "We do not want to leave our land": pacific ambassadors at the United Nations resist the category of "climate refugees". Geoforum 40:475–483
Mostafaeipour A, Mostafaeipour N (2009) Renewable energy issues and electricity production in Middle East compared with Iran. Renew Sust Energ Rev 13(6-7):1641–1645
Nicholson CTM (2014) Climate change and the politics of causal reasoning: the case of climate change and migration. Geog J 180(2):151–160
Nunn PD, Hunter-Anderson R, Carson MT, Thomas F, Ulm S, Rowland MJ (2007) Times of plenty, times of less: last-millennium societal disruption in the Pacific Basin. Hum Ecol 35 (4):385–401
Pattullo P (2000) Fire from the mountain: the tragedy of montserrat and the betrayal of its people. Constable and Robinson, London
Pelling M, Dill K (2010) Disaster politics: tipping points for change in the adaptation of sociopolitical regimes. Prog Hum Geog 34(1):21–37
Petersen W (1958) A general typology of migration. Am Soc Rev 2(3):256–266
Segalla SD (2012) The 1959 Moroccan oil poisoning and US Cold War disaster diplomacy. J North Afr Stud 17(2):315–336
Shotte G (2006) Identity, ethnicity and school experiences: relocated montserratian students in British schools. Refuge 23(1):23–39
Steel T (2011) The life and death of St. Kilda. The moving story of a vanished island community. HarperCollins, London
Tobin GA, Whiteford LM (2006) Community resilience and volcano hazard: the eruption of Tungurahua and evacuation of the Faldas in Ecuador. Disasters 26(1):28–48
UNISDR (2015) Sendai framework for disaster risk reduction. UNISDR (United Nations International Strategy for Disaster Reduction), Geneva
Warnaar M (2005) Shaken, not stirred: Iranian foreign policy and domestic disaster. Master's thesis, Universiteit van Amsterdam
Warnaar M (2013) Iranian foreign policy during Ahmadinejad ideology and actions. Palgrave Macmillan, New York, NY
Wickremesinghe R (2006) Peace process in Sri Lanka. South Asian Surv 13(1):5–15
Yamamoto L, Esteban M (2014) Atoll island states and international law: climate change displacement and sovereignty. Springer, Berlin

Chapter 13
Local Knowledge for Addressing Climate Change Risks at Local Level: A Case Study from Nepal

Sanjaya Devkota and Ajay Chandra Lal

1 Introduction

One of the underlying causes of global warming is an excessive emission of greenhouse gasses, primarily due to fossil fuel combustion and industrialization (Halsnaes and Sara 2009; ICIMOD 2008; IPCC 1996; Practical Action Nepal 2009; Upreti 2007). However, deforestation changes in land-use and urbanization also play a crucial role, resulting in variations in solar energy, temperature, and precipitation (Dore 2005; FAO 2014; ICIMOD 2008; IPCC 2007). Global warming effects are likely to amplify climate change-induced disasters which in turn give rise to an increase in the potential threat to lives and properties of people around the world as a whole. This is particularly felt however, in developing countries like Nepal, where local populations have extremely low immediate adaptation capacities (Huq et al. 2004; ICIMOD 2008; Mirza and Monirul 2003; OECD 2006). This is also because developing country economies generally rely on climate-sensitive resources (for example, forests, water resources, agriculture, and livestock) and are thus less able to cope with the impacts of climate change (Huq et al. 2004; Mirza and Monirul 2003; OECD 2006). Among the developing countries, Nepal is situated in the heart of a climate-sensitive zone, the Himalayas, where small increments in temperature change are augmenting adverse impacts (Mirza and Monirul 2003). Furthermore, Nepal is comprised of fragile ecosystems, complex topography, and a geological context that is very sensitive to slight changes in

S. Devkota (✉)
FEED (P) Ltd., Jhamsikhel, Lalitpur, Nepal

Institute of Engineering, Tribhuvan University, Pulchok Campus, Kathmandu, Nepal
e-mail: feed@wlink.com.np

A.C. Lal
Institute of Engineering, Tribhuvan University, Pulchok Campus, Kathmandu, Nepal

© Springer International Publishing Switzerland 2017

K. Sudmeier-Rieux et al. (eds.), *Identifying Emerging Issues in Disaster Risk Reduction, Migration, Climate Change and Sustainable Development*,
DOI 10.1007/978-3-319-33880-4_13

climate variability, thus placing Nepal in the fourth most vulnerable position in regards to climate change (Maplecroft 2011).

Although the history of recorded temperature and precipitation in Nepal is not exhaustive, analysis of available data has indicated consistent and continuous warming after the mid-1970s (Shrestha et al. 1999). Shrestha et al. (1999) also found that the national average temperature rise was 0.06 °C per year, while it was 0.04 °C and 0.08 °C per year respectively in the Terai and Himalayas. The warming rates follow the elevation gradient and are more pronounced in the Himalayan region (Agrawal and Berg 2002; Baidya et al. 2008; Bhusal 2014; Shrestha et al. 1999). This finding was further reinforced by the study carried on the north side of the Himalayas (that is, the Tibetan Plateau) by Liu and Chen (2000). Similarly, more erratic patterns of precipitation have been observed in all ecological regions of Nepal (Baidya et al. 2008; Baidya and Karmacharya 2007).

Several studies have shown that climate-induced disasters such as floods, landslides, intense rain, hailstorms, droughts, cold and heat waves, and vector-borne diseases have become more frequent (Colom and Pradhan 2013; IFAD 2013; Mirza and Monirul 2003). Of the country's 75 districts, 49 are prone to floods and/or landslides, 23 to wildfires, and one to wind storms that lead to impacts on livelihoods. This includes a low level of crop production, the retreating of drinking water sources, and low energy production from micro-hydropower plants (NAPA 2010).

To address the increasing impacts of climate change, the Government of Nepal has developed a policy known as "National Climate Change Policy-2011" (MoSTE 2011). The policy recognizes that changes in temperature and rainfall might cause hazard impacts such as landslides and flooding, and in some cases, Glacial Lake Outburst Floods (GLOFs). In line with the National Climate Change Policy (2011) the Government of Nepal prepared a National Adaptation Plan of Action (NAPA) in 2010 in order to formulate a national strategy and framework to deal with climate change issues. Additionally, NAPA (2010) has identified that climate-induced disasters such as floods, landslides, and droughts have killed more than 4,000 people over the last 10 years. Besides causing the loss of thousands of lives and associated economic losses, climate change has caused rural–urban migration along with losses of property and agricultural production totaling approximately US$5.34 billion (NAPA 2010).

Although we now better understand the multiple types of impacts that climate change has on livelihoods, there are still many underlying uncertainties that limit climate change adaptation (CCA) and disaster risk reduction (DRR) practices and policies (Prabhakar et al. 2009; Schipper and Pelling 2006). For example, hydro-climatic data that supports how the climate has been changing over a period at local levels is poorly understood in countries like Nepal. The Glacier Trust (GT), an organization based in Nepal with the objective of monitoring glaciers and glacial lakes, has mentioned that the official Mean Annual Temperature (MAT) data are not sufficiently accurate as weather stations are not as densely distributed as they ought to be (GT 2015). Such limitations have led us to explore people's local knowledge and observations to better understand climate change impacts at the local level and in order to formulate suitable adaptation and development plans.

Simultaneously, understanding climate change as observed by local people can help to better allocate scarce resources for mainstreaming climate change risks in development plans and projects.

Local knowledge and practices are increasingly recognized and used as valuable resources for planning climate change adaptation (IPCC 2007; UNFCCC 2013). Since adaptation is essentially a local response, locally developed mechanisms and practices are better options than external approaches to immediate climate risk response such as water shortages, crop failures, flood disasters, landslides, soil erosion, and infrastructure damages (MoSTE 2015; Nakashima et al. 2012). In Nepal, vulnerable communities use indigenous practices to plan adaptation and DRR activities at the local level such as crop diversification, conservation of springs and water sources, flood protection, and the building of infrastructure (trail and trail bridges) (MoSTE 2015). A combined approach drawn from modern science and traditional/local knowledge can therefore provide practical and cost effective options for better preparedness leading to a more resilient local community (Mukhopadhyay 2009; WB 1998).

This paper was a part of the assignment given to the authors by the National Academy of Science and Technology (NAST) of Nepal to prepare a Climate Risk Management (CRM) Toolkit to address climate change risks at the local level through the mainstreaming of development planning and projects. The study was carried out in three districts (Dolakha, Sindhuli, and Mohattari) covering all three major ecological regions of Nepal (Mountain, Hill, and Terai) in order to explore local knowledge and observations about how climate change risks have been mainstreamed in development plan and projects.

2 Overview of the Case Study Districts

Nepal's ~200 km wide north–south strip of land is divided into three ecological regions namely the Mountain, Hill, and Terai regions (CBS 2012). The three districts, Dolakha, Sindhuli, and Mahottari, were selected for the study to represent the entire ecological region in central Nepal (Fig. 13.1).

Dolakha District is located in the mountain to hilly ecological regions. The lowest elevation in the district is ~723 masl whereas Mt. Gauri Shankar is the highest peak at ~7,134 masl. Being a mountainous district, Dolakha is highly vulnerable to landslides as well as Glacier Lack Outburst Floods (GLOF) and other climatic phenomena such as lightning and cold waves (DesInventar 2012; NAPA 2010). The District Disaster Preparedness and Response Plan (DDPRP 2010) identifies that 22 out of 51 Village Development Committees (VDCs) (the smallest administrative unit) are at high risk of slope failure.

Sindhuli District is situated in the hill ecological region with two distinct physiographic regions (1) the mid-hill and (2) the lower hill. The mid-hill consists of steep slope mountain peaks while the lower hill is fragile Siwalik. The DDRPP (2010) of Sindhuli states that the district is highly susceptible to landslide and flood

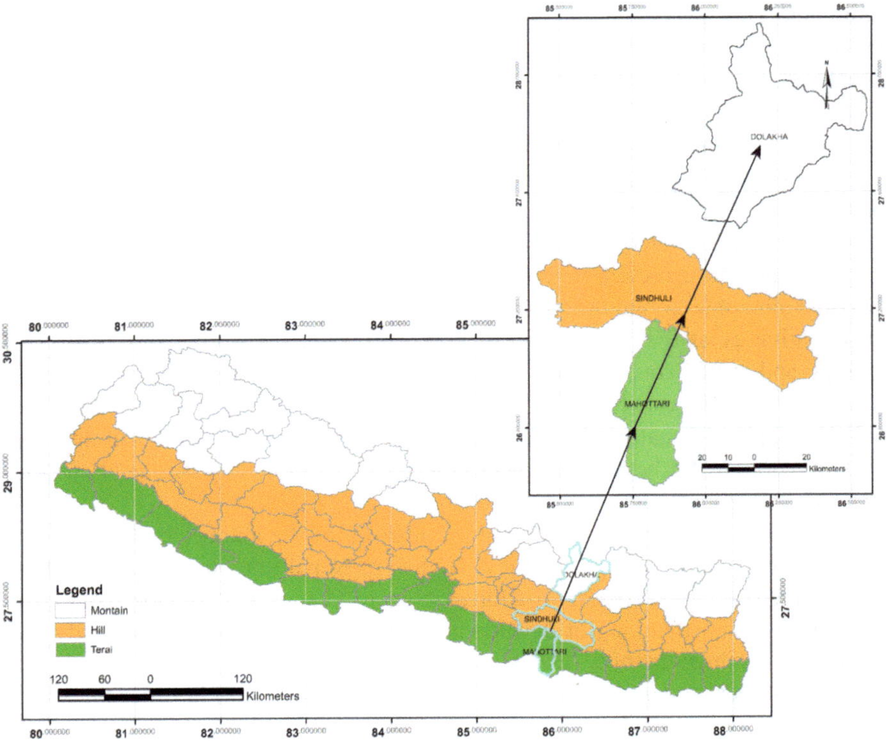

Fig. 13.1 Ecological map of Nepal showing study area

hazards. Flooding is the major challenge faced in the river valley, followed by landslides in the upper reaches of the hills. 14 out of 53 VDCs in this district are found to be at high risk of flooding, followed by 10 VDCs in a medium risk zone for different disasters (DDRPP 2010).

Mahottari district lies in the Terai region, with an altitude varying from ~70 m.a.s.l. in south and ~808 m.a.s.l. in the northern Siwalik region. Flooding is the major challenge in the Mahotari district and has caused significant impact to local livelihoods and property (NAPA 2010). The DDRPP identified that 34 out of 76 VDCs of the district are classified as being at high risk of flooding. In addition, floods, cold waves, and fires are some other extreme events observed in the district (DesInventar 2012).

3 Methods

Descriptive and exploratory methods were used for the study such as an extensive literature review of pertinent reports, publications, papers, and texts. The literature review was followed by consultation workshops at the NAST headquarters in

Kathmandu with government departments, national and international non-governmental organizations (e.g. Department of Forestry, Department of Environment, Ministry of Home Affairs, UNDP, DFID, WWF, IUCN, etc.) involved in the study of climate change and its implications. Consultation workshops finalized the field study approach and developed the most appropriate methodology to be adopted. Similarly, four key themes were identified to further streamline the study: (1) agriculture and food security, (2) forest and bio-diversity, (3) water resource and energy and (4) climate-induced disasters.

Long term climate variables, mainly precipitation and temperature of the districts, were collected from the Department of Hydrology and Meteorology (DHM 2012) and analyzed to see how such variables have changed over the last few decades. Maximum daily temperature data was analyzed to detect trends such as those discussed by the World Meteorological Organization (WMO 2009). Similarly, daily precipitation data were classified into two categories (a) monsoon seasonal rain and (b) other seasonal rain patterns according to monsoon season wet days. These two data sets were later analyzed to obtain a general picture of monsoonal rainfall intensity and variation in monsoonal rainy days, as 80 % of the annual rain occurs over a 4-month period (June–Sept) (e.g. Shrestha 2000). Understanding the monsoon season is important as it is during this time that most slope failure and flooding occurs (e.g. Gerrard and Gardner 2000).

The field investigation was based on consultative meetings, focus group discussions (FGDs), key informant surveys and site observations. The consultative meetings were held at district headquarters with government officials and other concerned stakeholders (e.g. NGOs professionals, social workers, and media persons) responsible for planning and executing associated development activities. The consultation workshop was organized to explore whether the issue of climate change and its associated risks were considered in formulating development plans and programs in order to mainstream impacts of climate change in four key themes mentioned earlier. It also aimed to identify the communities to be visited within the district for further consultation to understand the role of peoples' local knowledge and observations about climate change at the local level.

The consultative meeting was followed by FGDs, key informant interviews (mostly local senior citizens) and expert observations. Even though at first 1 FGD was to be organized per selected community in each district, there was a high demand by communities to participate (Dolakha = 65, Sindhuli = 45 and Mahottari = 35). Depending on the numbers, participants were divided into several groups in order to maintain 10–12 persons per group for effective discussion (Krueger 2002). It was attempted to create homogeneity in the participation of community members including women in each FGD considering ethnicity, economic activity, age, and education. Altogether, 13 FGDs were conducted (Dolakha = 6, Sindhuli = 4 and Mahottari = 3), where women's participation in Sindhuli was 12, followed by 8 in Dolokha and 4 in Mahottari.

As the study was focused on exploring local people's knowledge and observations about changing weather patterns and impacts (Devkota 2014; Dhakal et al. 2013), the FGDs were primarily focused on obtaining peoples' experiences and understandings about climate change impacts on their local livelihoods and surroundings. During the consultations and interviews, the phrase "climate change" was not used unless the informants themselves brought it up. In such cases, if informants themselves did not mention climate change by the end of the interview, they were asked if they had heard about it.

Finally, the data and information collected from workshops, field surveys (based on FGDs and interviews) and the literature review were analyzed. Interviews were conducted with local development planners and program implementing agencies (e.g., government officials and social workers) to understand what actions were being taken to mainstream climate change issues to mitigate climate risks.

4 Results

4.1 Climate Variables

Rainfall in Nepal varies greatly from place to place due to the sharp topographical variations as mentioned in Nayava (1980) and Baidya et al. (2008). As the rain bearing winds approach Nepal in the summer monsoon (June-September) from the southeast, heavier rainfall falls in the foothills of Siwalik range, increasing with the altitude in the windward side and sharply decreasing in the leeward (Nayava 1980). To understand the trend of monsoonal rainfall intensity, the number of wet days and annual maximum temperature recorded at three weather stations is provided in the table below (Table 13.1).

While undertaking this analysis, temperature recordings of the Sindhuli District in Sindhuli Gadi (Index No 1107) were found to be highly irregular and thus discarded from the analysis. The maximum daily temperature data of Jiri (1982–2012) and Janakpur (1969–2009) stations were analyzed with a simple regression technique. The fitted trend line showed weak R^2 (0.15), which is not statistically significant however it can be seen that the trend is upward in direction (Fig. 13.2) as this was also the case found by McSweeney et al. (2010).

Similarly, analyses of monsoonal rainfall intensity and monsoon season wet days also reveal that there is no strong increasing trend. However, the visible interpretation of plots in the figure (Fig. 13.3) indicate that monsoonal rain intensity

Table 13.1 Descriptions of the DHM stations included in this study

Station No	Station name	Elevation (masl)	Latitude (N)	Longitude (E)	Ecological region
1103	Jiri	2,003	27° 38′	86° 14′	Mountain
1107	Sindhuli Gadi	1,463	27° 17′	85° 58′	Hill
1111	Janakpur	90	26° 43′	85° 58′	Terai

Source: DHM 2012

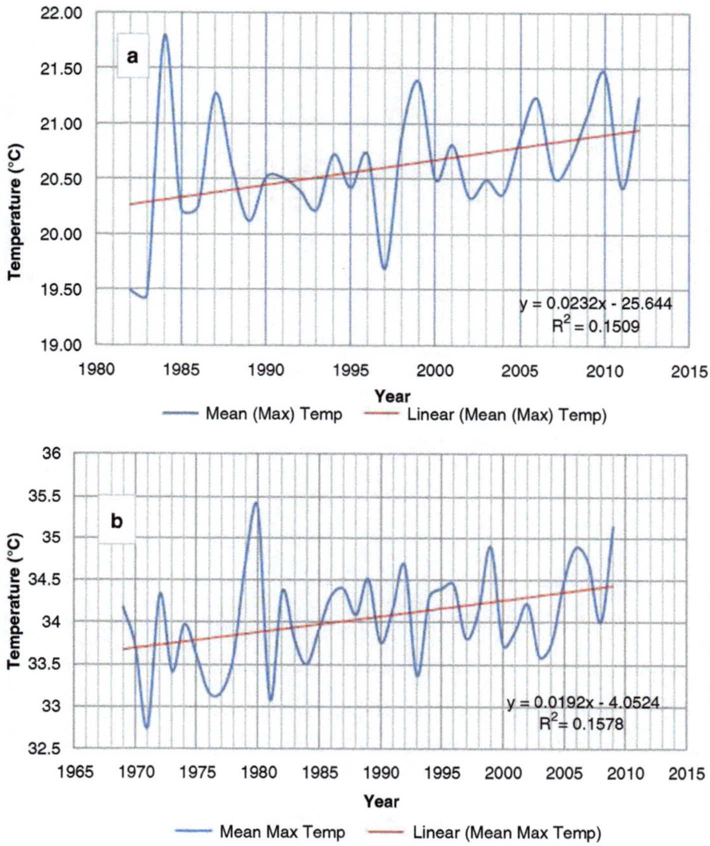

Fig. 13.2 Near surface maximum air temperature trends—(**a**) Jiri and (**b**) Janakpur

is increasing over the Terai ecological region with a slight decrease in the number of monsoonal wet days. Among the three ecological regions, the trend is relatively high in the Terai and it is much weaker as we move north (that is, in the hill and mountain ecological regions).

Shrestha (2000) and Nayava (1980) also found that the monsoonal rainfall intensity in higher altitudes of Nepal is low in comparison to the lower-lying lands as in the Terai.

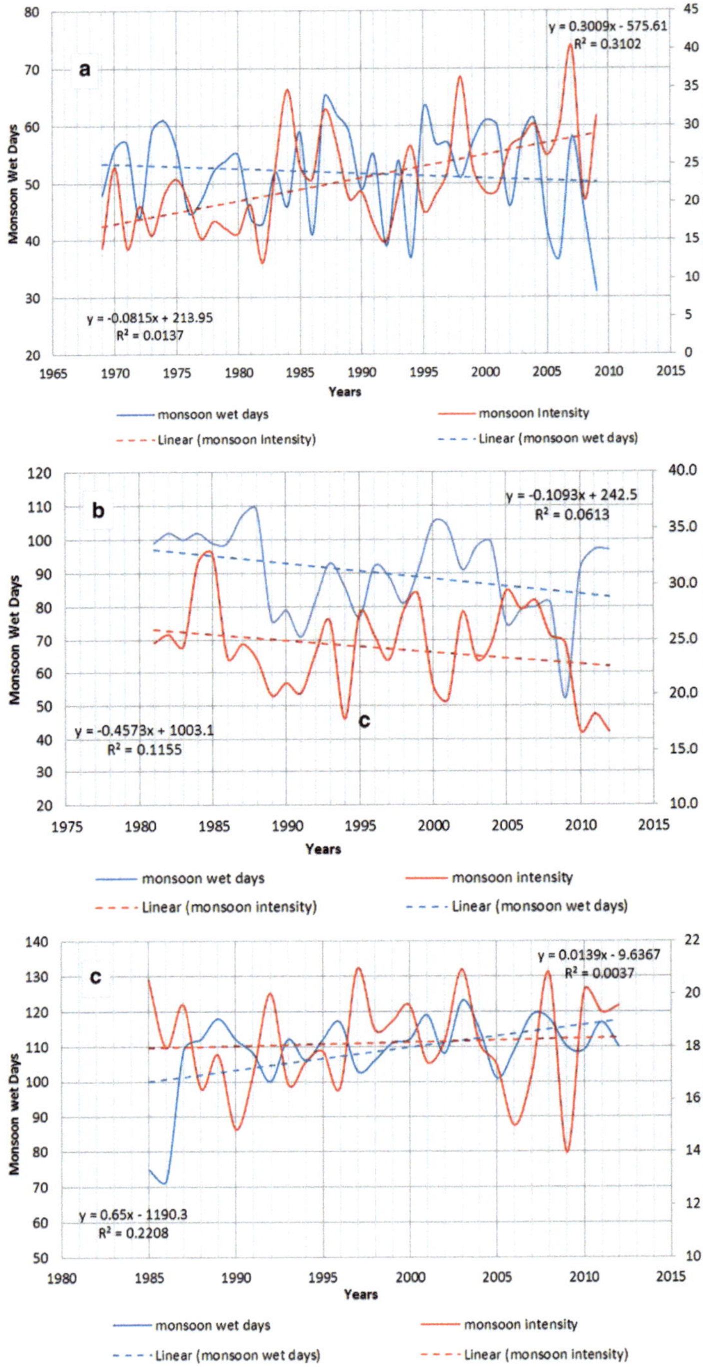

Fig. 13.3 Monsoonal precipitation trends—(**a**) Janakpur, (**b**) Sindhuli, and (**c**) Jiri

4.2 *Peoples' Knowledge and Observations About Climate Change at the District/Local Level*

4.2.1 Local Community People

For a farmer in the upper middle hills of Dolakha districts, climate change means less snow in the winter seasons, which often means poor quality potato production and the drying up of small springs. During the FGDs, people mentioned that some of the water sources (springs) dried up in the winter and resulted in a shortage of drinking water in mountain settlements. Some farmers claimed that vector-borne diseases, which used to be found primarily in the Terai region, are now also observed in mountainous areas.

In Sindhuli, participants observed more erratic, high-intensity, and unpredictable monsoons along with decreased rainfall, shorter winters and longer summers, presumably all associated with climate change. Senior people (>60 years of age) in Sindhuli have noticed that the intensity of monsoonal rain has increased in the last 10 years, leading to frequent flash flooding and sedimentation, which often damages existing cultivated land. People in the hill regions have also noticed early flowering of some of the plant species and the presence of new and invasive plant species, which have never been seen before.

For the people in the Mahottari district in the Terai region, climate change means winter cold waves, excessively hotter, longer, and drier summers (extended drought), and erratic and unpredictable, short but intense monsoonal rainfall that leads to frequent devastating floods. Consequently, the Mahottari District was identified as one of the most flood-prone districts in Nepal (NAPA 2010). During the consultations and FGDs, local people reported that the frequency of flooding has increased in the past 10 years. Delayed monsoon and frequent flood events have had several impacts (1) farmers cannot plant rice on time; (2) flood events damage the rice fields leading to severe food insecurity; and (3) delayed monsoons reduce the opportunity for those who usually work for wages, forcing workers to migrate to Indian border cities in search of employment. Table 13.2 summarizes the findings from the FGDs for each of the key themes mentioned above:

Key informant interviews were 50 % male and 50 % female (Dolakha = 12, Sindhuli = 8 and Mahottari = 6) and designed to explore observations about how climate conditions had changed over the past 10 years. The informants were asked open-ended questions without mentioning the term "climate change." Table 13.3 presents the results of the interviews and summarizes four main themes that describe changes in climate variables:

The data reveals that a majority of the population consulted during the field study had observed unusual weather patterns (e.g. short and warm winters, hot and longer dry summers, short but intense rainfall).

Table 13.2 Local knowledge and observations of CC impacts per themes

Sector	Districts		
	Dolakha	Sindhuli	Mahottari
Agriculture and food security	Low- and poor-quality production of potatoes and cereals	Low productivity of crops	Low production of rice and other cereals
	Increased insects	Increased insects and invasive plant species	Increased insects and invasive plant species
	Crop production is not enough	Crop production is just enough but mostly depends on monsoon	Crop production is mostly depends on monsoon
Forest and bio-diversity	Frequent forest fire	Increased forest fire	Sissoo (*Dalbergia latifolia*) are dying
	Early shedding of fodder tree leaves	Early flowering of fruits (e.g. mango)	Difficult in getting fuel wood
	Getting fuel wood is difficult	Extinction of black crow	Less wild animals
			Increased forest fire
Water resource and energy	Small springs are drying	Increased frequency of flash floods and sedimentation in river valley	Unexpected flood breaches the river bank
	Flash floods and sedimentation	No or less water availability for winter crops	Damaged or contaminated ground water sources
	Less production of energy from Micro-hydro Power Plant		Streams originating in Siwalik are rapidly drying as monsoon ends
Climate-induced disasters	Landslides frequently block rural roads	Frequently blocked rural road and damaged infrastructure (e.g. trail bridges, electric poles)	Frequent flash floods, river banks breached, the inundation of property and houses
	Landslide damaged agricultural land and water sources	Excessive soil erosion and soil losses from agricultural land	Sedimentation over the rice fields
			Damaged bridges, culverts, and electricity tower/pole

Table 13.3 Local people's observation on climate variables

S. N.	Description	Districts		
		Dolakha (%)	Sindhuli (%)	Mahottari (%)
1	Shorter and warmer winters	72	85	77
2	Hot and longer dry summers	66	76	83
3	Decreasing monsoonal rainy days	73	68	81
4	Increasing monsoonal intensity of rain	85	78	69

Source: Field Survey in April and August 2012: compiled by the Authors

4.2.2 District-Level Development Workers

Each of the district-level consultation workshops were attended by local government officials (e.g. Chief District Officer, Local Development Officer, District Engineer and Planner, District Agriculture and Livestock Officer, District Forest Officer and others), professionals from local NGOs, social workers, and the media. In general, this group was familiar with the term "climate change" and its impacts. Local populations, on the other hand, have noticed the changes in monsoonal rainfall levels and temperature but do not really know about the term climate change. What they do know however, is that something is going wrong within their surroundings and environment. Development workers have observed delays in monsoonal rainfall patterns in the Mahottari with much less or no snow in the upper-middle hills of the Dolakha District as observed by the local population. Rainfall was observed to be very erratic, unpredictable, and increased in frequency. The district-level workshops focused upon the four sectors that were discussed earlier that development planners and concerned stakeholders had identified following assessments of the impacts, planning tools and potential adaptation techniques for mainstreaming climate change climate resilience in local communities (Table 13.4). The four sectors discussed in the consultation workshop are climate-sensitive issues and directly relate to the sustainability of rural livelihoods in several ways.

It is clear that climate change is not only about changes in temperature and precipitation but also about the adverse impacts that have been depicted in Table 13.4. However, the impacts of climate change largely depend on access to education, infrastructure, socioeconomic activities, and households in particular (Wongbusarakum and Loper 2011). Such impacts are more severe in the poor communities as they have weaker adaptation capabilities. The cumulative impacts of climate change result in loss of livelihoods, which can trigger both seasonal and permanent migration (Werz and Conley 2012).

4.3 Climate Change and Migration

Migration has been perceived as a cumulative effect of climate change on individuals and society (Werz and Conley 2012). Although the magnitude and impact of migration has widely been acknowledged by local people and district-level professionals, it has only recently become evident that climate change has in fact contributed to higher overall levels of migration. For several decades, Nepal has experienced seasonal labor migration to neighboring countries of India and in the Middle East. However, rural–urban migration within the nation over the past decade has mainly resulted from the additional push factors caused by disasters and conflict (Upreti and Muller-Boker 2010).

Table 13.4 Planning and adaptation measures for climate change impacts in four climate-sensitive themes

Agriculture and food security	Forest and bio diversity	Water resource and energy	Climate induced disasters
Impacts	Impacts	Impacts	Impacts
New diseases and pest insects	Forest Fire, land-slide, and Flood result in decrease in bio diversity and genetic loss	A depletion of glaciers in the Himalayas	GLOF, Land Slide and Soil Erosion are major climate induced disaster
Very unpredictable rainfall and decreased numbers of rainy days	A decrease in forest coverage due to a high demand for agricultural land as well as timber.	An increase in the size and number of glacier lakes	Flood, Fire, Heat Wave and Cold wave along with and Soil Erosion are major climate induced disaster
Frequent flood and loss of agricultural land and investment	Vultures are not seen commonly which used to be common few decades ago	A depletion of dis-charge in rivers	Loss of soil
Decreased production	Few wild animals like red deer and birds like peacocks are not seen frequently	A depletion of the ground water table in the Terai	Loss of agricultural land and desertification
Food surpluses but heavy pressure on land for increasing population		A decrease in the dis-charge of rivers and streams and its effect on irrigation and drinking water sup-plies and reduction in hydropower generation	Loss of settlement
Desertification in flood plain of Rato and Jangha river in Mahottari, Tamakoshi in Dolakha and Kamala in Sindhuli Districts		An increase in the demand for water and energy	Loss of infrastructure like roads
Fire in settlement and Forest fire and pasture fire due to prolonged summer		A decrease in rainfall duration	Loss of forest and biodiversity
Loss of local crop variety		An increase in the demand of water for irrigation, drinking, and industrial uses	Loss of houses
Banmara (*Ageratina adenophora*) and other parasitic plants have been emerging		Dry ponds, tube-wells, and other source of water	

(continued)

Table 13.4 (continued)

Agriculture and food security	Forest and bio diversity	Water resource and energy	Climate induced disasters
Shift in habitat of Yak (*Bos grunniens and Bos mutus*) in higher altitude			
Planning Tools:	Planning Tools:	Planning Tools:	Planning Tools:
Directives and guidelines from Ministries, DDCs, and VDCs	Nepal Forest Conservation Act	Environmental Protection Act and Regulations	Environmental Protection Act and Regulations
	Demand from the local community	Water Resource Act	Water Resource Act
	Manual for FUGs	Electricity Development Act	Directives from the department and ministries
	Directives from Ministries and departments	EIA, IEE	IEE, EIA, SWOT Analysis
		Directives from Ministries and NPC	District Disaster Preparedness Plan
Adaptation	Adaptation	Adaptation	Adaptation
New varieties of crops	Increase in private orchards	Forestation at different location	Bio-engineering
Only two crops	Plantation of more tree than falling	Forest User Groups for forest conservation	Gabion and Stone wall side protection
Maize as a main crops where irrigation is not possible	Increasing use of gas and alternative energy for cooking	Soil Conservation Programmes	Soil Conservation programme
Increase in wheat farming	Development of alternative Construction Material	Encouragement for alternative and clean energy e.g. solar, micro-hydropower, bio-gas, improved cook stoves, etc.	Bank protection and Dyke construction
Construction of Dykes			
Vegetable farming			
Grazing (grass farming)			
Change in Food habit			
Conservation of water sources,			

EIA Environmental Impact Assessment, *IEE* Initial Environmental Examination, *SWOT* Strengths, Weaknesses, Opportunities, and Threats

Local people in Dolakha have reported that some of the landslides that occurred in the mountainous areas have forced people to migrate to district headquarters and other nearby towns (e.g. Banepa, Panuti, etc.) in 2006 and 2009. Similarly, it is widely perceived that the low levels of potato production in the mountain regions of

Dolakha have significantly reduced the incomes of farming families, therefore forcing youth to migrate abroad in search of employment. The exact migration data have yet to be surveyed, however the large number of vacant houses observed during the field survey reveal the breadth of migration. According to local populations in Sindhuli, out-migration is mainly due to disasters and the decreased productivity of agriculture with climate change as one of the main drivers. The Mahottari district in the southern low land of Terai used to have high rice production but is now facing a dilemma of labor shortages due to high levels of out-migration, largely caused by the annual flooding of the Rato River and the subsequent unpredictable monsoonal rains.

According to NPC (2009), about 2 % of the population in Dolkha and Sindhuli are absent while 2.5 % have out-migrated to the Mahottari District. Similarly, about 3 % of the population of the Dolkha and Mahaottari and 8 % in Sindhuli Districts are in-migrants, mainly as government employees and security personnel deployed in the district (DDC 2010; NPC 2009). Local people have reported that there have been high levels of in-migration to the Sindhuli District mainly from neighboring districts (e.g. Khotang, Okhaldhunga, and Solukhumbu) due to the remoteness and various other push factors such as a lack of access to roads, markets, education, and health. Local people and professionals in both Mahottari and Dolakha districts have observed that out-migration was mainly due to decreased crop production, disasters, and unemployment.

5 Discussion

5.1 Climate Change, Weather Variables, and Impacts

The local population in the study area may not understand the concept and causes of "global warming" as observed by Baul et al. (2013), but their understanding of incremental temperature and precipitation level change is clear and informed, based on their long-term experiences with the realities of their local environment. In parallel, local populations have observed a reduction in monsoon wet days despite their somewhat increased intensity, which is yet higher again in the Terai ecological region. These results are consistent with similar studies in different parts of Nepal (Baul et al. 2013; Colom and Pradhan 2013; Devkota 2014; Synnott 2012). In general, the experiences of local populations are congruent with scientific studies, such as those conducted by Kansakar et al. (2004) and Shrestha (2000), where an estimated rate of temperature increase of 0.41 °C per decade has been predicted, based on meteorological data of Nepal.

Local populations have observed more intense and erratic monsoonal precipitation in the recent years compared with the last 10 years. However, local knowledge about the total annual rainfall is not supported by official statistics as the total monsoonal rainfall has not changed but rather the intensity has increased as measured by the total number of wet days (Baul et al. 2013).

Other studies have also indicated that local people have noticed visible yet slow climate change and impacts. According to Colom and Pradhan (2013: 72) "Nepali's believe that temperatures have risen, rainfall has become less predictable and floods and droughts have increased during the last 10 years. People also feel that the environment has changed and nearly nine in ten say that insects and pests have increased."

As Nepal is an agriculturally based economy, greater climate variability is now seen to undermine the livelihoods of the rural poor. Monsoonal-based agriculture is no longer sufficient for the survival of rural people in all the ecological regions of Nepal as there is significant food insecurity (Synnott 2012). Furthermore, the impacts to other sectors are equally adverse. Biodiversity has suffered degradation due to the extinction of flora and fauna, and new invasive plant species have spread in many areas, leading to instability of the ecosystem. In hilly areas, some small water sources are retreating, and in the low lands of the Terai, flooding is contaminating water sources (for example, groundwater supplies), thus creating a scarcity of drinking water and insufficient energy generation from the micro-hydro power plants found in the area. The increase of forest fire, landslides, and excessive soil erosion are also destroying existing ecosystems and increasing the vulnerability of local communities.

As a response to increasing climate change, the majority of rural people are adjusting as best as they can on their own by adapting and diversifying crops (i.e. breeds of crops, cultivation of crops that require less water, conservation of water sources, ground water recharge, etc.) (Colom and Pradhan 2013; Synnott 2012). Some are however, migrating to other places in search of better opportunities (Colom and Pradhan 2013).

It is now common that adult males migrate in search of better incomes, either in domestic urban centers or abroad. Reasons for this migration were attributed to lower agricultural production and fewer opportunities for finding labor to fulfill the minimum livelihood needs. Migration of male members has added additional burden to the female members living in rural communities as they now have to travel a long way to fetch drinking water, collect firewood for fuel and fodder, and manage agricultural land (Lokshin and Glinskaya 2009; Synnott 2012). The study and subsequent consultation with local populations in the three ecological regions indicate that climate variables are changing at different scales in different regions and these changes amplify the impacts at different levels from ecosystems to livelihoods, ultimately making communities more vulnerable.

5.2 Mainstreaming of CRM in Development Planning

This study reveals that most of the mainstreaming of CRM is not yet practised in development work such as rural road construction and agriculture. One of the reasons behind soil loss and mass movement is due to the unplanned construction of rural roads and traditional agricultural development. This has caused a negative impact on drinking water sources and an increase in the desertification of the river valley as reported by local people. The majority of the population in the study areas still report to follow traditional agricultural practices, install traditional irrigation systems, and maintain livestock farming, which has led to a decrease in crop production and an increase in hardship in rural and remote parts of the country (DDC 2010). At the same time, increased seasonal and temporal migration has led to a lowering of the available workforce for agriculture, and this has meant an increase in barren land. The district-level consultation workshops (held with district-level government officials, planners and development workers, representatives from NGOs, social workers, and media people) highlighted the actions recommended for mainstreaming climate change in key government activities and sectors (Table 13.5).

Two main themes relating to the mainstreaming of climate change in development planning emerge from the proposed activities listed in Table 13.5:

- To reduce the risk posed by climate change by developing project activities with concerned stakeholders that can result in the safeguarding of national and local investments. This is also referred to as "climate-proofing."

Table 13.5 Actions to be considered for mainstreaming climate change in different government activities and sectors

Agriculture and food security	Forest and biodiversity	Water resource and energy	Climate induced disasters
Agriculture and food security policy	Soil conservation and forest management policy and programmes	National water resources and Energy Policy	Separate departments for Water Induced Disasters
New ponds for fishing and agriculture	Managing Forest fire	Water user group	National Climate Policy
A need for awareness and discussion among local people	REDD + Programme	Rotation system for water distribution among upstream and downstream population	Inclusion of disaster risk reduction in district planning
Water harvesting for small scale irrigation	Promotion of Community Forest Users Group	Protection and conservation of springs/water sources	Mobilization of local clubs/ institutions
Collection and conservation of extinct species of crops and vegetation	Reduce open grazing	Protection/conservation of vegetation around water sources	Discouraging of the haphazard construction of rural roads
Use of drip and sprinkle for alternative irrigation		Use and promotion of renewable energy	
Off-season and cash crop farming			

- To ensure that project activities reduce the vulnerability of target populations to climate change through interventions that in no way increase vulnerability as a result of projects.

Both of the objectives can be achieved at the local level by incorporating climate change issues in project management life cycles as discussed by EC (2004), also known as the mainstreaming of climate change risk in development planning.

6 Conclusions

In many developing countries, the process of mainstreaming is in its earliest stages and there is very little accepted doctrine on how the process should occur. According to ODI (2011), mainstreaming has been described in the context of climate change, as a "holistic" or "development-first" approach, whereby adaptation and mitigation objectives are integrated within development agendas. Strategic-level mainstreaming, as defined by CARE (2010), addresses the organizational environment in which policies and programs are planned and implemented. This can include activities such as building staff awareness and capacities, putting appropriate institutions or mechanisms in place and identifying entry points for adaptation action (Olhoff and Schaer 2010). According to our study, although many of the district-level development communities, including government officials are aware about climate change and its impacts, they are yet to be mainstreamed into development plans and projects. The most frequently cited reason behind the delay in mainstreaming CRM is weak governance, limited resources and commitment of the concern stakeholder (OECD 2006). The Sendai Framework of Action for Disaster Risk Reduction (UNISDR 2005) explicitly makes mention to the need for nations to mainstream disaster risk considerations into planning procedures for major infrastructure projects, including the criteria for design, approval, and implementation of such projects shall be based on social, economic, and environmental impact assessments.

There is growing knowledge about how people are adjusting to changing climate conditions, especially in the agricultural sector (Baul et al. 2013; Nakashima et al. 2012) without any intervention from government authorities (FAO 2014). Proper efforts to mainstream the concept of CRM into development work with the consideration and integration of local knowledge will contribute to more sustainable solutions and help make better use of resources leading to more resilient communities.

Although the government of Nepal has emphasized incorporating climate friendly development plans and policies, these have not been implemented thus far. Some of the plans and policies to address climate change impacts are already there but the problem is in the enforcement of these policies and the establishment of efficient monitoring mechanisms. The four major themes discussed in Table 13.5 above are the rural livelihood lifelines that must be incorporated in any development plan or program. However, in a country like Nepal where the enforcement of law and regulation is very weak and ignorance of responsibility in the government system is common, questions around climate change safety do indeed arise.

This study has led to several questions for further research: (1) How can government institutions effectively disseminate their knowledge, plans, and policies at the local level? (2) How can we bridge local experiences and knowledge in mitigating climate change risks? and (3) Is migration an adaptation strategy or is it the result of weak governance?

This research suggests that local populations have ample experience when it comes to knowledge about their land and surroundings. Local experiences therefore need to be incorporated within development plans and policies. People in low-lying lands have different experiences and problems than those living in higher altitudes. Developing countries with poor infrastructure and economy, have significant uncertainty in the prediction of climate change impacts at the local level as data are not readily available to the general public or even, in many cases, to researchers. The best we can do is to understand the potential climate-induced risks at the local level and advocate for development programs that address or reduce such risks by applying and integrating local and expert knowledge.

References

Agrawal S, Berg M (2002) Development and climate change project: concept paper on scope and criteria for case study selection, COM/ENV/EPOC/DCD/DAC (2002)1/final. OECD, Paris

Baidya SK, Shrestha ML, Sheikh MM (2008) Trends in daily climate extremes of temperature and precipitation in Nepal. J Hydrol Meteorol 5:38–51

Baidya SK, Karmacharya J (2007) Climate change and its impacts on Nepalese agriculture. J Agric Environ 9:62–71

Baul TK, Ullah KMA, Tiwari KR, MacDonald MA (2013) Peoples' local knowledge of climate change in Middle-Hills of Nepal. Indian J Tradit Knowl 12(4):585–595

Bhusal JK (2014) Climate change implication for Hydropower development in Nepal Himalayan region. Dir Res J Chem Mater Sci 2(2):27–35

CARE (2010) Climate vulnerability and capacity analysis hand book. CARE International. http://www.careclimatechange.org

CBS (2012) Geographical Setting of Nepal. Central Bureau of Statistic. Accessed on 18 January 2015. http://cbs.gov.np/wp-content/uploads/2012/Year%20Book%202007/frame_CBS/new3.html

Colom A, Pradhan S (2013) How the people of Nepal live with climate change and what communication can do. Whitehead S, Wilson D (eds.). BBC Media Action, Climate Asia. http://www.bbc.co.uk/mediaaction/climateasia

DDC (2010) District Development Committee Profile of Dolakha, Sindhuli and Mohattari Districts, Nepal. Central Bureau of Statistics, Government of Nepal

DDPRP (2010) District Disaster preparedness and Response Plan of Dolakha, Sindhuli and Mohatari, Nepal

Devkota RP (2014) Climate change: trends and people's perception in Nepal. J Environ Prot 5 (4):255–265

DesInventar (2012) National Society of Earthquake Technology, Kathmandu, Nepal. http://www.DesInventar.org/en/database

Dhakal CK, Regmi PP, Dhakal IP, Khanal B, Bhatta UK, Barsila SR, Acharya B (2013) Perception, impact and adaptation to climate change: an analysis of livestock system in Nepal. J Anim Sci Adv 3(9):462–471

DHM (2012) Department of hydrology and meteorology. Government of Nepal, Kathmandu

Dore HI (2005) Climate change and changes in global precipitation patterns: what do we know? Environ Int 31:1167–1181

EC (2004) Project Cycle Management Guideline. European Commission, European Aid Corporate office. http://europa.eu.int/comm/europeaid

FAO (2014) Managing climate risks and adapting to climate change in the agriculture sector in Nepal. Climate, energy and tenure division (NRC), Food and Agriculture Organization of the United Nations (FAO), Rome

Gerrard AJ, Gardner RAM (2000) Relationships between rainfall and landsliding in the Middle Hills, Nepal. Nor Geogr Tidsskr 54:74–81

The Glacier Trust, Online document. Accessed on 3 July 2015 http://theglaciertrust.org/climate-change/

Halsnaes K, Sara T (2009) Development and climate change: a mainstreaming approach for assessing economic, social, and environmental impacts of adaptation measures. Environ Manag 43:765–778. doi:10.1007/s00267-009-9273-0

Huq S, Reid H, Konate M, Rahman A, Sokona Y, Crick F (2004) Mainstreaming adaptation to climate change in Least Developed Countries (LDCs). Clim Pol 4:25–43

ICIMOD (2008) The Case of the Hindu Kush–Himalayas ICIMOD's Position on Climate Change and Mountain Systems. Mountain Research and Development Vol 28 No 3/4 Aug–Nov 2008

IFAD (2013) Nepal Environmental and Climate Change Assessment, Main Report prepared for IFAD's Country Strategic Opportunities Programme 2013–2018. International Fund for Agricultural Development

IPCC (1996) Climate change 1995: the science of climate change. In: Houghton JT, Meira LG, Callander BA, Harris N, Kattenberg A, Kaskell K (eds) Contribution of working group I to the second assessment report of the IPCC on Climate Change. Cambridge University Press, Cambridge

IPCC (2007) Climate change 2007: impacts, adaptation and vulnerability. In: Parry ML, Canziani OF, Palutikof JP, van der Linden PJ, Hanson CE (eds) Contribution of working group II to the fourth assessment report of the intergovernmental panel on climate change, Annex I. Cambridge University Press, Cambridge, 976 pp

Kansakar SR, Hanna DM, Gerrard J, Rees G (2004) Spatial patterns in the precipitation regime of Nepal. Int J Clim 24:1645–1659

Krueger RA (2002) Designing and conducting focus group interviews. University of Minnesota. http://www.eiu.edu/~ihec/Krueger-FocusGroupInterviews.pdf

Lokshin M, Glinskaya E (2009) The effect of male migration on employment patterns of women in Nepal. World Bank Econ Rev 23(3):481–507

Liu X, Chen B (2000) Climate warming in Tibetan Plateau during recent decades. Int J Climatol 20 (14):1729–1742

Maplecroft (2011) Climate change risk atlas. UK. https://maplecroft.com/themes/cc/

McSweeney C, New M, Lizcano G (2010) UNDP climate change country profiles, Nepal. Accessed on 15 July 2015. http://country-profiles.geog.ox.ac.uk/

Mirza M, Monirul Q (2003) Climate change and extreme weather events: can developing countries adapt? Clim Pol 3:233–248. doi:10.3763/CPOL.2003.0330

MoSTE (2011) National framework for local adaptation plan for action. Ministry of Science, Technology and Environment. Government of Nepal

MoSTE (2015) Indigenous and local knowledge and practices for climate resilience in Nepal. Mainstreaming climate change risk management in development. Ministry of Science, Technology and Environment (MoSTE), Kathmandu, 122 pp

Mukhopadhyay D (2009) Cultural values, indigenous knowledge for climate change adaptations in developing countries. Delhi University, Economics Department, NoidaIndia, 156 pp
Nakashima D, Galloway J, McLean K, Thulstrup HD, Ramos Castillo A, Rubis JT (2012) Weathering uncertainty: traditional knowledge for climate change assessment and adaptation, 120. Paris, UNESCO, and Darwin, UNU
NAPA (2010) National Adaptation Programme of Action (NAPA) to Climate Change; Ministry of Science, Technology and Environment, Government of Nepal
Nayava JL (1980) Rainfall in Nepal. The Himalayan Review-Nepal Geological Society 12
NPC (2009) Regional development strategy of Nepal. National Planning Commission publication under the support of Asian Development Bank. TA 4752-NEP. P 163
OECD (2006) Putting climate change adaptation in the development mainstream. Policy Brief. Organization for Economic for Economic Cooperation and Development (OECD), Paris
ODI (Overseas Development Institute) (2011) The 'Mainstreaming' approach to climate change adaptation: insights from Ethiopia's water sector. 12 pp
Olhoff A, Schaer C (2010) Screening tools and guidelines to support the mainstreaming of climate change adaptation in development assistance- a stock tacking report. UNDP, New York, NY http://www.fao.org
Prabhakar SVRK, Srinivasan A, Shaw R (2009) Climate change and local level disaster risk reduction planning: need, opportunities and challenges. Mitig Adapt Strat Global Change 14 (1):7–33
Practical Action Nepal (2009) Temporal and spatial variability of climate change over Nepal, ISBN: 978-9937-8135-2-5
Schipper L, Pelling M (2006) Disaster risk, climate change and international development: scope for and challenges to, integration. Disasters 30(1):19–38
Shrestha AB, Wake CP, Mayewski PA, Dibb JE (1999) Maximum temperature trends in the Himalaya and its vicinity: an analysis based on temperature records from Nepal for the Period 1971–94. J Clim 12:2775–2787
Shrestha ML (2000) Interannual variation of summer monsoon rainfall over Nepal and its relation to southern oscillation index. Meteorol Atmos Phys 75:21–28
Synnott P (2012) Climate change, agriculture, and food security in Nepal. Developing adaptation strategies and cultivating resilience. Report prepared for Mercy Corps Nepal
Upreti CR (2007) A technology for reducing methane from livestock's. Hamro Sampada, Indreni offset press, Kathmandu, pp 102–103
Upreti BN, Muller-Boker U (2010) Livelihood insecurity and social conflict in Nepal. South Asia Regional Coordination Office, Swiss National Centre of Competence in Research (NCCR) North-South, Kathmandu, 271 pp
UNISDR (2005) Hyogo Framework of Action (2005–2015), United Nations International Strategy for Disaster Reduction. http://www.unisdr.org
UNFCCC (2013) Best practices and available tools for the use of indigenous and traditional knowledge and practices for adaptation, the application of gender-sensitive approaches and tools. Technical Paper; FCCC7TP7/2013/11
WB-World Bank (1998) Indigenous knowledge for development: a framework for development. Knowledge and Learning Center, Africa Region, World Bank
Werz M, Conley L (2012) Climate change, migration, and conflict - addressing complex crisis scenarios in the 21st Century. Centre for American Progress. Accessed on 19 July 2015. http://www.americanprogress.org
WMO (2009) Guidelines on Analysis of extremes in changing climate in support of informed decisions for adaptation, Climate Data and Monitoring, World Meteorological Organization WCDMP-No. 72
Wongbusarakum S, Loper C (2011) Indictors to assess community level social vulnerability to climate change: an addendum to SoMan and SEM-Pasitika. Accessed on 18 July 2015 http://www.socmon.org/download.ashx?docid = 64623

Chapter 14
Building Farm Resilience in a Changing Climate: Challenges, Potentials, and Ways Forward for Smallholder Cocoa Production in Bolivia

Johanna Jacobi, Patrick Bottazzi, Maria Isabel Pillco, Monika Schneider, and Stephan Rist

1 Introduction

Smallholder farming families around the world are facing severe challenges, and more and more are migrating from their rural areas of origin to cities or to other rural areas. Challenges include changing environmental conditions, difficult socio-economic circumstances, and unfavorable institutional settings. In line with the Andean concept of *vivir bien*,[1] resilience is an important property of farming systems that enables farmers to create and maintain sustainable livelihood options.

Diversified and organic farming systems have been receiving increasing attention in this context (Altieri et al. 2014; Halberg and Muller 2013; IAASTD 2008; UNCTAD 2013). The Food and Agriculture Organization of the United Nations (FAO) and the International Fund for Agricultural Development (IFAD) found that organic farming systems, and the social organization of farmers often associated with it, contribute to improving ecological, as well as socioeconomic, aspects of farmers' livelihoods (IFAD 2012; Scialabba 2013). Nevertheless, the relevance of

[1] *Vivir bien* is a concept included in the Bolivian constitution that means "good life" or "living well"—and not "living better," as is often stressed (Choquehuanca 2009). The concept is widely understood as an alternative to Western growth-oriented principles.

J. Jacobi (✉) • S. Rist
Centre for Development and Environment, University of Bern, Bern, Switzerland
e-mail: johanna.jacobi@cde.unibe.ch

P. Bottazzi
Faculty of Geosciences and the Environment, University of Lausanne, Lausanne, Switzerland

M.I. Pillco
Faculty of Agronomy, Universidad Mayor de San Andrés (UMSA), La Paz, Bolivia

M. Schneider
Research Institute of Organic Agriculture, Frick, Switzerland

© Springer International Publishing Switzerland 2017

K. Sudmeier-Rieux et al. (eds.), *Identifying Emerging Issues in Disaster Risk Reduction, Migration, Climate Change and Sustainable Development*,
DOI 10.1007/978-3-319-33880-4_14

organic farming for sustainable development and sustainable livelihoods remains little understood (Halberg and Muller 2013). In this chapter we compile the most important results from a research project on the resilience of organic and nonorganic smallholder cocoa farms in tropical Bolivia, where Andean migrants have developed different livelihood strategies than local indigenous groups, but are facing similar challenges. We interpret the findings in the context of climate change and climate change adaptation, and derive conclusions and ways forward for fostering farm resilience in the study area and in similar contexts.

1.1 Social-Ecological Resilience of Farming Systems

Social-ecological resilience is a promising approach for analyzing livelihood systems based on agriculture, because it is capable of capturing the interface between social and ecological systems. The question that matters is, what makes smallholder farms resilient to multiple threats, and how can vulnerable farms become more resilient? Rather than equating "resilience" with "sustainable development," we understand it as a property of a sustainable social-ecological system. Resilience as such is not necessarily good, because undesirable systems may be highly resilient as well. However, combining the concept of resilience with a normative understanding of sustainable development as a goal, we argue that a system must be resilient to be able to resist and adapt to changes, maintain its functions, and develop in a sustainable way.

Resilience, when applied to ecosystems, is the ability to withstand stress factors and to reorganize after disturbance (Adger 2003; Folke et al. 2010). This concept has been further developed with a view to applying resilience theory to social-ecological systems (Carpenter et al. 2001; Nicholls et al. 2013). A social-ecological system is an interface of human–nature interactions, such as, for example, a farming system (Milestad 2003). The concept of social-ecological resilience combines social and natural aspects and dynamics, and it emphasizes the strengths of a system rather than focusing on its vulnerability (Obrist et al. 2010). For example, agroforestry practices can increase organic matter in the soil, help to retain water, to provide nutrients in a plant-available form, and to reduce erosion by stabilizing the soil structure. By doing so, they can increase the farm's resilience to the risk factors of drought and soil degradation at the same time (Borron 2006).

Compared to ecosystem resilience, social-ecological resilience is a more dynamic concept that highlights the ability for learning and adaptation to change (Folke 2006). In an attempt to make the idea of resilience more tangible, the more specific concept of *farm resilience* (Milestad 2003) distinguishes three main features of resilience. *Buffer capacity* is the capacity to absorb change (Folke 2006; Holling 1973; Nicholls et al. 2013) and refers to the state of, and access to, livelihood assets—in particular ecological ones, such as biodiversity (Milestad 2003). *Self-organization* refers to an ecosystem's capacity to reorganize after disturbances (Altieri et al. 2014), but also to social organization as well as shared

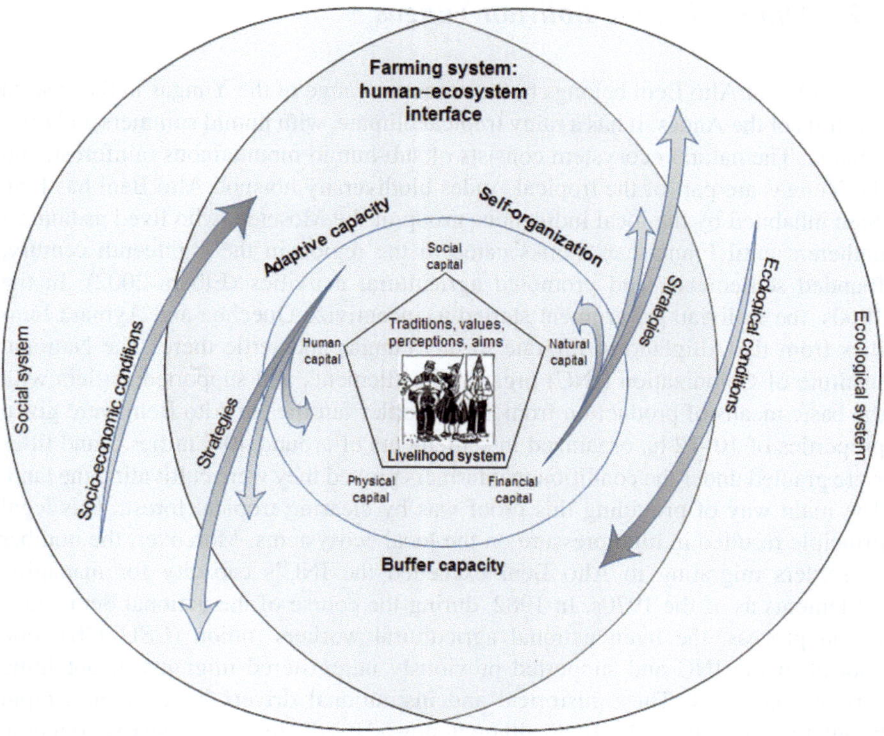

Fig. 14.1 Conceptual framework, author's design, central picture from Baumgartner et al. (2011)

norms and rules (Cabell and Oelofse 2012; Ostrom 1990). *Adaptive capacity* is based on the access to knowledge and learning opportunities, as well as on exchange of knowledge, social learning, and the combination of different types of knowledge, including traditional ecological knowledge (Boillat and Berkes 2013). These three features provide a starting point for developing indicators (Ifejika 2010; see also Cabell and Oelofse 2012).

We interpret the findings from the research project along the conceptual framework shown in Fig. 14.1, which combines aspects from resilience thinking (Carpenter et al. 2001; Milestad 2003), the sustainable livelihoods approach (DFID 1999), and rural livelihood systems (Baumgartner et al. 2011). Actors—in our case, farming families—base their decisions not only on external conditions (the natural environment and the social, political, and economic system), but also on inner realities (traditions, values, perceptions, and aims). Accordingly, they develop a set of strategies to mitigate perceived risks and increase their chances of making a living from their activities. These strategies have feedback effects on actors' livelihoods as well as on external influences, and they enhance or reduce the resilience of the social-ecological system in which the actors are operating.

1.2 *Alto Beni in the Bolivian Yungas*

The valley of Alto Beni belongs to the mountain range of the Yungas in the eastern foothills of the Andes. It has a rainy tropical climate, with humid summers and drier winters. The natural ecosystem consists of sub-humid mountainous rainforest, and the Yungas are part of the tropical Andes biodiversity hotspot. Alto Beni has long been inhabited by the local indigenous group of the Mosetén, who lived as hunter-gatherers until Franciscan monks came to the region in the eighteenth century, founded settlements, and promoted agricultural activities (Elbers 2002). In the 1960s, the Bolivian government started to incentivize Quechua and Aymara families from the Altiplano to migrate to the Yungas and settle there. The National Institute of Colonization (INC) organized settlements and supported settlers with the basic means of production from 1965. Settler families in Alto Beni were given properties of 10–12 ha organized in settlements of around 40 families. Land titles were granted under the condition that farmers proved they were cultivating the land. The main way of providing this proof was by clearing tropical forest. This legal principle resulted in high pressure on the local ecosystems. Moreover, the number of settlers migrating to Alto Beni exceeded the INC's capacity for managing settlements as of the 1970s. In 1982, during the course of the national democratization process, the main national agricultural workers' union (CSUTCB) took control of the INC and supported previously unregistered migrants in obtaining formal land titles. These historical and institutional drivers have favored rapid population growth in Alto Beni, although nowadays the region is also experiencing out-migration, with many young people moving to cities as in most rural areas in Bolivia (Brandt et al. 2013). Today, the Mosetén make up less than 10 % of the estimated 23,000 inhabitants of Alto Beni (von Stosch 2014). They were granted a common land title in 1996 which covers about half the area of Alto Beni, but land conflicts with settlers and other actors have persisted (Bottazzi 2008). Mosetén families combine agricultural activities with hunting and gathering, making use of their highly developed ecological knowledge (Bottazzi et al. 2013; von Stosch 2014).

Settlers from the Altiplano, however, brought with them strong social organizations and a greater capacity for linking up with national and international markets than the local indigenous groups have (Bottazzi 2008). Many settlers came from mining communities who had suffered liberalization measures and the collapse of the world tin market, or from farming communities in the Altiplano, whose ecosystem is very different from that of the Yungas. Accordingly, most of them had little knowledge of tropical plants and ecosystems (Bottazzi 2008). Most of today's agricultural practices in Alto Beni are poorly adapted to the ecosystem. Agriculture focuses on a few cash crops, which are mostly cultivated in monocultures. Shifting cultivation based on slash-and-burn forest clearing is common, and farmers are progressively reducing the length of fallow periods. This leads to heavy degradation of the area's fragile soils as well as its biodiversity. When their land becomes unproductive, many settlers move further into the Amazon rainforest to clear themselves a new plot.

Although Andean migrants and other nonindigenous settlers have been criticized for degrading fragile Amazonian ecosystems, social-ecological research has shown that farming families have a high capacity for adapting to changing climate conditions, thus building social-ecological resilience, and in fact, developing sustainable innovations (Adams et al. 2009). For many farming families in Alto Beni, cocoa has become a major source of income, and while local indigenous familiest end to grow cocoa landraces for their own use, settlers cultivate Trinitario and Forastero varieties for export markets.

Today, cocoa farmers in Alto Beni face a number of challenges. They mention heat waves, prolonged dry periods, heavy rains, hail storms, landslides, and floods as climate change-related impacts that represent a major risk for cocoa production and their livelihoods (Jacobi et al. 2013). Rains in the last years have been more intense and of shorter duration than usual. Increasing pressure on crops from pests and diseases has also been observed. Socioeconomic challenges include reduced working hours and force due to increasing temperatures, as well as fluctuating farm gate prices and limited access to markets. At the same time, a strong market development has been boosting demand for specialty and certified organic cocoa, fruit, timber, and other non-timber forest products (Jacobi et al. 2013).

The high level of social frictions, risks, and vulnerability, but also of opportunities in Alto Beni contribute to making it a "critical area" where threshold effects can spur rapid change in positive or negative directions. People in the study area show different reactions and adaptation strategies, some of which have completely opposite impacts on social-ecological resilience. Strategies include migrating to other recently colonized areas further down the road to Ixiamas and beyond, moving further into the remaining rainforest, or focusing on other sources of income. Increasing timber extraction is a common strategy among the Mosetén, whereas settlers may shift to coca (*Erythroxylum coca*) cultivation or diversify their crops and incorporate more trees in their land use systems. More than 1,200 families in Alto Beni cultivate certified organic cocoa. Most of them are settlers, but there are also a number of Mosetén families. Organic farming, in this case, is export-oriented and has to meet US and EU standards, which above all means not using a grochemicals. Families have developed various strategies to adapt their cultivation systems to climate change impacts, most of which include planting trees—as hedgerows or shade trees—or keeping large trees from the rainforest and growing crops below them. Families who adopt these strategies tend to make use of a high agricultural biodiversity and diversify their diets, keeping their level of dependence on external resources low (Jacobi et al. 2013).

Agroforestry systems have been promoted and implemented by development projects as locally adapted diversified farming systems that protect biodiversity and soils. In Alto Beni, agroforestry means diversifying cocoa plots by including trees and other crops (Fig. 14.2). Farmers in Alto Beni see certain benefits in such diversified agroforestry systems, including shade to work in, greater soil fertility, less pests and diseases, reduced water stress, less damage to crops from extreme

Fig. 14.2 Sketch of a diversified agroforestry system in Alto Beni (DED 1997)

weather events, and a diverse harvest (Jacobi et al. 2014). A globally relevant side effect of agroforestry systems is their high carbon sequestration potential (Nair 2011; Somarriba et al. 2013). Implementation of an agroforestry system in Alto Beni can be constrained by the unavailability of plants and seeds, equipment, or know-how needed for shade management, insufficient knowledge about tree and crop management, and the assumption that agroforestry systems require more work in the first years after they are established than monocultures (Jacobi et al. 2014).

Cocoa is a crop from the rainforest understory that grows naturally under shade, which makes it suitable for cultivation in an agroforestry system. Cocoa also grows in full-sun monocultures, but yields decrease after a few years, as plants have a shorter life cycle and are more susceptible to pests and diseases (Gyau et al. 2014; ICRAF 2014; Rice and Greenberg 2000). Nonetheless, monocultures are common practice in the tropics worldwide (Tscharntke et al. 2011), and Alto Beni is no exception, with both organic and nonorganic cocoa cultivation.

Research linking agroforestry and resilience to climate change has mainly focused on coffee, demonstrating, for example, that shade trees in coffee plantations mitigate microclimate extremes (Lin 2007). Verchot et al. (2007) link climate change adaptation and mitigation by means of coffee agroforestry, arguing that tree diversification reduces risks from climate change impacts while sequestering carbon. An extensive study in Nicaragua found that agroecological farms which included woody perennials in their cropping system (e.g., as living fences) suffered significantly less damage from hurricane Mitch than less diversified farms (Holt-Giménez 2002). Another study analyzing landscape resilience found similar connections, including that reducing the complexity of shade canopy on coffee farms increased the farm area affected by landslides (Philpott et al. 2008). Other authors found the impacts of soil erosion (Holt-Giménez 2002), water stress (ICRAF 2014; Moser 2010; Ruf and Schroth 2004; Verchot et al. 2007), heavy rains and hailstorms (Altieri and Koohafkan 2008; Holt-Giménez 2002), and pests and diseases (Holt-Giménez 2002) reduced on diversified farms. These findings point to the potential of agroforestry systems for reducing the risk of disaster while building resilience to a range of climate change impacts.

2 Comparing Farm Resilience Indicators on Organic and Nonorganic Cocoa Farms in Alto Beni: Approach and Main Findings

Our research project comparing farm resilience on organic and nonorganic cocoa farms lasted 3 years (2010–2012). Research was conducted by a Swiss–Bolivian interdisciplinary research team from the University of Bern, the Swiss Research Institute of Organic Agriculture (FiBL), and the La Paz Universidad Mayor de San Andrés (UMSA). Data gathering included sampling of resilience indicators that were determined together with cocoa producers (Jacobi et al. 2013), as well as interviews with cocoa producers and participant observation of cocoa producing families (Jacobi et al. 2015a). The research process and methods used were as follows: (1) defining, together with cocoa farmers, challenges (farmer and expert interviews, focus group discussions, literature) and then resilience indicators (in a participatory workshop, backed and complemented by literature: Adger 2000; Cabell and Oelofse 2012); (2) gathering data on these indicators from 52 cocoa farms by means of plot sampling and interviews (we selected 30 organic and 22 nonorganic farms with different degrees of diversification of their cocoa plot, plus five experts from local agricultural organizations); and (3) searching for perspectives and recommendations based on interviews and participant observation. Cocoa farmers' consent to participate was obtained in each case, and results for each farm were printed out, presented to, and discussed with the participating families.

From the focus groups and the participatory workshop, we derived resilience indicators for (1) buffer capacity: tree species and crop varieties diversity, diversity of ant species (as a proxy for wild species), Ah soil horizon, soil organic matter, soil bulk density, cocoa yields, the infestation of cocoa trees with "witches' broom disease" (*Moniliophthora perniciosa*), (the most severe cocoa disease in the region at the time of research), and the number of different sources of income; (2) for self-organization: affiliation to farmers' organizations, subsistence level of the farm, and the annual family income; and (3) for adaptive capacity: participation of family members in courses on cocoa cultivation and the number of different sources of information the family had. These indicators were later analyzed for significant differences between organic and nonorganic cocoa farms. The influence of organic certification and solidarity economy patterns on resilience building were evaluated using a qualitative approach and content analysis (Patton 2002). For details on the data sampling methods and the statistical analysis, see Jacobi et al. (2013) and Jacobi et al. (2015a).

2.1 Farm Resilience

Most indicators of buffer capacity, self-organization, and adaptive capacity were more favorable on organic farms than on nonorganic farms. Differences were highly significant for tree species diversity, infestation with "witches' broom disease" (*Moniliophthora perniciosa*), affiliation to farmers' organizations, and the participation in courses on cocoa cultivation. All indicators were higher on organic farms, with the exception of infestation with *Moniliophthora perniciosa*, which was higher on nonorganic farms. The spread of a newly arrived cocoa disease, *Moniliophthora roreri*, was not assessed in this study because research had already been concluded at the time it began to spread. Ant species diversity, the Ah soil horizon, cocoa yield, and annual family income were all significantly higher on certified organic farms. Soil organic matter, soil bulk density, crop diversity, the number of different income sources, subsistence level, and the number of information sources a family had did not differ significantly (Jacobi et al. 2013; Jacobi et al. 2015a). All of these indicators were highly influenced by agroforestry, and most organic farms managed an agroforestry system; the average organic farmer had 48.6 % of their farm area under a tree canopy, as opposed to only 18.7 % for the average nonorganic farmer. A reduced tree canopy may make the whole farm more susceptible to damage from heat and droughts, heavy rains, and hail storms. Agroforestry farms had a higher cocoa yield, fewer infested cocoa trees, and a higher subsistence level regarding food consumed that was produced on the farm. All farmers but one (a nonorganic monoculture farmer) used a mix of an average seven cocoa varieties (Trinitario and Forastero varieties, and in some cases *Cacao Nacional Boliviano*). Since only one farmer in our sample used one single variety (ICS-95), we cannot further explore the influence of genetic uniformity or diversity of cocoa varieties on pests and diseases.

Shade trees were mentioned as an important component by almost all families interviewed: 28 organic farmers and 20 nonorganic farmers stated that a diversity of well-adapted trees was key to more resilient cocoa cultivation systems because they create shade and balanced temperatures, reduce pest pressure, and provide a diversity of products (timber, fruit, fodder, construction material, natural medicine, and ornamental plants). However, most nonorganic farmers had done little to diversify their cocoa plots with shade trees, although half of them reported that they planned to do so. As reasons, they mentioned a higher labor input and a lack of equipment to prune the trees, a lack of market channels for agroforestry products, as well as a lack of knowledge on agroforestry management and plant species. The main reason was that they expected cocoa yields to be lower in agroforestry systems than in monocultures, an assumption that was not supported by our data. In fact, yields in our sample were about 35 % higher under agroforestry than in monocultures, which had to do with better management and greater knowledge among those farmers who practised agroforestry (Jacobi et al. 2013). Our qualitative analyses showed that processes related to knowledge and social networks are important in building resilience, as discussed below.

2.2 *Solidarity Economy: The Example of El Ceibo*

The importance of farmers' organizations was expressed by IFAD (2012:2): "*Farmers' organizations will make or break the success of small organic farmers.*" The reason for this, according to IFAD, is multidimensional. Farmers' organizations often enable their members to participate in economies of scale through collective marketing; they provide training in organic production; and they reduce certification costs for the individual member, among other benefits. This matches our findings with regard to training in cocoa cultivation and making markets and certification accessible to smallholder farmers. In order to find out more, we analyzed a farmers' organization from our study area: the umbrella organization of organic cocoa cooperatives El Ceibo.

El Ceibo was founded in 1977 as an umbrella organization of cocoa cooperatives in Alto Beni, and obtained organic certification in the 1980s as the world's first exporter of certified organic cocoa. With 59 cooperatives, more than 1,200 families certified, and around 200 in the transition phase, the organization has become a showcase in the history of Bolivian cooperatives. At the time of research, El Ceibo had an educational center that ran a radio program on cocoa cultivation and postharvest treatment. They also operated a forest seed bank and a tree nursery, a cocoa tree nursery with a collection of cocoa varieties and an elite tree selection program, and a microcredit system. They provided technical assistance to their members, and they had an internal system of social insurances. El Ceibo was in charge of organizing organic cocoa and fair trade certification, as well as of providing technical assistance and capacity building on sustainable cocoa cultivation. Most positions were filled by cocoa producers and rotated every 2–4 years to keep the connection between the umbrella organization and the farmers as close as possible. An advantage of this strategy is surely their emic perspective, but the concept may also have disadvantages, including a lack of skilled staff for certain tasks and a lack of continuity in the work being done. The organization mitigated these drawbacks by hiring external experts for some positions.

The cooperatives had norms and regulations that influenced members' livelihoods. These included a ban on the use of agrochemicals, the requirement for the implementation of an agroforestry system (not required by certification standards but recommended and supported by El Ceibo's technical staff), participation in the cooperatives' gatherings and general assemblies, a following of postharvest management guidelines, and a sense of loyalty to the cooperative, that is, not to sell cocoa to other buyers who do not invest in long-term improvements of local livelihoods.

An important feature of social-ecological systems' adaptive capacity is social connectedness, as described by Pretty and Smith (2004). Our interviews showed a strong perception of togetherness and solidarity among cooperative members (Jacobi et al. 2015a), as well as a strong ethical commitment to sustainable cocoa production and the cooperative's social and marketing networks. This was true for both settlers and Mosetén families. Mosetén families cultivated mainly cocoa landraces (*Cacao Nacional Boliviano*), which have a substantially higher quality and different aroma profile compared to the other genetic material used in the region, but produce less quantity. The affiliated farmers were not just delivering primary material; they were owners and took part in the whole value chain, for example as decision-makers in the assemblies or as staff in the different stages of processing. El Ceibo was started by local farmers, which may explain members' strong identification with the organization to some extent. Some young people from El Ceibo families moved to the cities as well, but according to our expert interviews, Alto Beni experienced less out-migration compared to other rural areas in Bolivia because young people had fairly good work and income perspectives within El Ceibo and because cooperative members had a sense of stronger social connectedness.

Studies have long stated that El Ceibo has had a strong influence on the region. Benefits include overall higher cocoa prices, market and transport infrastructure, creation of jobs, a store with good equipment and organic products, and a bank with low interests on loans for agricultural purposes (Bebbington et al. 1996; Hillenkamp 2006). Another indirect effect is the emergence of new cooperatives following the example of El Ceibo with a focus on other products, such as organic bananas (Garming et al. 2011).

Institutional challenges faced by El Ceibo include a high entry price for new members, who have to buy a share of El Ceibo. Many nonaffiliated families mention this high entry cost as the reason for not having organic certification and the other benefits of El Ceibo membership. Starting similar organizations may be a possible approach to this issue, and may also help to overcome certain problems in the El Ceibo administration, whose vertical structures were described and criticized as inflexible and persistent in our expert interviews. Some interviewees complained about a lack of willingness to innovate within the organization. The relatively high cost of membership in the organization and the lower levels of connectedness among nonmembers may be reasons for the low adoption of agroforestry practices outside El Ceibo. Migration further down into the lowlands was found to be a way not only of colonizing new physical spaces for production but also of creating new political spaces in order to attain a higher position in sociopolitical structures (Bottazzi 2014). Displacement and the expansion of social-ecological degradation can therefore be considered a consequence of a lack of institutional support to organic agroforestry practices at the local, national, and international levels. The main institutional challenges at this point might be (1) to find ways of keeping farmers' organizations such as El Ceibo accessible to resource-poor farmers and

(2) to revise the conditions of access to land and productive assets in colonized areas by defining sustainable land use as a condition instead of land clearing. The support and valuation of agroforestry products should be extended to other non-timber forest products and to products that are not only for export, but also increasingly for local markets.

A major challenge in rural Bolivia is to support and restore the value of rural livelihoods so that young people can consider farming a viable livelihood strategy. Rural–urban migration of young people in Bolivia has been found to decrease ethnobotanical knowledge and the valuation of plants (Brandt et al. 2013). In this way it might also have a negative effect on the resilience (in particular the adaptive capacity) of farming systems. In this context, strong local farmers' organizations may contribute to decelerating young people's outmigration. El Ceibo, for example, supports perennial land use systems, thereby providing a long-term prospect. Indeed, the El Ceibo families we interviewed expressed no intentions to move further down into the rainforest, in contrast to several non-affiliated families. Temporal migration of family members was common, but it was also common for young people to return after finishing their education, for example in agronomy, forestry, or accounting—that was often funded through a scholarship from El Ceibo. Remittances flowed from the cocoa farm to family members in the city rather than vice versa. Farmers' organizations like El Ceibo can thus create perspectives for rural livelihoods combined with temporal migration, provided that they develop the strong social networks and social security that is required for *vivir bien*. Such temporal migration within a rural livelihood must not be understood as a linear movement but rather as a circular one that enables a diversification of activities, education, and household income sources.

2.3 The Role of Organic Certification

Organic certification, according to minimal standards of the EU and the USA, may not necessarily enhance on-farm diversity, but the associated premium price was the main reason for cocoa producers to join El Ceibo, which then provided support for diversification, agroforestry, and postharvest management. The common perspective of organic production also provided a basis for defining additional common goals going beyond agriculture—for example, concerning environmental or sociocultural issues—and for developing an ethical understanding of agroecological systems and human–nature interactions. The organization had therefore adopted rules that were stricter than the basic organic regulations, following suggestions from the International Federation of Organic Agriculture Movements (IFOAM) regarding external inputs and on-farm resource management.

Certification schemes have been criticized for letting price premiums stagnate while increasing requisites, and for reproducing unjust trade structures (Hillenkamp 2006). Organic and fair trade certification may also have disadvantages such as tying up financial and human resources, leaving insufficient time for technical assistance and capacity building activities (El Ceibo pers. comm.). A study on the effect of fair trade certification on the extent of coffee areas in Costa Rica found that fair trade certification alone was insignificant in encouraging farmers to cultivate coffee, but that associated shade tree diversification greatly diminished costs and risks, and adopters kept more coffee under cultivation during the coffee crisis than non-adopters (Babin 2014). Participatory Guarantee Systems (PGS)[2] are alternative low-cost verification processes where farmers, consumers, and other stakeholders participate and benefit. PGSs serve smallholder farmers and local markets and are on the rise in Bolivia, as local and national markets for organic products are growing.

Settlers in our study cultivated Trinitario and Forastero cocoa varieties and produced larger quantities of dry grain than local indigenous Mosetén families, who tended to cultivate the sought-after *Cacao Nacional Boliviano* that has a higher quality but produces smaller quantities. This pattern offers different strategies according to the two groups' different social-ecological systems; intensive and diversified agroforestry systems might be a strategy for settlers to slow down slash-and-burn shifting cultivation, diversify production, and adapt to climate change, but *Cacao Nacional Boliviano* combined with replanting of depleted timber species might be a strategy for Mosetén families, as they hold a large community land title. El Ceibo and other buyers have developed market channels for both, for national as well as international markets.

2.4 Sharing Knowledge for Agricultural Diversification

Resilience to climate change and disasters is closely linked to on-farm biodiversity, and the only means for disaster risk reduction in marginal environments are often based on inventive self-reliance, experimental knowledge, and locally available resources (Altieri and Nicholls 2013). El Ceibo courses, technical assistance, and workshops on cocoa management and diversified agroforestry are open to all, but are almost exclusively frequented by members (settlers as well as Mosetén families). Experts interviewed stated that there was a general openness to learning and exchanging knowledge. All events we visited had many participants, even in remote areas. Because of their acceptance and broad network throughout Alto Beni, El Ceibo had an important function as a knowledge resource center—a hub where knowledge became accessible (through capacity building) and applicable (through technical support)—even if this function was somewhat limited to its members. A further way of building resilience to climate change impacts in well-organized farmers' groups like El Ceibo might be to bring together traditional (endogenous) ecological knowledge and scientific (exogenous) ecological

[2] IFOAM homepage for more information: http://www.ifoam.org.

knowledge (Borron 2006). El Ceibo could emphasize this point as they link settlers up with Mosetén families who have highly developed traditional ecological knowledge. This may lead to the coproduction of new, specialized knowledge about climate change adaptation and risk mitigation (see Rist 2006).

Our interviews showed that knowledge was still not accessible enough. Producers demanded more courses on agricultural practices and agroforestry (Jacobi et al. 2014). Constraints on attending courses in cocoa cultivation included lack of time, long distances, and transport problems. Accordingly, we consider essential that local institutions are encouraged and supported in establishing farmer to farmer learning systems. Another constraint stated in the expert interviews was a lack of cooperation between the various organizations in the region, despite a great potential for synergies and mutual benefits (Jacobi et al. 2015a).

2.5 *Agroforestry and Solidarity Economy for Resilience Building in Alto Beni*

After comparing key parameters of farm resilience, we conclude that diversified agroforestry systems for cocoa production support resilient social-ecological systems and can be a viable alternative to the widespread slash-and-burn practices of annual crops or cocoa monocultures in Alto Beni. Given that adaptation to climate change is one mayor challenge for smallholder agriculture in Alto Beni, the role of organizations such as El Ceibo who foster agroforestry and agricultural diversification cannot be overestimated.

Organic cocoa farms that were affiliated with El Ceibo had the prospect of being more resilient to stress factors than nonaffiliated, nonorganic cocoa farms. This was because organic farmers were more likely to implement agroforestry and were socially better connected to one another than those who were not members of El Ceibo or a similar organization. El Ceibo, as a local farmers' organization, enhances the process of integrating families—settlers and Mosetén alike—into cooperatives and promotes the adoption of organic principles, with important implications for farm resilience. Cocoa cultivation in agroforestry systems helps families benefit from environmental services to face climate change impacts and adapt to changing conditions. In addition, it provides a multitude of diverse products and has a potential for increasing farming families' self-sufficiency and diversifying their sources of income.

El Ceibo is a good example when it comes to understanding the socioeconomic implications of organic agriculture, for it offers long-standing experience in terms of how farmers' organizations can profit from organic cultivation through strong social networks. Because El Ceibo is an organization founded and managed by local farmers, it enjoys a high local acceptance and is likely to persist in the region.

The concept of farm resilience, with its components of buffer capacity, self-organization, and adaptive capacity, seems useful in assessing interfaces between environmental services and socioeconomic trade-offs—as in this case, where we used it to understand the role of organic certification and cooperatives in building farm resilience. We found that knowledge management is a key factor for sustainable cocoa cultivation under agroforestry and hence for resilience building to climate change impacts (Jacobi et al. 2015a).

We conclude that strong farmers' organizations play a crucial role in achieving *vivir bien* and sustainable livelihoods, as well as in providing young people with a perspective for the future—which is part of the "inner realities" mentioned in Fig. 14.1. They may also be able to slow down migration to the ever-growing twin cities of El Alto and La Paz, and further down into the primary rainforest.

3 Conclusions and Way Forward

Results from the project indicate that cocoa agroforestry systems in Alto Beni may be a strategy to reduce climate change impacts and disasters in the social-ecological system of cocoa-based livelihoods. Farming families were aware of benefits such as shade to work in, greater soil fertility, less pests and diseases, reduced water stress, less damage to crops from extreme weather events, and a diverse harvest (Jacobi et al. 2014). To enhance agroforestry management, farmers and experts in Alto Beni demand more support for local smallholder farmers' organizations from development projects, in the form of expertise and financial support for Mosetén families (e.g., in the field of specialty cocoa cultivation and quality management). Current challenges regarding commercialization include linking the diverse products from agroforestry systems to markets, promoting better postharvest techniques, and supporting the production of high-value specialty cocoa such as *Cacao Nacional Boliviano* varieties. Agroforestry might require additional economic incentives, with REDD+ projects being critically discussed elsewhere (Bottazzi et al. 2014; Hufty and Haakenstad 2011; Pokorny et al. 2013). The Bolivian alternative "Joint Mitigation and Adaptation Mechanism for theIntegrated and Sustainable Management of Forests" under the Framework law of the Rights of Mother Earth will hopefully build such incentives in the future. Access to cooperatives should be made easier for new cocoa producers, for example by starting new organizations following the example of El Ceibo.

Projects should focus more on coproduction and an exchange of knowledge among cocoa farmers and technical staff skilled in agroforestry, with a view to improving the accessibility and applicability of knowledge about organic cocoa cultivation and the use of a diverse range of agroforestry products. Additionally, local organizations like El Ceibo could emphasize their role as knowledge resource centers even more. Cocoa producers—including from other regions—should be encouraged to interact with researchers and agricultural consultants (extension services) of other local and external organizations as teachers and trainees at the same time. Moreover, combining Mosetén families traditional ecological knowledge with the Andean settlers' technical knowledge on cocoa cultivation could enhance knowledge coproduction in Alto Beni. A priority should therefore be to increase the level of interaction among local organizations as well as between them and also with and between external organizations such as the Faculty of Agronomy of the public university of La Paz (UMSA) and its research center in Alto Beni. This would help to build a supportive multi-scalar institutional structure for cooperatives that enhances social networks that supports and promotes resilient and diversified farming systems.

References

Adams C, Harris M, Murrieta R (2009) Amazon peasant societies in a changing environment. Springer, Sao Paolo

Adger N (2003) Building resilience to promote sustainability. Update IHDP, Newsletter of the International Human Dimensions Programme on Global Environmental Change 2(3)

Adger WN (2000) Social end ecological resilience: are they related? Progr Hum Geogr 24(3):18

Altieri M, Nicholls C (2013) The adaptation and mitigation potential of traditional agriculture in a changing climate. Clim Change:1–13

Altieri MA, Koohafkan P (2008) Enduring Farms: Climate Change, Smallholders and Traditional Farming Communities. Third World Netw Environ Dev Ser 6:72

Altieri MA, Koohafkan P, Nicholls CI (2014) Strengthening resilience of modern farming systems: a key prerequisite for sustainable agricultural production in an era of climate change. Third World Netbook Briefing Paper 70:8

Babin N (2014) The coffee crisis, fair trade, and agroecological transformation: impacts on land-use change in Costa Rica. Agroecol Sust Food Syst 39(1):99–129

Baumgartner R, Hoegger R, Rist S (2011) Hacia estrategias de vida sostenibles—culturas, recursos y cambios en India y Bolivia. DSpace Repository, La Paz

Bebbington A, Quisbert J, Trujillo G (1996) Technology and rural development strategies in a small farmer organization: lessons from Bolivia for rural policy and practice. Public Admin Dev 16:195–213

Boillat S, Berkes F (2013) Perception and interpretation of climate change among Quechua farmers of Bolivia: indigenous knowledge as a resource for adaptive capacity. Ecol Soc 18(4):21

Borron S (2006) Building resilience for an unpredictable future: how organic agriculture can help farmers to adapt to climate change. FAO, Rome

Bottazzi P (2008) Linking 'Socio-' and 'Bio-' Diversity: The Stakes of Indigenous and Non-indigenous Co-management in the Bolivian Lowlands. In: Galvin M, Haller T (eds) People, Protected Areas and Global Change: Participatory Conservation in Latin America, Africa, Asia and Europe. Perspectives of the Swiss National Centre of Competence in Research (NCCR) North-South. Geographica Bernensia, Bern, pp 81–110

Bottazzi P (2014) Une écologie politique des territoires tsimane' d'Amazonie bolivienne: "notre grande maison". Karthala/Graduate Institute Publications, Paris, Genève

Bottazzi P, Crespo D, Soria H, Dao H, Benavides JP, Swarzer S, Rist S (2014) Carbon sequestration in community forests: trade-offs and synergies, multiple outcomes, and institutional diversity in the Bolivian Amazon. Dev Change 45(1):105–131

Bottazzi P, Reyes-García V, Crespo D, Marthez-Stiefel S-L, Galvarro HS, Jacobi J, Clavijo M, Rist S (2013) Productive diversification and sustainable use of complex social-ecological systems: a comparative study of indigenous and settler communities in the Bolivian Amazon. Agroecol Sust Food Syst 38(2):137–164

Brandt R, Mathez-Stiefel SL, Lachmuth S, Hensen I, Rist S (2013) Knowledge and valuation of Andean agroforestry species: the role of sex, age, and migration among members of a rural community in Bolivia. J Ethnobiol Ethnomed 9(1)

Cabell JF, Oelofse M (2012) An indicator framework for assessing agroecosystem resilience. Ecol Soc 17(1):18
Carpenter S, Walker B, Anderies JM, Abel N (2001) From metaphor to measurement: resilience of what to what? Ecosystems 4(8):765–781
Choquehuanca D (2009) El Vivir Bien como Respuesta a la Crisis Global. Ministerio de Relaciones Exteriores, La Paz
DED (1997) Guía para el Establecimiento de Sistemas Agroforestales. Servicio Alemán de Cooperación Social-Técnica (DED), La Paz
DFID (1999) Sustainable livelihoods guidance sheets. UK Department for International Development, London
Elbers J (2002) Agrarkolonisation im Alto Beni Landschafts- und politisch-ökologische Entwicklungsforschung in einem Kolonisationsgebiet in den Tropen Boliviens. Doctoral thesis, Heinrich-Heine-University, Duesseldorf
Folke C (2006) Resilience: the emergence of a perspective for social–ecological systems analysis. Glob Environ Chang 16:253–267
Folke C, Carpenter SR, Walker B, Scheffer M, Chapin T, Rockstrom J (2010) Resilience thinking: integrating resilience, adaptability and transformability. Ecol Soc 15(4)
Garming H, Guardia S, Pocasangre L, Staver C (2011) Farmers' community enterprise for marketing organic bananas from Alto Beni, Bolivia: impacts and threats. Enterprise Dev Microfinance 22:210–224
Gyau A, Smoot K, Kouame C, Diby L, Kahia J, Ofori D (2014) Farmer attitudes and intentions towards trees in cocoa (Theobroma cacao L.) farms in Côte d'Ivoire. Agroforest Syst 88:1035–10451
Halberg N, Muller A (2013) Organic agriculture, livelihoods and development. In: Halberg N, Muller A (eds) Organic agriculture for sustainable livelihoods. Earthscan, London, pp 1–20
Hillenkamp I (2006) Central de Cooperativas El Ceibo: Un Estudio con Enfoque de Economía Solidaria. PhD, Universidad de Ginebra, Geneve
Holling CS (1973) Resilience and stability of ecological systems. Annu Rev Ecol Syst 4:1–23
Holt-Giménez E (2002) Measuring farmers' agroecological resistance after Hurricane Mitch in Nicaragua: a case study in participatory, sustainable land management impact monitoring. Agr Ecosyst Environ 93(1–3):87–105
Hufty M, Haakenstad A (2011) Reduced emissions for deforestation and degradation: a critical review. Consilience J Sust Dev 5(1):1–24
IAASTD (2008) Agriculture at a crossroads. Synthesis report. Washington, DC
ICRAF (2014) Trees for life: creating a more prosperous future through agroforestry. World Agroforestry Centre, Nairobi
IFAD (2012) Organic agriculture: a promising alternative for small farmers Thematic evaluation
Ifejika Speranza C (2010) Resilient adaptation to climate change in African agriculture. DIE studies. Deutsches Institut für Entwicklungspolitik, Bonn
Jacobi J, Andres C, Schneider M, Pillco MM, Calizaya P, Rist S (2014) Carbon stocks, tree diversity, and the role of organic certification in different cocoa production systems in Alto Beni, Bolivia. Agroforest Syst 88:1117–1132
Jacobi J, Schneider M, Bottazzi P, Pillco M, Calizaya P, Rist S (2013) Agroecosystem resilience and farmers' perceptions of climate change impacts in cocoa farms in Alto Beni, Bolivia. Renew Agric Food Syst. Online first
Jacobi J, Bottazzi P, Schneider M, Huber S, Weidmann S, Rist S (2015a) Social–ecological resilience in organic and non-organic cocoa farming systems in Bolivia. Agroecol Sust Food Syst 39:798–823
Jacobi J, Zanella MA, Mariscal M, Lucana AC, Rist S (2015b) Alternatives for the sustainable development of Alto Beni, Bolivia. In: Rosendahl J, Weigelt J, Durand JM, Zanella MA (eds) Pro-poor resource governance under changing climates. IASS, Potsdam
Lin BB (2007) Agroforestry management as an adaptive strategy against potential microclimate extremes in coffee agriculture. Agric For Meteorol 144(1–2):85–94

Milestad R (2003) Building farm resilience—challenges and prospects for organic farming. Doctoral thesis, Swedish University of Agricultural Sciences, Uppsala
Moser G (2010) Response of cocoa trees (Theobroma cacao) to a 13-month desiccation period in Sulawesi, Indonesia. Agroforest Syst 71:171–187
Nair PKR (2011) Carbon sequestration studies in agroforestry systems: a reality-check. Agroforest Syst 86(2):243–253
Nicholls CI, Ríos Osorio LA, Altieri MA (eds) (2013) Agroecología y resiliencia socioecológica: adaptándose al cambio climático. Red Iberoamericana de Agroecología para el Desarrollo de Sistemas Agrícolas Resilientes al Cambio Climático (REDAGRES), Red Adscrita al programa Iboamericano de Ciencia y Tecnología para el Desarrollo (CYTED), Sociedad Científica Latinoamericana de Agroecología (SOCLA), Medellín
Obrist B, Pfeiffer C, Henley R (2010) Multi-layered social resilience: a new approach in mitigation research. Progr Dev Stud 10(4):283–293
Ostrom E (1990) Governing the commons. The evolution of institutions for collective action. Cambridge University Press, Cambridge
Patton MQ (2002) Qualitative research & evaluation methods. Sage, Beverly Hills, CA
Philpott SM, Lin BB, Jha S, Brines S (2008) A multi-scale assessment of hurricane impacts on agricultural landscapes based on land use and topographic features. Agr Ecosyst Environ 128:12–20
Pokorny B, Scholz I, de Jong W (2013) REDD+ for the poor or the poor for REDD+? About the limitations of environmental policies in the Amazon and the potential of achieving environmental goals through pro-poor policies. Ecol Soc 18(2)
Pretty J, Smith D (2004) Social capital in biodiversity conservation and management. Conserv Biol 18(3):631–638
Rice R, Greenberg A (2000) Cacao cultivation and the conservation of biological diversity. AMBIO J Hum Environ 29(3):167–173
Rist S (2006) Natural resources, sustainability and social learning processes—pathways towards co-production of knowledge for sustainable development. Habilitation, University of Bern, Bern
Ruf F, Schroth G (2004) Chocolate forests and monocultures: a historical review of cocoa growing and its conflicting role in tropical deforestation and forest conservation. In: Schroth G, Fonseca G, Harvey C, Gascon C, Vasconcelos HL, Izac AMN (eds) Agroforestry and biodiversity conservation in tropical landscapes. Island Press, Washington, DC, pp 107–134
Scialabba N (2013) organic agriculture's contribution to sustainability. Paper presented at the USDA organic farming systems research conference, Washington, DC
Somarriba E, Cerda R, Orozco L, Cifuentes M, Davila H, Espin T, Mavisoy H, Avila G, Alvarado E, Poveda V, Astorga C, Say E, Deheuvels O (2013) Carbon stocks and cocoa yields in agroforestry systems of Central America. Agr Ecosyst Environ 173:46–57
Tscharntke T, Clough Y, Bhagwat SA, Buchori D, Faust H, Hertel D, Hölscher D, Juhrbandt J, Kessler M, Perfecto I, Scherber C, Schroth G, Veldkamp E, Wanger TC (2011) Multifunctional shade-tree management in tropical agroforestry landscapes—a review. J Appl Ecol 48:619–629
UNCTAD (2013) Wake up before it is too late: make agriculture truly sustainable now for food security in a changing climate. Trade and environment review
Verchot LV, Van Noordvik M, Kandji S, Tomich T, Ong C, Albrecht A, Mackensen J, Bantilan C, Anupama KV, Palm C (2007) Climate change: linking adaptation and mitigation through agroforestry. Mitig Adapt Strat Glob Chang 12:901–918
von Stosch K (2014) Indígenas y Campesinos en Alto Beni: Diferentes visiones en torno a tierra, territorio y recursos naturales. Fundación Tierra, La Paz

Chapter 15
The Role of Traditional Knowledge to Frame Understanding of Migration as Adaptation to the "Slow Disaster" of Sea Level Rise in the South Pacific

Keith Morrison

1 Introduction

This chapter seeks to outline a conceptual model framing the role of traditional knowledge to guide proactive transformation of social-ecological structures, wherein migration figures as a significant adaptation. The entry point is the supposed triggering of migration through climate change induced sea-level rise in the South Pacific. Sea level rise can be characterized as a slow-onset hazard for many low lying countries, including for several in the South Pacific where they face the hazard of being inundated under common scenarios produced by the Intergovernmental Panel on Climate Change (IPCC) (Parry 2007). Political rhetoric from leaders of several of these countries, in particular Kiribati[1] and Tuvalu,[2] portray the slow-onset hazard as national disasters requiring an international response. But, given the well-accepted distinction between hazard (e.g., sea level rise) and disaster (i.e., inability to manage the hazard), a question has to be asked about the human activities in the South Pacific nations themselves: How are activities increasing the likelihood that this hazard becomes a disaster? That sea-level rise is a slow-onset hazard only emphasizes the need to use this time to avoid the creation of a disaster in the making.

[1] The president of Kiribati, Anote Tong, has likened Kiribati to the canary being sacrificed (Tong 2014; Zakaria 2014).

[2] The Tuvalu prime minister, Enele Sopoaga, has likened climate change to a weapon of mass destruction (ABC News 2014).

K. Morrison (✉)
Pacific Centre for Environment and Sustainable Development, The University of the South Pacific, Suva, Fiji
e-mail: keithdmorrison59@gmail.com

© Springer International Publishing Switzerland 2017
K. Sudmeier-Rieux et al. (eds.), *Identifying Emerging Issues in Disaster Risk Reduction, Migration, Climate Change and Sustainable Development*,
DOI 10.1007/978-3-319-33880-4_15

The question can therefore be rephrased to consider the efficacy of adaptation to sea-level rise. In other words, any characterization of climate change and consequent sea-level rise as a disaster begs the question about what is failing in the adaptation process. But maybe this is just splitting hairs between what is a hazard and what is a disaster—because what is really meant by this political rhetoric constitutes a challenge to the international community. The challenge is to recognize how low-lying island countries are being forced into a situation of needing to adapt on a large scale. The rhetoric may simply be a means by which to try to leverage funds to assist the adaptation process. But underlying such a move are nevertheless serious issues that deserve attention, no matter how true it is that the political rhetoric has other agendas. One is that many low-lying countries are potentially being faced with an inordinate economic challenge to adapt to sea-level rise. The second is that it is possible that migration is not seen as a type of adaptation by many in these low-lying countries, but rather as a failure of adaptation. The central interest of this chapter is to explore how framing the issue through the lens of cultural traditions enables practical adaptive processes, dynamics, and options beyond the political rhetoric. While there has been much rhetoric about, and considerable research outlining the perceptions of South Pacific communities on how climate change migration affects their cultures, there has not been as much effort put into understanding how traditional knowledge works to practically facilitate and construct adaptive measures. The aim of this chapter is to redress this lack.

The hypothesis of the chapter is that migration is often an effective adaptation in the South Pacific when guided by traditional knowledge, but that migration can also lead to maladaptation due to assimilation of indigenous people into modernized society rather than their successful adaptive acculturation of useful aspects of modernization and globalization. While at the same time, migration is sometimes interpreted as a failure of adaptation: traditional knowledge can help avoid maladaptation. To be able to introduce the conceptual model used to analyze the case study, short-term and long-term resilience are first introduced and distinguished. Second adaptation and maladaptation are defined, and then the drivers of migration are outlined to complete the framing of the issue.

1.1 Short-Term and Long-Term Resilience

"Short-term resilience" is defined as reversible adaptation within a particular social-ecological structure, and "long-term resilience" is defined as irreversible adaptation incorporating the need for proactive transformation of social-ecological structures. Insight into the distinction between short-term and long-term resilience is given by a related distinction between adaptation for resilience and adaptation through transformation (Pelling 2011). Adaptation for resilience refers to ensuring situations under stress are able to "bounce back" to how they were before the stress (Moser 2010), whereas adaptation through transformation has the sense of a change brought either by stress or by preparation for stress.

Transformation is about seeking alternatives (O'Brien 2012). Further insight is provided through the distinction between "engineering resilience" and "ecological resilience" (Gunderson et al. 2003). Whereas "engineering resilience" refers to "the ability to return to the steady state following a perturbation," "ecological resilience" refers to "the magnitude of disturbance that can be absorbed before the system redefines its structure" (Gunderson et al. 2003: 530). The key point is that resilience can be proactively maintained long-term if a social-ecological system undergoes transformation before it is forced onto the system. Therefore, whereas the short-term resilience concept appears to be a very appropriate concept to guide adaptation in face of climate variability, it is nevertheless less useful to ensure long-term resilience to climate change and sea-level rise. Transformations of social-ecological structures are required in a process of coevolution between human social structures and the environment, for example in face of anthropogenic climate change bringing sea-level rise.

Similarly, there appears to be a mistaken belief in the ability to contain and accept the inevitability of risks by controlling the environment and resources through technological and financial means (Beck 1992). What this "normalization of risk" strategy overlooks is that the ability to increase short-term resilience can occur at the expense of long-term resilience. Modernization and globalization, which favor short-term resilience to high frequency low impact events, can come at the expense of decreasing long-term resilience to low frequency high impact events (Bunce et al. 2009; Lauer et al 2013; Maru et al. 2014; McCubbin et al. 2015). By contrast, there is evidence that traditional social-ecological systems are able to recognize ecological resilience and so deliberately transform themselves to engage in proactive adaptation to climate change for long-term resilience (Maru et al. 2014; Mercer 2010).

1.2 Adaptation and Maladaptation

To clarify the root causes of migration experienced by many societies as a failure to adapt, a useful starting point is once again Pelling (2011), where he argues that the root causes of anthropogenic climate change are to be found in maladaptive social-ecological structures, which encourage sustaining the wrong type of adaptive measures. Zografos et al. (2013) refers to maladaptation, as adaptation that decreases security for one segment of the population, while perhaps increasing it for others. The implication is that such social-ecological structures need to be transformed to ensure the goals or ideals of social equity, justice and inclusivity. One example of this is the need to ensure land tenure and access to natural resources, which become pertinent issues when adaptation involves migration.

To recognize the difference between adaptation and maladaptation requires first to appreciate a somewhat ironical feature of adaptation: Adaptation is a result of changes needed to maintain elements of a system (e.g., a society or household) that ought be kept constant, such as community relationships (Rappaport 1999), whereas maladaptation results from mistaken adaptation of elements of a system that stayed constant rather than evolving, such as inappropriately built houses through capital investments in a floodplain. Adaptation is about exploring strategic alternatives that maintain the highest priorities. This facet of adaptation refers primarily to long-term resilience and proactive transformation of social-ecological systems. To date there are however few conceptual models to guide proactive transformation of social-ecological structures. It is the hypothesis of this chapter that one reason is because the role of culture, and traditional knowledge in particular, has been overlooked as a useful framework for guiding successful adaptation. One strategic adaptation where the role of traditional knowledge can clearly be seen is how migration is successfully carried out under the guidance of cultural traditions.

1.3 The Drivers of Migration

For migration to be a successful adaptation strategy to climate change, it has to either transform the maladaptation wrought by modernization and globalization, or avoid maladaptation by avoiding pressures created by globalization and modernization, through transforming socio-ecological systems toward long-term resilience. Analysis is however required to determine the leverage points where cultural traditions can guide successful migration. Black et al. (2011) defined five key types of drivers for migration: economic, political, social, demographic, and environmental. There are two key features to Black et al.'s (2011) analysis. One is that drivers are both positive and negative, for example environmental drivers include both ecosystem services as well as hazards. The second is that underlying environmental degradation as a push factor for migration is largely unrepresented in standard theories about migration. Explanation is provided by other researchers (McCubbin et al. 2015). They argue that environmental effects are very complex, and that even if people are responding to environmental stressors, this is usually mediated through non-climatic dimensions, for example overcrowding, urbanization, economic choices, changing land use and shifting cultural values. Cultural traditions can therefore be expected to have a widespread influence through the complex social interactions at play with environmental drivers.

What has distracted analysis away from the complex sociocultural influences, through oversimplifying the complexity, has been sensationalizing of environmental degradation due to climate change (Farbotko 2005). Seminal research on the interaction between culture and the environment, for example by Rappaport (1968) in Papua New Guinea, and work on migration within the traditional cultural context in Fiji (Bedford 1976), became eclipsed by an exclusive focus on potential climate change effects of global warming. In the 1980s the first prominent South Pacific regional report on the environmental impacts of climate change was produced (ASPEI 1988), and quickly taken up by politicians in the international arena

(Paeniu 1991). Reports and political actions focusing almost exclusively on environmental effects of climate change impacts have continued unabated until today.

There is similar little appreciation of the degree of complexity when the economic drivers for migration are analyzed. Betts (2011) for example has suggested a framework through an analysis of the intersection of governance and the economic drivers of international migration. He characterized two emergent types of migration. One type consists of private goods due to employment through either skilled or season unskilled migration, producing remittances from both, and resulting in emergent bilateral or regional cooperation for the mutual benefits gained between nations. There is a wealth of information about this type of economy as a contemporary driver for migration in the South Pacific region (Campbell and Warrick 2014; Connell 2004, 2006), as well as a traditional driver prior to European colonization (Hau'ofa 1993). The problem is as much about immobility, exacerbated by the contemporary nation state, as it is about migration (Findlay 2011), but the role of the South Pacific traditional local economies are seldom considered (Morrison 2008). Yet further complexities arise with the second type framed by Betts (2011), namely migration as a global public good where the need for refugee migration is recognized as beneficial for all. A contemporary legal question, much debated, is whether or not this applies to environmental refugees due to climate change (McAdam 2012), but the reasons for the lack of resonance of the concept within local communities seldom focused upon (Farbotko and Lazrus 2012).

2 Toward a Conceptual Model: The Role of Cultural Traditions to Guide Migration and CCA

That there are influences by cultural traditions on migration decisions in the South Pacific is well documented. For example, Campbell and Warrick (2014) in a review of literature indicate that the main concerns when making migration decisions are about "loss of identity, culture, family ties and community for communities leaving homelands, and conflict and governance issues" (Campbell and Warrick 2014: 25). On the other hand, it is also recognized that people do nevertheless successfully maintain culture and communities when migrating, even across what are now national borders, and have been doing so for centuries in the South Pacific as part of a regional economy connecting a "sea of islands" (Farbotko and Lazrus 2012; Hau'ofa 1993). What is missing in the literature is an analysis of what enables culture and communities to be maintained after migration. Farbotko and Lazrus (2012) argue that such analysis should be a point of departure for research on the topic. What is also abundantly clear from the literature is that in the South Pacific, environmental concerns, including climate change, are seldom the reason for migration, and are often resisted (Campbell and Warrick 2014). Therefore adaptation through migration is likely to be proactive for benefits gained, while maintaining and extending a sense of home and community. This research proposes a framework for understanding how traditional knowledge provides keys to do so, through the capacity to avoid maladaptation by maintaining what is essential for successful adaptation. There are generally recognized social-ecological processes structuring cultural traditions and activities, which can serve as a starting point for an analysis of what the keys are. Understanding them helps determine leverage points where cultural traditions proactively transform social-ecological structures to ensure successful adaptation.

According to Rappaport (1999), a key feature of cultural traditions is a hierarchy of types of knowledge, all of which operate simultaneously. Rappaport (1999: 263) terms the highest level "Ultimate Sacred Postulates." They are invariant, and so they are the ultimate priority that drives adaptation within a culture. Their invariance is however at the expense of not having specific content. They refer mostly to attitudes and ethical orientations (Morrison and Singh 2009). The term "horizon" is often used to describe this level of knowledge. Rappaport (1999: 263–268) terms the next level down "cosmological axioms," which are worldviews associated with values, such as hospitality. They do adapt, but slowly and most often are not experienced as changing. Rules forming the structure of a society are the third level. They are experienced as adapting. The lowest level refers to contextually dependent implementation of rules, which change and adapt even if social-ecological structures do not change. Rappaport (1999: 267) emphasizes that a consequence of the simultaneous use of these four levels of knowledge is to make cultural traditions "highly flexible." With an invariant ethical horizon of eternal infinite hope beyond death to direct adaptation, rules are left free to adapt as necessary.

By contrast, the horizon of modernization and globalization is not invariant, nor is it used simultaneously with other types of knowledge. Rather the horizon of modernization and globalization is the evolving technological knowledge about how to control nature. Instead of a hierarchy of simultaneously operating types of knowledge, the power to control nature by technology is simply used back to structure society. This conflating of the cultural hierarchy has been termed "reflexive modernity" (Beck et al. 1994) by those analyzing risk. In contrast to the traditional cultural horizon beyond death, where risk is absent, modernization limits the ability to minimize risks associated with material events, through the control of material reality. The lack of an invariant and ethical priority to direct adaptation in modernization can be expected to not only produce maladaptation, but also to lack adaptive flexibility for transforming social-ecological structures.

According to adaptation theory, there are two facets to the experience of adaptation guided by a traditional culture (Rappaport 1999: 406–437). First, the invariant horizon and slowly adapting worldview provides a framing of parameters or rules to guide appropriate responses. This is partly captured by the "Law of Requisite Variety," which describes the need to be able to respond to significant influences, such as "king tide"[3] levels and rainfall patterns (Bossel 1998; Morrison

[3] "King tides" is a South Pacific colloquialism referring to the highest tides occurring twice or so a year.

and Singh 2009). Second, to determine these significant influences requires a more ethical orientation, including care and concern for others and the maintenance of reciprocity within an economy and the flow of ecosystem goods and services (Adger et al. 2009; Morrison 2008; Morrison and Singh 2009; Reenberg et al. 2008).

Cultural traditions therefore fulfill the role of providing a horizon and values to help integrate and make sense of a complex set of drivers—both pushes and pulls—for short-term and long-term resilience. Migration can enable both types of resilience: short-term resilience through circular migration; and long-term resilience through migration to become part of new societies with new and mobile social identities and roles. Once again this is provided by an eternal horizon and worldview of values of traditional cultures, which maximize freedom in relation to adapting to and developing new social rules and roles, as well as the implementation of new rules and roles in new contexts. Figure 15.1 summarizes the conceptual model proposed here.

For example, to ensure circular skilled migration achieves the function to benefit the Pacific Island communities the migrants travel from, they must send home remittances, and eventually return home to provide their professional services to their home country. Otherwise this migration becomes an example of a "brain drain," and the development process that trained the professionals does not add to the resilience of their home country. But for professional migrants to maintain their links to their communities they need to maintain the communitarian values of their cultural traditions. They must resist assimilation into the individualism of the modernization process, with which their training inevitably brings them into contact. This is similarly the case for seasonal unskilled migrants on work programs. Even though their migration is regulated to ensure it is cyclical, unless they become "overstayers," they have to equally resist the consumerism values of the culture and economy in which they work overseas. They are hard pressed not to spend more than they earn on the lowest of wages, even less to save money and send home remittances. Only strongly held communitarian values of their cultural traditions ensure that their migration is functional and adds to the resilience of their home and host communities.

3 Research on Cultural Traditions and Migration

This research on the role of cultural traditions on migration is a sub-component of a broader climate change adaptation research program primarily funded by the European Union - Global Climate Change Alliance (EU-GCCA) at The University of the South Pacific. The main component is concerned with the sociocultural aspects of climate change adaptation. Migration is a subcomponent of this work.

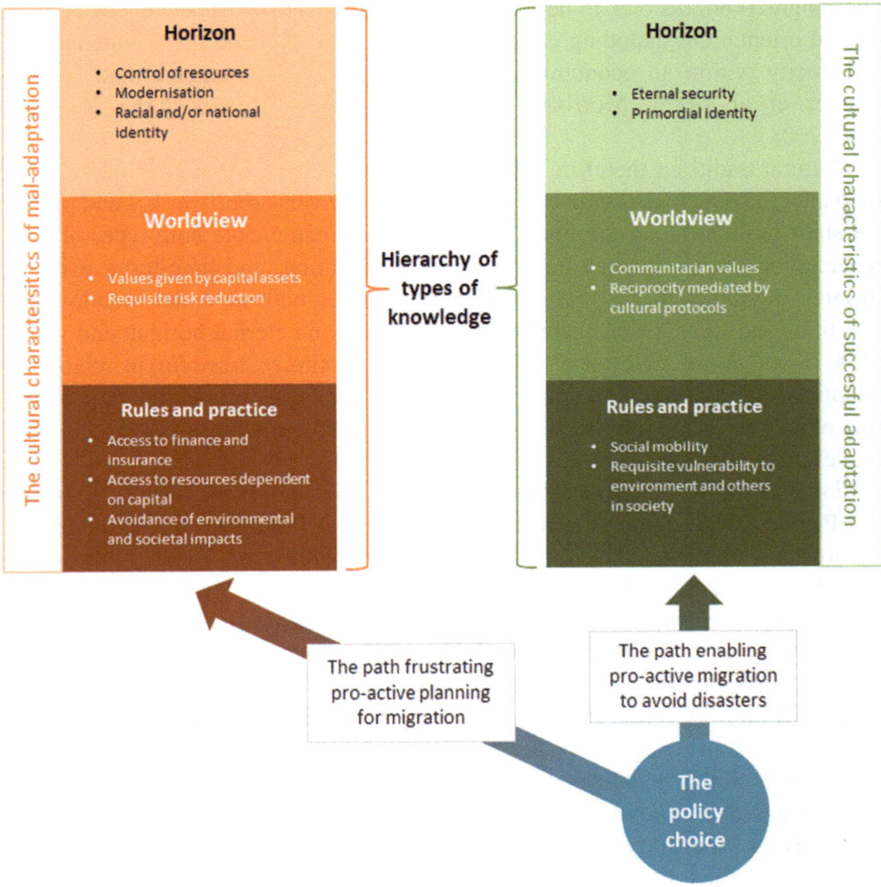

Fig. 15.1 Conceptual model for the role of cultural traditions to guide migration and CCA

The EU-GCCA funded the work through a postgraduate program in climate change adaptation, where community-based action research was utilized as a means by which postgraduate students experientially learn with the primary researcher and author of this chapter. Participatory action research, which has much in common with the community-based action research carried out, is recognized as an effective way for learning about adaptive capacity and associated social vulnerability (Fischer et al. 2013). Indigenous research methodology was utilized to carefully avoid framing the community-based action research in a way that is culturally different to the everyday practices of the community (Smith 1999). Standard qualitative methods were utilized, in particular in-depth interviews and "focus groups," but only as part of conversations that emerged out of everyday normal contexts and in accordance with cultural protocols (Teaiwa 2007; Wulff 2010). The cultural protocols and processes participated in were those of traditional indigenous Fijian culture.

The first step was to seek traditional approval to engage with the village through the *sevusevu*[4] protocol. A gift of *yaqona*[5] was made by the author to the *turaganikoro*[6] of the urban indigenous Fijian village of Kalabu, in the most densely populated district of Fiji, near the capital city, Suva. The gift was accepted and the *turaganikoro* then sought permission from the chief of Kalabu. Approval was given to meet with the village community and the communities of the informal settlements on Kalabu land. Four *talanoa*[7] discussions were held over a period of 4 months. The author, postgraduate students and the community participated, and in one *talanoa* personnel from governmental departments also participated. The *talanoa* were called by the *turaganikoro*, and they proceeded according to traditional protocols. Five in-depth interviews were carried out by the author during the *talanoa*. The interviewees for all interviews presented themselves during the *talanoa*, and the interviews were held as the *talanoa* were coming to a closure. There were also five "tours" undertaken around the Kalabu land, by men and women of Kalabu to inform the author and postgraduate students about the issues they wanted assistance with.

3.1 The Study of Horizons

Conceptually, the role of eternal horizons is apex to the adaptive capacity facilitated by cultural traditions. Moreover, the technological system that substitutes for eternal horizons in modernization frames the bounded short-term resilience found in modernized societies. Therefore an objective of reflection upon what was learnt through the community-based action research was to discover the role of horizons in providing adaptive capacity. Associated with this was however also a focus on social vulnerability because conceptually requisite social vulnerability provides the sensitivity and responsiveness necessary for adaptive capacity (O'Brien 2013), and is nurtured by traditional horizons and worldviews (Moser 2013). The empirical component of this research sought to explore adaptive capacity in societies that are both influenced by cultural traditions and modernization, namely as related to the management and degradation of natural resources and how this affects migration. This was done by discovering the horizons, or attitudes and ethical orientations of

---

[4] The *sevusevu* protocol involves a ceremonial gift of a local intoxicating plant product to the village.

[5] *Yaqona* is the Fijian name for what is more commonly termed "kava" (*Piper methysticum*). *Yaqona* is of cultural significance throughout the South Pacific. It has marked medicinal psychological effects, in particular overcoming anxiety. *Yaqona* is usually drunk in coconut shell cups dipped into a large carefully carved communal bowl, termed a *tanoa*.

[6] The *turaganikoro* is the local headman or mayor of a community. The title is not hereditary, and usually only given for a limited duration.

[7] A *talanoa* is a "focus group" type discussion according cultural protocols, which include respectful speaking and the communal drinking of *yaqona* out of the community *tanoa*.

various stakeholder groups in comparison with those of the researcher through participation in joint activities (Kearney 1995). The activities jointly participated in were data collection and discussions with the community along with personnel from government departments, so as to assist Kalabu village prepare for and to engage with government to address concerns they have about environmental degradation of their land, especially the pollution of the Lagere River flowing through their land. The process was found to take time and occurred as mutual understanding emerged; this was found to be mutually empowering to both the communities and postgraduate students involved.

The role of horizons in migration was teased out through two follow-up in-depth interviews carried out by the author, to further explore concepts that had arisen out of the 2014 phase of the community-based action research with Kalabu. The first was held in January 2015 with an elder in South Tarawa on Tarawa atoll in Kiribati, to explore what was meant by previously recorded references to "eternal safety," "moral safety," and "spiritual safety," and that the loss of this was what was most feared from climate change. The elder emphasized that the terms were references to eternal and universal values of humanity, and refer to a "primordial identity" rather than merely a social identity. Given that the loss of this eternal horizon was what was most feared, it appears that the adaptation by migration to another physical place and home is acceptable, just as long as the absolute horizon of hope, or "eternal safety," is maintained in the new home. The plausibility of this interpretation was borne out by a second follow up in-depth interview in February 2015, in Suva Fiji, with an elder from Tuvalu. He referred to how primordial identity in Tuvalu metaphorically alludes to intense experiences of a type of tree, referring to early childhood experiences of local vegetation. But he also made the point that the use of this metaphor no longer only applies to the ancestral island and islet, because so many are now brought up elsewhere. Nowadays it also refers to the place and home where a person is brought up.

What is therefore sought to be maintained through migration is the eternal horizon, and what is most feared in migration is the loss of the eternal horizon due to cultural assimilation into the horizons of "development," or in other words the maladaptation of modernization. Even though this was also described in terms of fearing a loss of cultural identity, the cultural identity referred to was a primordial identity. Just as adaptation is paradoxically maintained by what is kept constant, a primordial identity which is valued as the cultural identity provides flexibility to adapt and modify social identities and roles when migrating. There was the fear that migration to more "developed" countries could threaten this and hence their "eternal safety."

One further point was strongly emphasized by the elder in Kiribati, namely how a primordial identity provides personal agency and gives the right to everyone to proactively transform social structures they find themselves within. Even though an eternal horizon and primordial identity was recognized as providing long-term resilience and adaptive flexibility, one of the things feared from migration was the loss of social mobility. Nevertheless there was hope that their social freedom would be retained after migration as long as they kept their eternal horizon and primordial identity.

3.2 Findings: The Role of Informal Settlements and Migration

Kalabu lies within the greater Suva city boundary of 208,808 people (according to Fiji Government official 2014 statistics), out of the total Fiji population of 887,000. Urbanization in Fiji is occurring at a rapid rate. 53.4 % of Fijians lived in urban areas in 2014, up from 41.6 % in 1990. Suva is in Viti Levu Island, which is the largest of the 106 inhabited islands of Fiji. Fiji islands have a total land area of 18,272 km^2, in an ocean of 1,290,000 km^2, and are strategically placed in the South Pacific, east of Australia, north of New Zealand, and south west of Hawai'i.

Kalabu has 300 inhabitants (according to the 2007 census), owning 2,850 hectares (ha), of which 1,874 (ha) are under lease and managed by the iTauke Land Trust Board (TLTB),[8] with the Kalabu villagers receiving the rent. The 1,874 ha are leased as housing estates, industrial areas including the Kalabu Tax-Free zone, commercial agriculture and a quarry. However, this leaves Kalabu with only 976 ha to share with several thousand informal migrant settlers (1733 according to the 2007 census). The migrants do so as part of a traditional *vakavanua*[9] arrangement whereby the village landowners allow most people who respectfully seek somewhere to live, to live rent free, to build a house and grow food, just as long as they comply with traditional cultural protocols of the village and respect the authority of the traditional leadership. These informal settlers are however frowned upon by the TLTB as the settlers are outside of their jurisdiction, as well as outside of most regulatory processes of the formal governance of Fiji. They fall almost completely only under the authority of village regulations based on cultural traditions. The migrants include indigenous Fijians from other villages and islands who are refugees for various official and unofficial reasons, including environmental impacts, for example from the Lau group of islands in the Tovata settlement, and a Rabi community in the Savutalele settlement. Interspersed in the informal settlements are Solomon Islander descendants of indentured laborers "blackbirded" to Fiji during the British colonial era, along with Indian-descent Fijians who are also descendants of indentured laborers.

The Rabi migrants are refugees from Banaba Island, in present day Kiribati, relocated in 1945 to the then uninhabited Rabi Island in Fiji. Banaba Island was mined for phosphate by the British colonial government from 1900 to 1979. The indigenous population had long been an awkward hindrance to the mining operation, but a pretext to relocate them was only given after World War II. The invading Japanese forces during World War II had already moved the Banaba population into camps on the islets of Tarawa atoll in Kiribati (then part of the Gilbert and Ellice

[8] The iTauke land Trust Board (TLTB) is the legal guardian for all indigenous owned land in Fiji. They are the formal body who negotiates leases with those who wish to use the resources owned by indigenous Fijians. No land under trust to the TLTB is legally able to be sold.

[9] The *vakavanua* arrangement involves the *sevusevu* protocol offering of *yaqona* to the village, as used also by the research team to initiate the community-based action research.

Islands—now Kiribati and Tuvalu), and so after the war the British colonial government advised the Banaba population that they could not return and relocated them instead onto relatively isolated Rabi island. There are differing accounts of what happened, but the narrative maintained by the Rabi community is that they were tricked (McAdam 2013). After a court case against the British Government they obtained some redress, but continue to feel aggrieved. The population of Rabi Island was approximately 5,000 in 2014, with 95 % being descendants of the Banaban refugees.[10] Many Rabi islanders have now begun to migrate into greater Suva in Fiji, as part of the urbanization in Fiji, with one of the communities in Savutalele informal settlement.

Insight into what has driven the migration of the informal settlers to "squat" on Kalabu land, and what would drive their future migration to move on, emerged through the community-based action research *talanoa* about the apparent lack of security that the informal settler migrants have. The communities made it clear that the lack of a formal lease or ownership of land was not a major issue for them. Their security is rather found in being able to "live freely" and also "safely, quietly and in community." The free life referred to is not principally in relation to not having to pay rent but rather in relation to not having to comply with regulations other than the traditional cultural processes maintaining community in the village. Most also see their stay in the informal settlement as a transition to enable them to become successfully established into the wider society of urban Fiji. It is a means by which their migration to the capital city can successfully maintain and extend their community integration. Indeed commonly, as one generation moves on from the settlements after their children have become educated in local schools and obtained employment, another family from their home island or community will migrate to reside in their house in the informal settlement. Other reasons given for migration to the informal settlements are that they are able to grow their own food whilst being able also to work in the town, with their children able to go to good urban schools and have access to health care.

The migration-to-dwell according to the *vakavanua* arrangement appears to be an example of a successful transformative adaptation, based on traditional cultural understanding and practices. The challenge is that the formal governance of the land by the TLTB is trying to stop the process, using arguments such as the need to ensure landowners obtain market rates for leases and to ensure development is in accordance with regulatory processes. But when asked if they would consider being relocated by the government to rural lifestyle blocks where they can grow their own food they said they would consider it, but not to a (sub)urban subdivision. To "live freely" also means to be able to interact with the natural environment and rhythm of

[10] The forced migration from Banaba Island to Rabi Island can be contrasted to voluntary migration in 1947 from Vaitipu Island in present day Tuvalu onto Kioa Island neighboring Rabi Island. The migration to Kioa Island also occurred after World War II but was proactively planned. A notable consequence is that the "Tuvaluan" community of Kioa Island do not feel aggrieved with their immigration to Fiji and identify far more with being Fijian than do the "Rabi" (McAdam 2013).

ecological processes; an important driver as Kalabu village seeks to restore their river.

Another strongly expressed reason why migrant informal settlers stay in a mangrove wetland in Veiraisi is because of links they have through marriage or have developed through friendships with neighboring Kalabu village. They feel safe and connected not only to the natural environment but also the Kalabu community. When asked what they will do if they have to move due to climate sea-level rise, they say they plan to migrate to another informal settlement in the hills around Kalabu so that the connection with Kalabu community is maintained.

3.3 The Role of the Traditional Worldview

On the other hand, while both the migrants and the landowner leaders of Kalabu highly value their freedom from formal regulatory processes, they are also very aware and insistent on the need for the formal role of the government to be fulfilled, especially in relation to environmental regulations. The community-based action research involved the author and postgraduate students to work with Kalabu village to try to restore the health of the Lagere River flowing through their land. The river was once a prime source of food, and also contains Kalabu's totem, a fresh water shellfish (Fig. 15.2).

The village leaders are very aware of the demarcation of their authority and the government's enforcement authority. They commonly express how they feel betrayed by the government allowing the environment to become so degraded, and

Fig. 15.2 The very heavily polluted Lagere river flowing through Kalabu land (Shamita Chandra 2014)

have for several decades been trying to seek redress. The village leadership very strongly assumes proactive adaptive comanagement as the norm. Thus the community-based action research was a contribution to facilitating linkages between sympathetic personnel within key government ministries and the village leadership.

The community-based action research process also brought engagement with migrant informal settlers at Veiraisi along the coast in a mangrove wetland, and informal settlements in the hills around Kalabu village itself. Talking with the informal settlers about their migration there, and also the threat of a further need for migration for those living in the mangrove wetland at Veiraisi, brought some insights into what the drivers for migration are for them and how their horizons and worldviews operate to guide the process. The village leadership and community of landowner villagers and informal migrants clearly recognize the need for the traditional worldview to guide long-term resilience, but equally see the need for formal governmental processes to take responsibility to regulate the development processes of modernization. They clearly see formal regulatory processes as a necessary complement to their traditional processes. They assume and demand that a proactive adaptive comanagement process is carried out. Proactive adaptive comanagement facilitated by traditional processes is not however a gentle rational dialogue.

The traditional perspective recognizes and prepares traditional leadership to test governmental officials. The posturing, rhetoric, demands and debating by traditional leadership are however commonly misinterpreted by government officials, donors and their consultants. It is commonly heard from government officials, donors and consultants that traditional leadership is unthankful, demanding, rude and uncooperative. This interpretation appears to arise due to a lack of understanding of the traditional perspective, and even romanticism of it. The traditional perspective is not naive, it includes protocols to explicitly challenge to test visitors and potential partners, to ensure they are sincere and have commitment. Traditionally this was a key role of chiefly orators. For example the Kalabu village leadership has successfully taken government ministries to court and is in constant debate with TLTB and government ministries. Far from this showing a lack of willingness to engage in proactive adaptive comanagement, it actually proves the commitment of the traditional leadership to the process. Authentic trust and working relationships are formed through challenges and learning how to respond to prove the honesty of intentions and the real capacity to be able to help and to work together. If anything, there is a failure of government officials, donors and their consultants to know how to effectively and constructively engage in the necessary debates, which also makes them equally reluctant to engage fully in scientific debates about what needs to be done. The apparent uncooperativeness of traditional leadership actually belies a far stronger commitment and capacity to face up to maladaptation and the need for transformative alternatives. The need for effective migration policies to enable successfully transition to settlement in urban areas in Fiji is a case in point.

4 Conclusions

After first reviewing the issue of the need for long-term resilience and avoidance of maladaptation, and how these drive and affect migration, the chapter proposes a conceptual model incorporating a key feature of cultural traditions, namely how a hierarchy types of knowledge are used simultaneously as a means for which to provide high adaptive flexibility. At the apex of the hierarchy of types of knowledge are eternal horizons of absolute hope where there is no risk, as the horizon sees beyond death and material nature, and slightly below, worldviews comprising of communitarian values. These two highest levels provide the context in which adaptation of social structure, roles and rules and their specific implementation are carried out with great flexibility. Next the chapter explores the hypothesis that cultural traditions provide the basis for successful migration due to enabling long-term resilience. Key facets of the proposed conceptual model are addressed. Clarification is achieved about cultural identity and how adaptation of social roles influence migration decisions, affirming the hypothesis that cultural traditions nurture adaptive flexibility to successfully migrate, by providing the insight that social identity and roles are not fixed when guided by cultural traditions. The main case study provides further insights about adaptive flexibility and social mobility, i.e., cultural traditions enable long-term resilience of migration. These include proactive and assertive engagements between various levels of governance structures (i.e., village and government) to seek transformation through comanagement of natural resources which were manifesting dysfunctionality as demonstrated by the failure by the government to stop environmental degradation. Follow-up research to consider other necessary social transformations in relation to education, health care, and employment deserves attention, including key drivers for urban migration, education and health of children, as well as employment. What has been made clear from the initial results is that the eternal horizon and communitarian values of traditional worldviews are key to empowering communities to freely maintain social mobility, improve their situation, and make their migration successful, but that there is also need for balance through comanagement between the two options outlined in Fig. 15.1.

One of the challenges of climate change adaptation is to ensure transformation of social structures through flexibility in social rules and roles. Such long-term resilience through adaptive flexibility is missing in modernized and globalized societies where rigid land tenure arrangements emphasize capital accumulation rather than community well-being. This is manifest in increasing low-frequency high magnitude disasters which are a driver for migration, including slow-onset ones like sea-level rise. On top of this, in the South Pacific context, migration creates the fear of becoming socially immobile without a home and community in a modernized society, without the capacity to adapt and improve livelihoods. The capacity to maintain social mobility needs to be addressed if migration is to be successful. Nonetheless there is keenness to migrate, including into urban settings, for the benefits that can be gained. But it will not be done willingly unless a cultural community is able to be maintained. Perhaps paradoxically, this appears to be assisted by extant cultural communities, even of different cultural traditions and ethnicities, in the host country. Traditional knowledge appears to avoid the fraught racial and national identity politics. The reason appears to be due to a common horizon driving adaptation among cultural traditions, namely the centrality of community and the reciprocity of relations, including with the natural environment.

Because maladaptation is associated with assimilation into modernization and globalization, cultural traditions were hypothesized as potentially providing a role to guide the process of successful migration. This appears to be verified from our research which used a traditional worldview conceptual model to frame the issue of adaptation and migration. For cultural traditions to guide the transformation of social ecological processes, they need however to be better understood as solutions rather than barriers to adaptation and migration. The lack of adequate conceptual models to guide proactive transformation of social-ecological structures for long-term resilience is a limiting factor. Most models focus on short-term resilience operating within the existing social-ecological structures of modernization and globalization.

Even though the main case study is an example of how cultural traditions guide migration for long-term resilience in internal migration within Fiji, the concepts explored are in principle applicable also for international migration. It is worth noting that internal migration within Fiji is often between islands, involving distinct dialects and cultural traditions, including from the low-lying islands of what are now other nations threatened by sea-level rise. What the case study suggests is that the public good of recognizing environmental refugees could be realized if the immigrants are welcomed to become part of existing communities maintained by cultural traditions. A type of incubator is required, where the extant cultural adaptive capacity present in the cultural traditions of the immigrants is supported so that the immigrants in their own time adapt to accept and to develop new social rules and roles in their new environment to maintain their social mobility. Such an approach will help communities objectively threatened by the slow onset hazard of sea-level rise to proactively migrate, not only to avoid impending disaster but also to benefit from new resources and opportunities.

Acknowledgements The author wishes to thank Kalabu and the USP 2014 classes of EV414 and EV425 for the adventure of working with them, and the taxpayers of the European Union for funding the work.

References

ABC News (2014) Tuvalu Prime Minister Enele Sopoaga says climate change ‘like a weapon of mass destruction’. http://www.abc.net.au/news/2014-08-15/an-tuvalu-president-is-climate-change-27like-a-weapon-of-mass-/5672696

Adger WN, Dessai S, Goulden M, Hulme M, Lorenzoni I, Nelson DR, Wreford A (2009) Are there social limits to adaptation to climate change? Clim Change 93(3-4):335–354

ASPEI (1988) Potential impacts of greenhouse gas generated climatic change and projected sea-level rise on Pacific Island States of the SPREP region: preliminary report. South Pacific Regional Environment Programme, Noumea

Beck U (1992) Risk society: towards a new modernity. Sage, London

Beck U, Giddens A, Lash S (1994) Reflexive modernity: politics, tradition and aesthetics. Stanford University Press, Stanford

Bedford R (1976) Kabara in the 1970s Project Working Report No. 3, Man and the Biosphere (MAB) Project 7. Ecology and rational use of island ecosystems. Population and environment project: Eastern Islands of Fiji. UNESCO/UNFPA, Suva

Betts A (ed) (2011) Global migration governance. Oxford University Press, Oxford, pp 1–33

Black R, Adger WN, Arnell NW, Dercon S, Geddes A, Thomas D (2011) The effect of environmental change on human migration. Glob Environ Chang 21(1):S3–S11

Bossel H (1998) Earth at a crossroad: paths to a sustainable future. Cambridge University Press, Cambridge, UK

Bunce M, Mee L, Rodwell LD, Gibb R (2009) Collapse and recovery in a remote small island—a tale of adaptive cycles or downward spirals? Glob Environ Chang 19(2):213–226

Campbell J, Warrick O (2014) Climate change and migration issues in the Pacific. UNESCAP, Fiji

Connell J (2004) The migration of skilled health professionals: from the Pacific Islands to the world. Asian Pac Migr J 13(2):155–177

Connell J (2006) Migration, dependency and inequality in the Pacific: old wine in bigger bottles? (Part 2). In: Firth S (ed) Globalisation and governance in the Pacific Islands (No. 1). ANU E Press, Canberra, pp 81–106

Farbotko C (2005) Tuvalu and climate change: constructions of environmental displacement in the Sydney Morning Herald. Special Issue: Islands: Objects of Representation:. Geogr Ann Ser B Hum Geogr 87(4):279–293

Farbotko C, Lazrus H (2012) The first climate refugees? Contesting global narratives of climate change in Tuvalu. Glob Environ Chang 22(2):382–390

Findlay AM (2011) Migrant destinations in an era of environmental change. Glob Environ Chang 21(1):S50–S58

Fischer AP, Paveglio T, Carroll M, Murphy D, Brenkert-Smith H (2013) Assessing social vulnerability to climate change in human communities near public forests and grasslands: a framework for resource managers and Planners. J Forest 111(5):357–365

Gunderson L, Holling CS, Pritchard L, Peterson PD (2003) Resilience. In: Mooney HA, Canadell JG (eds) The Earth system: biological and ecological dimensions of global environmental change, of Encyclopedia of Global Environmental Change, vol 2. Wiley, New York, pp 530–531

Hau'ofa E (1993) Our sea of islands. In: Waddell E (ed) A new Oceania: rediscovering our sea of islands. University of the South Pacific, Suva, pp 21–16

Kearney R (1995) Poetics of modernity: towards a hermeneutic imagination. Humanities Press, New York

Lauer M, Albert S, Aswani S, Halpern BS, Campanella L, La Rose D (2013) Globalization, Pacific Islands, and the paradox of resilience. Glob Environ Chang 23(1):40–50

Maru YT, Smith MS, Sparrow A, Pinho PF, Dube OP (2014) A linked vulnerability and resilience framework for adaptation pathways in remote disadvantaged communities. Glob Environ Chang 28:337–350

McAdam J (2012) Climate change, forced migration, and international law. Oxford University Press, Oxford

McAdam J (2013) Caught between homelands. http://insidestory.org.au/caught-between-homelands

McCubbin S, Smit B, Pearce T (2015) Where does climate fit? Vulnerability to climate change in the context of multiple stressors in Funafuti, Tuvalu. Glob Environ Chang 30:43–55

Mercer J (2010) Disaster risk reduction or climate change adaptation: are we reinventing the wheel? J Int Dev 22(2):247–264

Morrison K (2008) Indigenous entrepreneurship in Samoa in the face of neo-colonialism and globalization. J Enterprising Commun People Places Global Econ 2(3):240–253
Morrison KD, Singh SJ (2009) Adaptation and indigenous knowledge as a bridge to sustainability. In: Begossi A, Lopes P (eds) Current trends in human ecology. Cambridge Scholars Publishing, Newcastle upon Tyne, pp 125–155
Moser SC (2010) Now more than ever: the need for more societally relevant research on vulnerability and adaptation to climate change. Appl Geogr 30:464–474
Moser SC (2013) Navigating the political and emotional terrain of adaptation: community engagement when climate change comes home. In: Moser SC, Boykoff MT (eds) Successful adaptation to climate change: linking science and policy in a rapidly changing world. Routledge, London
O'Brien K (2012) Global environmental change II: from adaptation to deliberate transformation. Progr Hum Geogr 36(5):667–676
O'Brien K (2013) The courage to change: adaptation from the inside out. In: Moser SC, Boykoff MT (eds) Successful adaptation to climate change: linking science and policy in a rapidly changing world. Routledge, London
Paeniu B (1991) Address by the Rt Hon Bikenibeu Paeniu, Prime Minister of Tuvalu. In: Jäger J, Ferguson HL (eds) Climate change: science, impacts and policy. Proceedings of the second world climate conference. Cambridge University Press, Cambridge, pp 527–530
Parry ML (ed) (2007) Climate Change 2007: impacts, adaptation and vulnerability: contribution of Working Group II to the fourth assessment report of the Intergovernmental Panel on Climate Change, vol 4. Cambridge University Press, Cambridge
Pelling M (2011) Adaptation to climate change: from resilience to transformation. Routledge, London
Rappaport RA (1968) Pigs for the ancestors: ritual in the ecology of a New Guinea people. Yale University Press, New Haven
Rappaport RA (1999) Ritual and religion in the making of humanity. Cambridge University Press, Cambridge
Reenberg A, Birch-Thomsen T, Mertz O, Fog B, Christiansen S (2008) Adaptation of human coping strategies in a small island society in the SW pacific—50 years of change in the coupled human–environment system on Bellona, Solomon Islands. Hum Ecol 36(6):807–819
Smith LT (1999) Decolonizing methodologies. Zed Books, London
Teaiwa KM (2007) Indigenous encounters: reflections on Relations between people in the Pacific. Occasional papers no. 43/Center for Pacific Islands Studies Occasional paper series/Center for Pacific Islands Studies, School of Hawaiian, Asian, and Pacific Studies, University of Hawai'i at Mānoa
Tong A (2014) Tag archives: Kiribati president Anote Tong. http://www.climate.gov.ki/tag/kiribati-president-anote-tong/
Wulff D (2010) Unquestioned answers: a review of research as ceremony: indigenous research methods. Qual Rep 15(5):1290–1295
Zakaria F (2014) On GPS: a sinking island nation. http://cnnpressroom.blogs.cnn.com/2014/06/08/president-of-kiribati-anote-tong-on-climate-change-its-too-late-for-us-on-cnns-fareed-zakaria-gps/
Zografos C, Goulden MC, Kallis G (2013) Sources of human insecurity in the face of hydro-climatic change. Global Environ Change. http://dx.doi.org/10.1016/j.gloenvcha.2013.11.002

Chapter 16
Conclusions: Linking Sustainable Development, Disaster Risk Reduction, Climate Change Adaptation, and Migration—Policy Implications and Outlook

Karen Sudmeier-Rieux, Manuela Fernández, J.C. Gaillard, Michel Jaboyedoff, and Ivanna M. Penna

2015 was the year when three major international frameworks were negotiated: the Sendai Framework for Disaster Risk Reduction 2015–2030 (SFDRR), the United Nations Framework Convention on Climate Change (UNFCCC), and the Sustainable Development Goals (SDGs). With 232 million people on the move in 2011 (UNDESA 2015), migration is also one of the most important cross-cutting themes—albeit contentious—in all three international agreements. How the above issues are linked is complex and the topic of this book: it sets out to address an array of migration issues that are linked to environmental issues, notably disasters which can become aggravated due to climate change. Thus, the chapters selected in this volume present a range of issues and geographical areas focusing on a number of sub-topics related to disaster risk reduction (DRR), climate change adaptation (CCA), and inter-linkages with migration and sustainable development. This final chapter revisits main emerging themes concluding with a number of policy recommendations that were developed throughout the book.

K. Sudmeier-Rieux (✉) • M. Fernández • M. Jaboyedoff
University of Lausanne, Lausanne, Switzerland
e-mail: karen.sudmeier-rieux@unil.ch

J.C. Gaillard
School of Environment, The University of Auckland, Auckland, New Zealand

I.M. Penna
Geological Survey of Norway, Trondheim, Norway

© Springer International Publishing Switzerland 2017

K. Sudmeier-Rieux et al. (eds.), *Identifying Emerging Issues in Disaster Risk Reduction, Migration, Climate Change and Sustainable Development*,
DOI 10.1007/978-3-319-33880-4_16

1 Disaster Risk Reduction, Migration, and Sustainable Development

The premise of this book is to seek greater conceptual clarity and present case studies about how DRR strategies are affected by migration patterns and vice versa, rather than the commonly used notion of climate change as a main driver of migration. In most countries around the world, DRR continues to operate in a short timeframe of immediate relief and not toward addressing the "underlying risk" factors that create disasters. The social construction of disasters (Alcántara-Ayala and Oliver-Smith, Chap. 7) unfolds through time and is deeply embedded in specific historical contexts that inevitably shape the unequal distribution of risk and vulnerability. Thus, disasters are rooted in the social, economic, and environmental history of the societies in which they take place (Cardona 2003; Garcia Acosta 2005; Hewitt 1995; Lavell 2000; Maskrey 1993; Oliver-Smith 2002). Therefore, most chapters of this book considered various aspects of DRR and how migration and sustainable development interact with risk, with migration having both the potential to increase and decrease vulnerabilities and exposure to risk. Except for the chapter on "Disaster Diplomacy" (Kelman, Chap. 12), the book does not to a great extent address mass migration due to conflict, which is a very critical type of disaster but one that fell outside the scope of this book. It also did not address the issue of so-called climate refugees directly but rather questions whether this term represents an oversimplification of the myriad of push and pull factors and multiple additional obstacles and pathways or intervening factors leading to migration.

As we consider how migration is linked to DRR from several different angles, we refer to the well-tested pressure and release (PAR) model (Wisner et al. 2004), which resonates with many of the contributions of this book. It describes how people's vulnerabilities progress from root causes, anchored in limited access to power structures and resources (Hewitt, Chap. 3; Jahn et al., Chap. 10), are aggravated by dynamic pressures such as rapid population changes (Upreti and Shrestha, Chap. 9; Devkota and Chandra Lal, Chap. 13), rapid urbanization (Texier and Edelblutte, Chap. 6), armed conflicts (Kelman, Chap. 12), decline in soil productivity (Schwilch et al., Chap. 11), and lack of appropriate human resources. These factors lead to poor livelihoods, including unsafe conditions such as people living in dangerous places (Pigeon, Chap. 5), in unprotected buildings, and low income levels. The PAR also highlights the lack of public actions and institutions, such as disaster preparedness (Alcántara-Ayala and Oliver-Smith, Chap. 7) or many other public precautions for keeping people safe. These are the factors that create disasters and within which migration can play a key role, both as a *positive* factor for releasing pressure as populations move from dangerous places, acquire new skills, and contribute to diversifying livelihoods and local economies (Upreti and Shrestha, Chap. 9), and at times as a *negative* factor when out-migration causes labor shortages, loss of local knowledge (Jacobi et al., Chap. 14), and more vulnerable conditions in the place of origin as migrants congregate in new places at risk, such as urban slums along rivers (Texier and Edelblutte, Chap. 6). Migration is also considered at various scales (household to national) and geographical contexts (North America, South America, Asia, Europe, and Pacific).

Thus, another strong theme that emerged is the question of migration and risk transfers. Using the UNISDR (2009) terminology, we recall that disaster risk is the combination of exposure (or the number of people and amount of time spent in a dangerous place), vulnerabilities (economic, social, and physical state of a population), and the hazard event (e.g., sudden hazards such as cyclones, earthquakes or slow on-set hazards such as drought, slow moving landslides). Several of the contributions pointed to a transfer of risk from rural to urban areas in Jakarta and Mumbai (Texier and Edelblutte, Chap. 6; Jahn et al., Chap. 10) or from mountainous places to low-lying areas of Nepal (Schwilch et al., Chap. 11; Devkota and Chandra Lal, Chap. 13). Let us remember that exposure and hazards can be high in both the place of origin and at times even higher in the place of destination, whether in rural or upstream areas affected by landslides and flash floods or low-lying urban slums, which often become hotspots of disaster risk. However, by moving to places with better economic opportunities, what people are clearly seeking to reduce is their vulnerability while enhancing their capacity to deal with future shocks and setbacks, even if this means being forced to move to a place with high exposure (Gaillard and Cadag 2009; Nathan 2008).

In the place of destination, migrants may be unaware of local environmental conditions, early warning signs, or evacuation routes, making DRR even less effective. On the other hand, many migrants originate from areas experiencing disasters and may bring with them innovative and new ways of coping with disasters and creating resilience, or the ability to "bounce-forward" (Manyena et al. 2011). Hence with continued exposure to dangerous physical factors, what migrants might hope to reduce are their levels of vulnerability through better access to resources, employment, education, and markets or the livelihood conditions expressed in the PAR framework. According to Guadagno (Chap. 2, p:18), "mobility-based choices are part of the strategies households can pursue in order to obtain well-being outcomes, in a context of more or less limited choices and significant tradeoffs."

> "People, throughout history, have frequently moved for numerous environmental and social reasons, including responding to short-term and long-term environmental changes, responding to short-term and long-term social changes, education, joining family, adventure, and seeking different livelihoods. Such migration is sometimes entirely voluntary, is sometimes entirely forced, and is most frequently somewhere along the voluntary-forced continuum, with conflict not necessarily being the norm but with cooperation and interaction usually emerging."
>
> (Kelman, Chap. 12, p:208).

The chapters provided different illustrations about linkages between migration, disasters, and climate change impacts in the place of origin: less available manpower, a loss of local knowledge, land abandonment, reduced soil fertility, balanced with increased remittances, returnees arriving with new trade skills, and cases of greater female empowerment. For the place of origin, migration was often depicted as a needed coping strategy but one that gnawed on the social fabric, creating inequalities between the receivers and non-receivers of remittances, while hallowing local food security as food consumption patterns shift from local to imported foods (Schwilch et al., Chap. 11; Upreti and Shrestha, Chap. 9).

The positive and negative aspects of migration were illustrated throughout the book: in Nepal, the quest for greater mobility has led to a boom in rural road construction, a positive development for mobility but a negative one due to the exponential increase in erosion and shallow landslides along the roads (Devkota and Lal Chandra, Chap. 13; Schwilch et al., Chap. 11). As expressed by Guadagno (Chap. 2, p:15), "it is therefore difficult to single out "positive" and "negative" forms of mobility; mobility choices should rather be investigated for their positive and negative effects, trying to understand how these latter are distributed and how they interplay with existing patterns of vulnerability and risk. The relevance of mobility as an attribute of resilience should not be underplayed." This includes the juxtaposition between household gains from migration (Upreti and Shrestha, Chap. 9; Schwilch et al., Chap. 11) versus the public capacity and responsibility to safeguard populations from disasters and climate change impacts but also from negative effects of migration (Hewitt, Chap. 3; Alcántara-Ayala and Oliver-Smith Chap. 7; Devkota and Chandra Lal, Chap. 13). Yet, other than contributing to household resilience, the extent to which migration might contribute to individual or collective efforts to prevent, mitigate, or better prepare for disasters is less conclusive.

Positive examples of migration and DRR included the post-Haiti earthquake of 2010 (Jahn et al., Chap. 10) where a potentially explosive land tenure conflict was addressed through preventive conflict resolution. The case study also illustrated the fuzzy zone between humanitarian action and development planning. While the total lack of sustainable development through land-use planning, building codes, and zoning laws inherently created the huge humanitarian crisis that was triggered by the earthquake, it was through humanitarian intervention that the land tenure conflict was finally resolved. Kelman (Chap. 12) describes a rare positive example of "disaster diplomacy" in the aftermath of the Indian Ocean tsunami in 2004, where a huge tragedy was turned into an opportunity for the internally displaced population of Banda Aceh as the resolution of a decades-long conflict was catalyzed.

Finally, Collins (Chap. 8) evokes the important issue of migration as a rights-based approach to DRR, highlighting the need for improved governance of migration, including safeguarding migrants during and after disasters. Furthermore, "rights, well-being aspirations and sustainable development are core principles for interpreting the application of DRR to unstable migration settings" (Collins,

Chap. 8, p:127). This point has a number of important policy implications which are highlighted below.

2 Climate Change Adaptation and Migration

As mentioned above and as stated by Kelman (Chap. 12), a persistent theme of this book is that migration is not and never has been confined to contemporary climate change, despite much populism over climate change causing migration. Certainly climate-induced changes such as the sea-level rise affecting island nations such as Kiribati in the Pacific Ocean (Morison, Chap. 15) are challenging many island nations' capacities to adapt and may cause the need to seek higher ground. Yet as Morrison points out, the recent climate-induced migration is ultimately rooted in an inadequate economic and political base. In such cases, it is difficult to clearly distinguish between migration as a response to complex socioeconomic interactions with climate change as an additional trigger or possible scapegoat for structural inequalities that are more difficult to tackle than claims for refugee status abroad.

This being said, there is no denying that climate change effects are acting as triggers or push factors for short- and long-term migration in many instances, as documented by many of the chapters in this volume. In addition to the example given from Kiribati, climate change is being attributed to the spread of invasive species on land being abandoned by out-migrants in Nepal (Schwilch et al., Chap. 11). In Dakar, Senegal (Pfeifer et al., Chap. 4), climate change effects have led to increased drought in the Sahel and a movement toward the city, which is now subsiding due to overextraction of groundwater. Rising sea levels, combined with more extreme periodic rainfall events, have paradoxically led to increased incidents of flooding. Several of the chapters (Jacobi et al., Chap. 14; Devkota and Chandra Lal, Chap. 13 and Upreti and Shrestha, Chap. 9) note that climate change is leading to more uncertainty for farmers in many places as the normal growing season and zones are changing, in addition to more extreme weather events. In the face of such uncertainty and pulled toward cities and foreign destinations with the promise of more modern lifestyles, young rural people are increasingly seizing the opportunity to move out of agriculture (Upreti and Shrestha, Chap. 9). In this situation, migration may increase resilience of remaining rural populations by providing a more regular stream of remittance income while at the same time undermining human resources and local knowledge at the place of origin.

As for DRR, stories about how remittances might directly support climate change adaptation are less conclusive. In Nepal, Devkota and Chandra Lal (Chap. 13) described a program intended to strengthen climate change governance in Nepal by providing guidelines to local government for adaptation; yet the extent to which the uptake will be successful is yet to be seen, as the region is also highly affected by high mobility. Jacobi et al. (Chap. 14) described organic agro-forestry systems in Bolivia that are being developed as a sustainable adaptation measure to climate change and one that is also affected by migration to cities in both positive and negative ways. Finally, Morrison (Chap. 15) described the traditional way of thinking of the people of Kiribati as a positive way of adapting to changing environmental conditions. In the end, perhaps the most telling example of CCA is migration itself, again as a final tipping factor on top of the abovementioned economic, social, and power root causes that create the need for mobility.

3 Linking Sustainable Development, DRR, CCA and Migration, Policy Implications

Utopic, out-of-fashion, or still of utmost importance to addressing underlying disaster risks and adaptation challenges, sustainable development continues to spark debate, as does the role of migration as part of sustainable development. The Sustainable Development Goals (SDGs) were finally established with 17 goals that are both lauded as highly ambitious and consensual and criticized as "sprawling and misconceived" (The Economist 2015). They include a little of everything, including a number of goals that mention disasters and environment. Migration was included in the SDGs in goal 10.7.: "Facilitate orderly, safe, regular and responsible migration and mobility of people, including through the implementation of planned and well-managed migration policies" (OWGGASDGs 2015: 18). Migration is also mentioned in the introduction, as a criterion for data disaggregation and a specific issue for the respect of worker's right. This is certainly an improvement from the Millennium Development Goals which did not include any mention of human mobility. However, it remains to be seen to what extent the 17 "sprawling" SDG goals will become an effective guidance for benchmarking progress on sustainable development. They do reflect the point of this book: that migration, DRR, CCA, and sustainable development are all closely interlinked. "Addressing poverty, disaster risk reduction, sustainable development and climate change adaptation together is largely the conflation of aspects of the same agenda of survival" (Collins, Chap. 8, p:139).

Despite its shortcomings, notably with regard to achieving more clearly defined targets, sustainable development, DRR, and CCA are clearly interlinked in the post-2015 Sendai Framework for Disaster Risk Reduction (SFDRR) 2015–2030, and the role of migration in DRR has been elevated. References were clearly made to linking the post-2015 development agenda, financing for development, climate change, and DRR goals, to the extent that some of the original SFDRR text with provisions on loss and damages was scaled back so as not to set a precedent before the upcoming UNFCCC Climate Change Conference of Parties in Paris, 2015. Whether or not the SFDRR text will be translated into realizable and accountable indicators is another debate.

Migration was one of the main points of contention during the SFDRR 2015–2030 negotiations, with some countries opposing the mention of "preventing and preparing for displacement," "facilitating remittance transfers to build the resilience of affected households," and limiting the scope of human mobility to only disaster-related human mobility. Otherwise, migration was highlighted as a positive factor in building resilience of home and host countries, and governments were encouraged to reduce risks created by migration while facilitating mobility to strengthen the resilience of affected people and promoting safer displacements in response to disasters. Other policy implications that were not directly mentioned in the SFDRR but deserve consideration in national and local policies related to migration and DRR and CCA are as follows:

- Considering the high degree of uncertainty and instability which condition the application of DRR to migration, there is a need to engage localized strategies and sustainable development outcomes for migrants (Collins, Chap. 8).
- DRR policies also need to consider migration as a rights-based approach to addressing issues of poverty and environmental sustainability (Collins, Chap. 8).
- More adequate DRR in both the place of origin and destination, especially with regard to expected more extreme and smaller scale events by providing access to adequate information to migrants about local hazards (Alcántara-Ayala and Oliver-Smith, Chap. 7; Pigeon, Chap. 5; Hewitt, Chap. 3).
- Urban planning and management of hotspot slum areas that are highly exposed to hazards so that migrants are not further marginalized (Texier and Edelblutte, Chap. 6).
- Policies encouraging easy transfers of remittances and facilitating public-private investments in DRR, early warning, or retrofitted housing (Le De et al. 2013).
- More attention to understanding the dynamics between out-migration, land abandonment, land tenure, and food insecurity (Jahn et al., Chap. 10; Schwilch, et al., Chap. 11; Jacobi et al., Chap. 14).
- Creating an enabling environment for returnee migrants in terms of skill development and entrepreneurship, especially targeting female skill development for women in the place of origin (Upreti and Shrestha, Chap. 9).
- Facilitating migration and migrant rights through awareness campaigns on both negative and positive effects of migration (Upreti and Shrestha, Chap. 9).
- Facilitating sharing and dissemination of research-based evidence and local knowledge about disaster and climate risks is crucial for research being effective to address social problems and issues and develop collective ownership (Devkota and Lal Chandra, Chap. 13).

4 Way Forward and Questions for Further Reflection

Decisions about whether to stay or move are most often made by individuals, households, or communities in order to improve opportunities and in the hope of reducing a variety of risks: economic, environmental, physical, and social. As this book has illustrated, migration plays an important role in shaping disaster risk as both a risk reducer and risk creator. This is why policies to facilitate positive outcomes of migration, while reducing migration as creator of risk, are extremely important, and also why it is extremely important to recognize the important socioeconomic and environmental factors that both push and pull migration trends, with climate change acting as an amplifier for migration in many cases.

As this book comes to an end, it is always useful to leave some thoughts for reflection on new channels of research and policy that can be developed. We focus on two main themes: mobility and risk transfers and mobility as a resilience-building strategy. Many of these research questions were first brought up by Guadagno (Chap. 2) and were addressed in this chapter.

1. Mobility and risk transfers
   - What is the overall impact of mobility in terms of risk reduction and risk creation? How can costs and benefits be measured and compared?
   - Can migration lead to a food security crisis when less food is produced locally and replaced with imported food? Hence, is migration increasing or decreasing resilience in rural areas over the long run in case of lower demand for migrant labor in cities and abroad?
2. Mobility as resilience-building strategy
   - Under what conditions is mobility's risk reduction potential best harnessed? What should be avoided in order to reduce its negative impacts on those moving, as well as those in host and home communities?
   - Does human mobility challenge or reproduce existing conditions of vulnerability? Is mobility a viable risk reduction strategy for the most vulnerable?
   - To what extent are remittances used for DRR/CCA and what are the incentives for households and communities to invest more in these activities?

Finally, one of the main barriers to furthering our understanding about the inter-linkages between population dynamics, environment, disasters, and climate change is the silo approach (Gall et al. 2015) with which we address such issues. We hope that this book has provided a nuanced and more comprehensive perspective on the complex inter-linkages between migration, disasters, and climate change than the one we are often fed by the media and its scaremongering tactics announcing overwhelming masses of climate refugees. Now it is time that this scientific debate be incorporated in the public agenda in order for policy makers at the global and local levels to provide adequate guidance, protection, and an enabling environment for producing the positive outcomes that migration can procure while ensuring maximum human security against disaster and climate risks whether for people in their places of origin or new places of settlement.

References

Cardona OD (2003) The need for rethinking the concepts of vulnerability and risk from a holistic perspective: a necessary review and criticism for effective risk management. In: Bankoff G, Frerks G, Hilhorst D (eds) Mapping vulnerability: disasters, development and people. Earthscan, London

Economist (March 28, 2015) The 169 commandments. The proposed sustainable development goals would be rose than useless. http://www.economist.com/news/leaders/21647286-proposed-sustainable-development-goals-would-be-worse-useless-169-commandments

Gaillard JC, Cadag JRD (2009) From marginality to further marginalization: experiences from the victims of the July 2000 Payatas trashslide in the Philippines. Jàmbá J Disaster Risk Stud 2 (3):195–213

Gall M, Nguyen KH, Cutter SL (2015) Integrated research on disaster risk: Is it really integrated? International Journal for Disaster Risk Reduction, 2015. 12 (June 2015):255–267

Garcia Acosta V (2005) El riesgo como construcción social y la construcción social de riesgos. Desacatos 19:11–24

Hewitt K (1995) Excluded perspectives in the social construction of disaster. What is a disaster. pp 75–91

Lavell A (2000) Desastres y Desarrollo: Hacia un Entendimiento de las Formas de Construcción Social de un Desastre: El Caso del Huracán Mitch en Centroamérica. In: Garita N, Nowalski J (eds) Del desastre al desarrollo sostenible: huracán Mitch en Centroamérica. San José, BID-CIDHS

Le De L, Gaillard JC, Friesen W (2013) Remittances and disaster: a review. Int J Disaster Risk Reduct 4(2013):34–43

Manyena SB, O'Brien G, O'Keefe P, Rose J (2011) Disaster resilience: a bounce back or bounce forward ability? Local Environment 16(5):417–424

Maskrey A (1993) Vulnerabilidad y mitigación de desastres. In: Maskrey A (ed) Los desastres no son naturales. La RED, Bogota, pp 93–110

Nathan F (2008) Risk perception risk management and vulnerability to landslides in the hill slopes in the city of La Paz, Bolivia. A preliminary statement. Disasters 32(3 Autumn):337–357

Oliver-Smith A (2002) Theorizing disasters: nature, power and culture. In: Hoffman SM, Oliver-Smith A (eds) Catastrophe & culture. School of American Research Press, Santa Fe, pp 23–47

OWGGASDG (2015) https://sustainabledevelopment.un.org/owg.html. Accessed 14 July 2015.

UNDESA (2015) Focusing on new trends in migration. http://www.un.org/en/development/desa/news/population/cpd46-trends-in-migration.html. Accessed 5 July 2015.

UNISDR (2009) Terminology on DRR. https://www.unisdr.org/we/inform/terminology.

Wisner B, Blaikie P, Cannon T, Davis I (2004) At risk, natural hazards, people's vulnerability and disasters, 2nd edn. Routledge, London and New York

Index

© Springer International Publishing Switzerland 2017

K. Sudmeier-Rieux et al. (eds.), *Identifying Emerging Issues in Disaster Risk Reduction, Migration, Climate Change and Sustainable Development*,
DOI 10.1007/978-3-319-33880-4

MIX
Papier aus verantwortungsvollen Quellen
Paper from responsible sources
FSC® C105338

If you have any concerns about our products,
you can contact us on
ProductSafety@springernature.com

In case Publisher is established outside the EU,
the EU authorized representative is:
Springer Nature Customer Service Center GmbH
Europaplatz 3, 69115 Heidelberg, Germany

Printed by Libri Plureos GmbH
in Hamburg, Germany